Structural and Biological Applications of Schiff Base Metal Complexes

Coordination compounds have been well-known for their wide variety of applications for over a century, as well as enhancing the researcher's interest and concern in evaluating their action mechanism. It is certainly one of the most intensely discussed research topics. Coordination compounds involve different metal-ion-ligand phenomena. The involved metal ions play a significant role in the structural association and functioning of several processes in the genetic and metabolism system.

In recent years, Schiff base ligands have gained significant interest and received keen interest from many researchers. Schiff's base ligands have been recognized to hold a wide variety of biological and medicinal activities due to the presence of donor atoms. They have proven exceptional pharmacological actions, such as antimicrobial, anti-tuberculosis, antiplatelet, antidiabetic, antiarthritis, antioxidant, anti-inflammatory, anticancer, antiviral, antimalarial, and analgesic qualities. These biologically active Schiff base ligands have also been shown to inhibit enzyme mobilization and, when bound to a metal ion, exhibit enhanced biological activity, making them useful in a number of fields. As a result, metal complexes of Schiff base ligands are gaining popularity due to their unique properties and functionalities. Schiff base complex-based research for educational and industrial purposes is booming, and the number of publications is gradually increasing. Despite these interests, there is currently no detailed book on Schiff base metal complexes that covers the structures, biological activities, and other non-biological perspectives.

This book delves into the structures of Schiff base metal complexes, which are critical in assessing the biological viability of any complex. It also highlights their biological significance in pharma and drug discovery like their antibacterial, antifungal, anticancer, anti-inflammatory, antiarthritis, antidiabetic, antioxidant, anti-proliferative, antitumor, anticancer, and antiviral properties. The fundamentals of metal complexes are described, as well as an up-to-date outline of developments in synthesis, characterization methods, and properties: chemical, thermal, optical, structural, and applications. This book also discusses the other applications of Schiff base metal complexes: as sensor (luminescent, electrochemical, and biosensor), as pigments in dying and paint industries, as photocatalyst to improve the degradation rate.

Features:

- This book would be useful for academia, researchers, and engineers working in the area of Schiff base and their metal complexes.
- It will give an in-depth account of the properties of Schiff base and their metal complexes.
- It will discuss the details of synthesis methods for Schiff base and their metal complexes.
- It will cover emerging trends in the use of Schiff base metal complexes in the industry.
- Finally, it will provide an overview of the wider biological applications of Schiff base metal complexes.

Structural and Biological Applications of Schiff Base Metal Complexes

Edited by
Pallavi Jain and Prashant Singh

CRC Press
Taylor & Francis Group
Boca Raton London New York

CRC Press is an imprint of the
Taylor & Francis Group, an **informa** business

First edition published 2023
by CRC Press
6000 Broken Sound Parkway NW, Suite 300, Boca Raton, FL 33487-2742

and by CRC Press
4 Park Square, Milton Park, Abingdon, Oxon, OX14 4RN

CRC Press is an imprint of Taylor & Francis Group, LLC

ISBN: 978-1-032-27125-5 (HB)
ISBN: 978-1-032-27120-0 (PB)
ISBN: 978-1-003-29145-9 (EB)

DOI: 10.1201/9781003291459

Typeset in Times
by Newgen Publishing UK

Dedicated to

---

late Shri Munna Lal Rajak
(M.Sc. in Agriculture)

for

his encouraging words to Understand Society
& a Vision to Live a Better Life

Yours Affectionate,

Prof. Prashant Singh

Contents

Organization of the Book

The book is organized into 12 chapters. A brief description of each of the chapters follows:

Chapter 1 focuses on the insight into the importance of Schiff bases and their metal complexes along with their preparation, optical and thermal properties.

Chapter 2 discusses the introduction to diabetes mellitus, Schiff base and Schiff base metal complexes. Additionally, it gives an insight into some interesting Schiff base metal complexes that have been synthesized and investigated for their antidiabetic activity in recent years.

Chapter 3 focuses on recent research studies carried out on the Schiff base metal complexes as anticancer agents. It has been found that some metal complexes exhibited fascinating anticancer properties against a wide range of cancer cell lines (MCF7, HepG2, HCT-116, HeLa, etc.), even higher than the approved standard drugs.

Chapter 4 discusses various naturally occurring and synthetic Schiff base complexes that have potential against bacterial fungal and many other diseases caused by microorganisms.

Chapter 5 describes the favorable usage of metal-based Schiff bases as antiviral and anti-parasitic drugs and further explores to acquire a better understanding of their effectiveness.

Chapter 6 concentrates on the synthesis of a few specific Schiff bases and their metal complexes possessing therapeutic effects, mainly anti-inflammatory properties.

Chapter 7 focuses on Schiff base ligand appended with different pharmacophores like coumarin, isatin, thiazole, benzimidazole, triazine, pyrazolone, and so on, or using different metal ions, emphasizing the potential candidature of Schiff base metal complexes in the quest of new anticancer agents with improved activity and poor toxicity.

Chapter 8 aims to describe the development in the field of metal complexes of Schiff bases capable of interacting with DNA. Experimental techniques employed to study their interactions and illustration of the type of binding as intercalation or groove style are also included.

Chapter 10 summarizes the sensor application of Schiff base metal complexes with a special emphasis on electrochemical sensing, fluorescent and colorimetric chemosensors, and optical chemical sensors.

Chapter 11 highlights a broad study on the biological activity of Schiff bases and their metal complexes, along with some miscellaneous applications.

Chapter 12 focuses on the latest research in medicinal and pharmaceutical chemistry. Complexes of molecularly designed Schiff bases show valuable applications in these fields. Molecular docking and virtual screening of these derivatives are crucial in predicting their properties in order to apply them in drug design.

Dr. Pallavi Jain
Prof. Prashant Singh

Editor Biographies

Dr. Pallavi Jain has a very luminous academic and professional career. She is currently working as Associate Professor in the Department of Chemistry, SRM Institute of Science and Technology Delhi-NCR Campus, Modinagar, Ghaziabad, India. Dr. Jain received her Ph.D. degrees in Chemistry from Banasthali Vidyapith, Rajasthan, India in 2017. She has extensive teaching experience of about 16 years. Her research interest focuses on the development of transition metal complexes incorporating Schiff base ligands for their biological importance in drug delivery. Her research interests also focus on the synthesis of metal complexes as effective biosensors. She has authored and co-authored over 45 publications in journals of international repute, and contributed more than 50 book chapters.

Dr. Prashant Singh has an M.Sc. in Chemistry from the Indian Institute of Technology, Delhi in 2005. After completion, he joined the research group of the late Dr. N. N. Ghosh, at Dr. B. R. Ambedkar Centre for Biomedical Research, University of Delhi, India. He received his Ph.D. in January 2010. He obtained the position of Assistant Professor at Atma Ram Sanatan Dharma (ARSD) College, University of Delhi in 2006 and published about 150 research articles/ reviews/ chapters in journals of international repute. His work has been cited in more than 2400 articles and publications. He has been awarded UGC postdoctoral research fellowship and also offered postdoctoral fellowship from North-West University, South-Africa.

Contributors

Khursheed Ahmad
Department of Chemistry,
National Institute of Technology,
Patna, India

Komal Aggarwal
Department of Chemistry,
Sri Venkateswara College,
University of Delhi, India

Divyam Bansal
Department of Chemistry,
Atma Ram Sanatan Dharma College,
University of Delhi, Delhi, India

Seema Gautam
Department of Chemistry,
Deshbandhu College,
University of Delhi, Kalkaji

Archana Gautam
Department of Chemistry,
S.G.T.B. Khalsa College,
University of Delhi, Delhi, India

Pallavi Jain
Department of Chemistry,
SRM-IST, Delhi-NCR Campus,
Modinagar, Ghaziabad

Shaina Joarder
Department of Chemistry,
Atma Ram Sanatan Dharma College,
University of Delhi, Delhi, India

Pankaj Khanna
Department of Chemistry,
Acharya Narendra Dev College,
University of Delhi, Kalkaji, New Delhi-110019, India

Leena Khanna
University School of Basic and Applied Sciences,
Guru Gobind Singh Indraprastha University,
New Delhi-110078, India

Kamlesh Kumari
Department of Zoology,
University of Delhi, Delhi

Sumit Kumar
Department of Chemistry,
Gurukula Kangri (Deemed to be University),
Haridwar (U.K.)

Rajesh Kumar
Department of Chemistry,
R.D.S. College, B.R.A.,
Bihar University, India

Sunny Manohar
Department of Chemistry,
Deen Dayal Upadhyaya College,
University of Delhi, New Delhi-110078, India

Namita Misra
Silver Oak University,
Ahmedabad

Amani F.H. Noureldeen
Biochemistry Department,
Faculty of Science,
Ain Shams University, Cairo, Egypt

Amreeta Preetam
Department of Chemistry,
University of Delhi, New Delhi, India

Ramadan M. Ramadan
Chemistry Department,
Faculty of Science,
Ain Shams University, Cairo, Egypt

Sapna Raghav
Science Block, NBGSM College,
Sohna, Gurugram University, Gurugram,
122103, India

Jyoti Raghav
Schools of Engineering and Applied Science,
Bennett University, Greater Noida, India

Madhur Babu Singh
Department of Chemistry,
Atma Ram Sanatan Dharma College,
University of Delhi, Delhi, India
And
Department of Chemistry,
SRM Institute of Science and Technology,
Delhi-NCR Campus, Uttar Pradesh, India

Jyoti Singh
Department of Chemistry,
Deen Dayal Upadhyaya College,
University of Delhi,
New Delhi-110078, India

Jagvir Singh
Department of Chemistry,
Nehru Mahavidyalaya, Lalitpur,
Uttar Pradesh

Monica Tyagi
Department of Chemistry,
Monad University, Hapur,
Uttar Pradesh, India

Sangeetha Talwar
Department of Chemistry,
Deen Dayal Upadhyaya College,
University of Delhi, New Delhi-110078, India

Prashant Tevatia
Department of Chemistry,
Gurukula Kangri (Deemed University),
Haridwar (U.K.)

Priyansh Kumar Utsuk
Department of Chemistry,
Gurukula Kangri (Deemed University),
Haridwar (U.K.)

Mansi
University School of Basic and Applied Sciences,
Guru Gobind Singh Indraprastha University,
New Delhi, India

Alka
Department of Chemistry,
SRM-IST, Delhi-NCR Campus,
Modinagar, Ghaziabad
and
Department of Chemistry,
Deshbandhu College,
University of Delhi, Kalkaji

Anuradha
Department of Zoology,
Raghuveer Singh Government P.G. College,
Lalitpur, Uttar Pradesh

Preface

Coordination compounds have been well-known for their wide variety of applications for over a century, as well as enhancing the researcher's interest and concern in evaluating their action mechanism. It is certainly one of the most intensely discussed research topics. Coordination compounds involve different metal-ion-ligand phenomena. The involved metal ions play a significant role in the structural association and functioning of several processes in the genetic and metabolism system.

In recent years, Schiff base ligands have gained significant interest and received keen interest from many researchers. Schiff's base ligands have been recognized to hold a wide variety of biological and medicinal activities due to the presence of donor atoms. They have proven exceptional pharmacological actions such as antimicrobial, anti-tuberculosis, antiplatelet, antidiabetic, antiarthritis, antioxidant, anti-inflammatory, anticancer, antiviral, antimalarial, and analgesic qualities. These biologically active Schiff base ligands have also been shown to inhibit enzyme mobilization and, when bound to a metal ion, exhibit enhanced biological activity, making them useful in a number of fields. As a result, metal complexes of Schiff base ligands are gaining popularity due to their unique properties and functionalities. Schiff base complex-based research for educational and industrial purposes is booming, and the number of publications is gradually increasing. Despite this interest, there is currently no detailed book on Schiff base metal complexes that covers the structures, biological activities, and other non-biological perspectives.

This book delves into the structures of Schiff base metal complexes, which are critical in assessing the biological viability of any complex. It also highlights their biological significance in pharma and drug discovery like antibacterial, antifungal, anticancer, anti-inflammatory, antiarthritis, antidiabetic, antioxidants, anti-proliferative, antitumor, anticancer, and antiviral properties. The fundamentals of metal complexes are described, as well as an up-to-date outline of developments in synthesis, characterization methods, and properties: chemical, thermal, optical, structural, and applications. This book also discusses the other applications of Schiff base metal complexes: as sensor (luminescent, electrochemical, and biosensor), as pigments in dying and paint industries, as photocatalyst to improve the degradation rate.

1 Schiff Base Metal Complexes

Structure and Chemical, Thermal, Optical Properties and Mechanical Structure-Property Relationships

*Divyam Bansal[1], Shaina Joarder[1], Madhur Babu Singh[1,2], and Kamlesh Kumari[3]**
[1] Department of Chemistry, Atma Ram Sanatan Dharma College, University of Delhi, Delhi, India
[2] Department of Chemistry, Faculty of Engineering and Technology, SRM Institute of Science and Technology, Delhi-NCR Campus, Uttar Pradesh, India
[3] Department of Zoology, University of Delhi, Delhi, India
*Corresponding author
Email: biotechnano@gmail.com

CONTENTS

1.1 INTRODUCTION

Schiff bases (SB), named after a German chemist Hugo Schiff, are prepared when primary amine combines with a ketone or aldehyde under certain circumstances (Scheme 1.1). SB are aldehyde or ketone analogues of nitrogen containing compounds in which the carbonyl has been changed to an azomethine or imine group. Due to the numerous ways that functional groups of SBs might coordinate, SBs are the subject of a lot of studies. Due to their availability, SBs are organic compounds which are extensively used in coordination chemistry. They readily create mono, di, tri and multinuclear complexes with d-block metal ions [1]–[3]. Metal complexes can serve as models for biologically significant species due to their chelating properties and biological applications. A variety of biological effects including antifungal, anti-inflammatory, antitumor, antiproliferative, antidiabetic, anticorrosive, anticancer, and antibacterial properties have been demonstrated by a number of SBs that possess the imino functionality [4], [5]. The ability to change the amines and aldehydes allows for the simple preparation and wide variety of SBs. Mono, di, tri, tetra, Penta, hexa, and multidentate chelating SB ligands were created based on the binding environments of metal ions. The following techniques can be used to prepare SBs and their complexes:

DOI: 10.1201/9781003291459-1

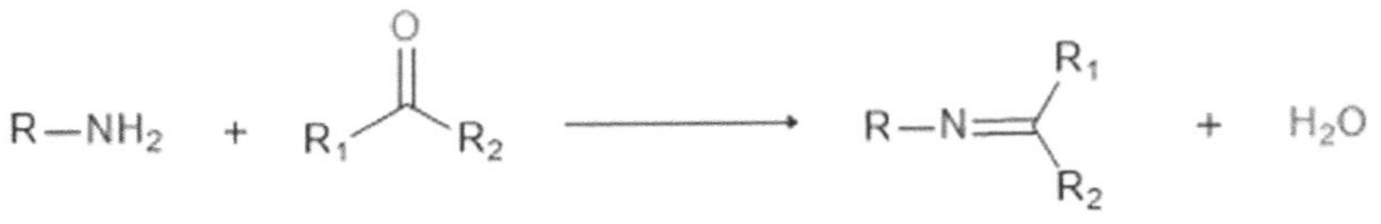

SCHEME 1.1 General scheme for formation of SB.

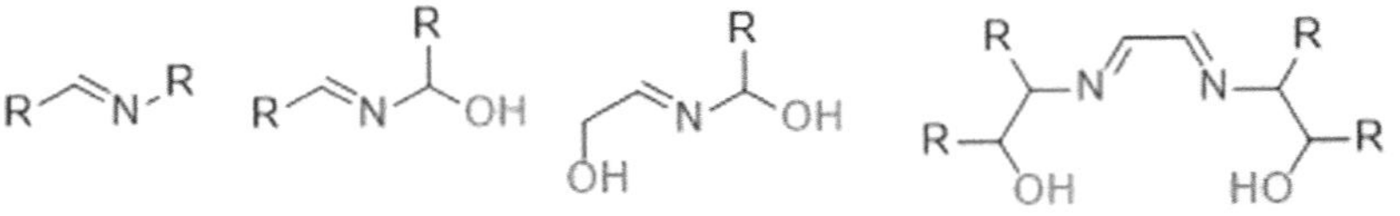

FIGURE 1.1 SBs of varying denticity; monodentate to tetradentate [6].

a. *Direct Ligand Synthesis and Complexation*: Prior to complexation, this process performs the separation and purification of SBs. After that, the metal ion and SBs are treated to prepare the complexes.
b. *Template Synthesis*: With this approach, complexes are synthesized in a single step by reacting an aldehyde, an amine, and a metal compound without the need to isolate SBs. By serving as a reaction template, the metal ions catalyze the process.
c. *Rearrangement of Heterocycles* (oxazoles, thiazoles etc.): This process involves combining the metal complex of one of the initial components (amine or aldehyde) with the other one to prepare the SB as its metal chelate [6].

The existence of donor sites affects the denticity of SB as shown in Figure 1.1. The majority of SB have nitrogen present as donor atoms, even though SB ligands containing donor atoms other than nitrogen can act as bi, tri, tetra, penta, or multidentate mixed donors. The primary amine/diamine and the kind of aldehyde/ketone used typically dictate the donor nature of the ligands [6].

1.2 STRUCTURE AND CHEMICAL PROPERTIES

Andiappan et al. created a series of SB complexes as shown in Scheme 1.2 via condensation reaction with 3,4-diaminopyridine and anthracene-9-carbaldehyde. They then added Pr, Er, and Yb (rare earth metals) to the resulting N2, N3-bis (anthracen-9-ylmethylene) pyridine3,4-diamine [7].

A new SB ligand was created by Karaomglan et al. through the condensation of 4-fluorobenzaldehyde and 1,10-phenanthrolin-5-amine and under the right circumstances, a unique series of its metal chelates (M = Ni(II), Co(II), Zn(II), and Cu(II)) were also prepared (Scheme 1.3) [8].

Dong et al. synthesized three open-chain ether SB complexes $[ZnDyL(NO_3)_2(OAc)(CH_3OH)]\cdot(CH_3OH)_3$ (Complex 1), $[DyH_2L(NO_3)_3]\cdot(CH_3CN)$ (Complex 2) and $[Zn_2DyL_2(OAc)(nicotinate)](NO_3)$ (Complex 3) ((H_2L = bis(5-bromine- 3-methoxysalicylidene) -3-oxapentane-1,5-diamine) Interfacial diffusion method was used to prepare these complexes (Scheme 1.4) [9].

By condensation reaction of 2-acetylthiophene and 2,6-diaminopyridine in ethanol at a 1:2 molar ratio, Ayoub et al. created a novel SB ligand named N, N-bis (1-(thiophen-2-yl) ethylidene) pyridine-2, 6-diamine (L1) (Scheme 1.5) [10].

Mohamed et al. prepared a tridentate SB ligand using the condensation of the drugs 2,6-diaminopyridine and isatin at a ratio of 1:1 M. (L). The complexes of following metals have been created: Mn(II), Co(II), Cr(III), Ni(II), Fe(III), Cd(II), Cu(II), and Zn(II) (Scheme 1.6) [11].

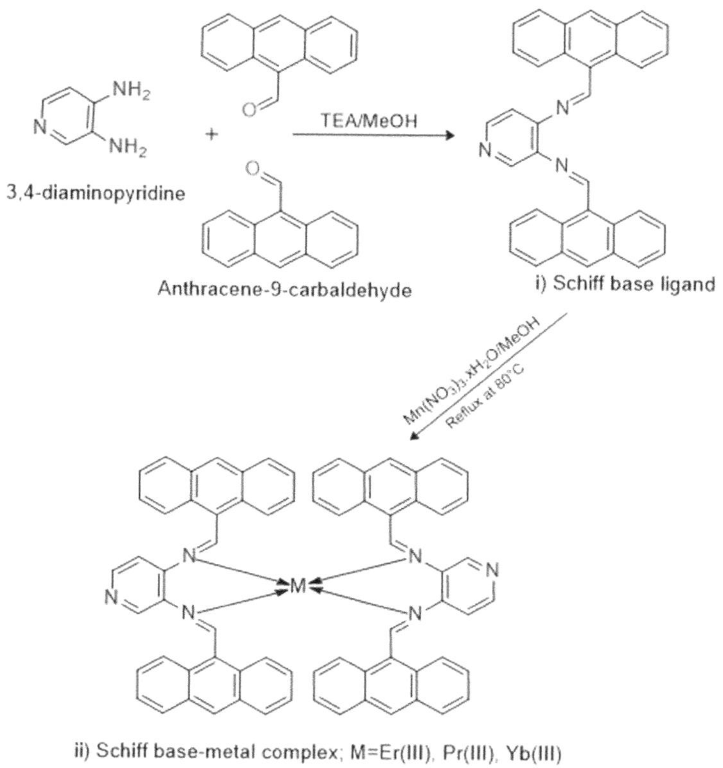

SCHEME 1.2 Synthesis of SB ligand and its rare earth metal complex [7].

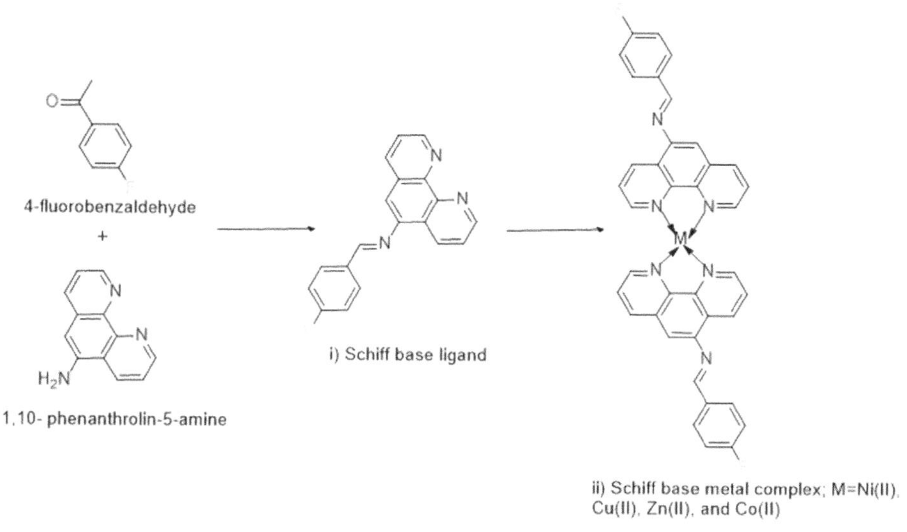

SCHEME 1.3 Condensation reaction of 4-fluorobenzaldehyde with 1,10-phenanthrolin-5-amine and formation of its metal complex [8].

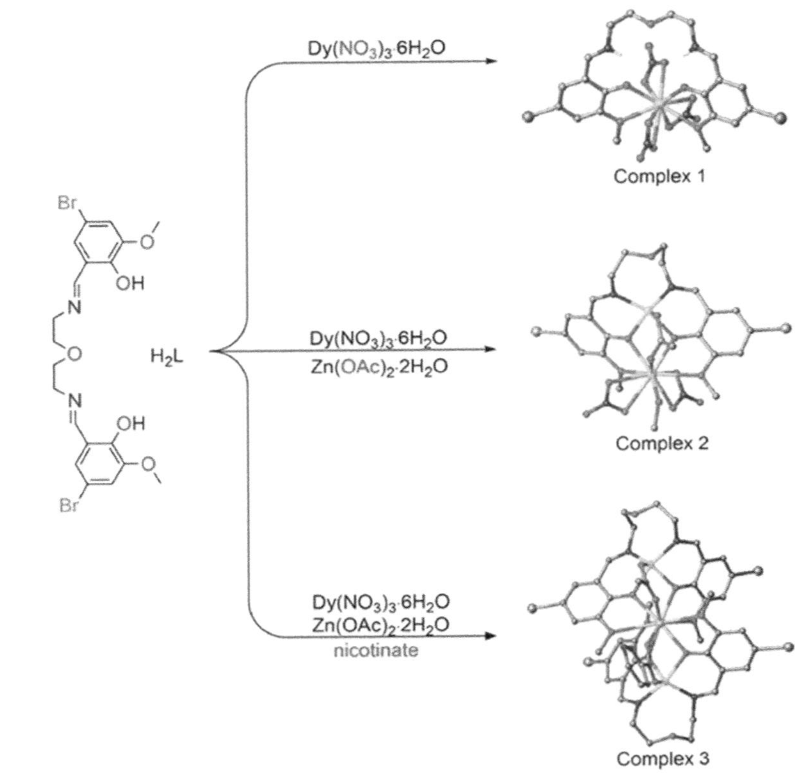

SCHEME 1.4 Synthetic method for preparation of complex 1, complex 2 and complex 3 [9].

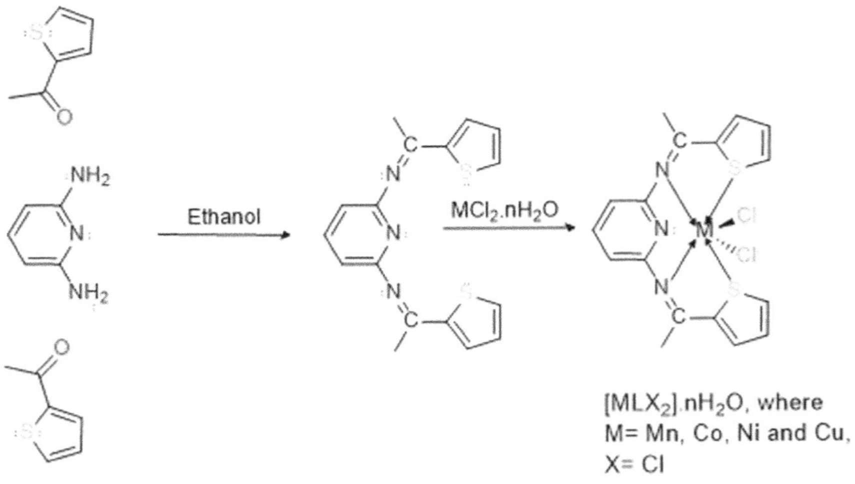

SCHEME 1.5 Preparation of SB ligand L1 and its chelates [10].

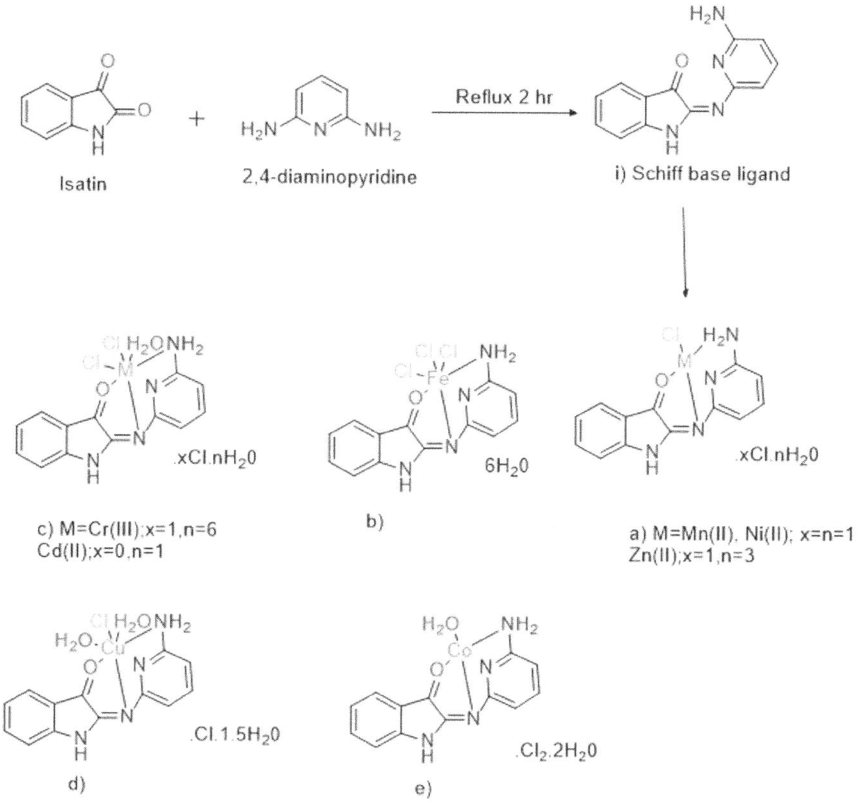

SCHEME 1.6 Condensation of 2,4-diaminopyridine and Isatin to form SB ligand and its complexes [11].

1.3 OPTICAL PROPERTIES

Andiappan et al. synthesized (Scheme 1.2) and studied the optical properties of SB metal complexes of anthracene-9-carbaldehyde and pyridine-3,4-diamine including Yb, Er and Pr metals. UV-vis absorption spectra of SB and their metal complexes are shown in Figure 1.2. SB displays keen absorption peaks at 314 nm and 276 nm arising from characteristic n-π^* and π-π^* carbonyl transitions of anthracene-9-carbaldehyde. The π-π^* and n-π^* transition peaks of SB-Pr, SB-Yb, and SB-Er complexes displaced a bit, and less intensity peaks occurred between 400 and 450 nm. Due to the ^{2}B1g- ^{2}A1g transition, metal complexes revealed curves indicating square planar shape for Pr, Yb, Er and ions. The obtained results are in line with the literature available [12], [13], [7].

Karaoglan et al. prepared (Scheme 1.3) and examined the UV-Vis absorption peaks of the ligand (E)1-(4-fluorophenyl)-N-(1,10-phenanthrolin-5-yl)methanimine (L2) and its metal chetals of Ni(II), Zn(II), Co(II) and Cu(II) in ethanol (Figure 1.3). The absorption bands that were observed between 287 nm and 297 nm in SB complex spectra related to the transition $\pi \rightarrow \pi*$. The absorption bands that were observed between 356 nm and 369 nm correspond to the transition π- $\pi*$ between the metal ion and the ligand, as well as the transition n $\rightarrow \pi*$ and $\pi \rightarrow \pi*$ of the -CH=N- group. The absorption peak of the metal chelates was found to be shifted to longer wavelengths in comparison to the absorption peak of the ligand. The wavelengths of the n- $\pi*$ and π- $\pi*$ transitions of SB complexes altered by roughly 24–37 nm and 13–23 nm, respectively. These changes confirmed that ligand was coordinated with metal ion [8].

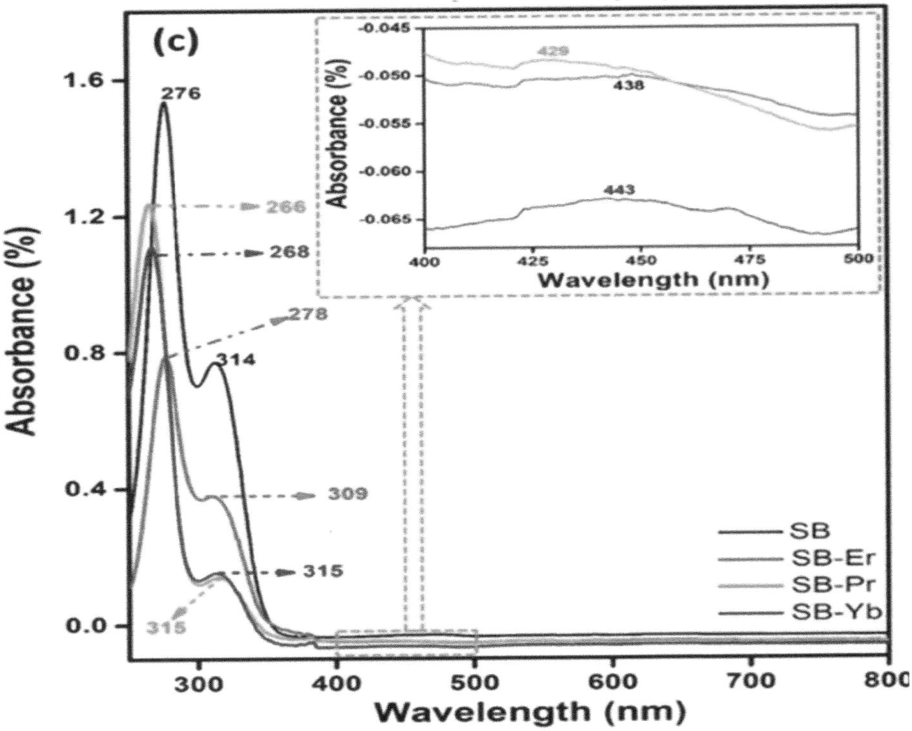

FIGURE 1.2 UV-vis spectra of SB metal complexes of anthracene-9-carbaldehyde and pyridine-3,4-diamine [7].

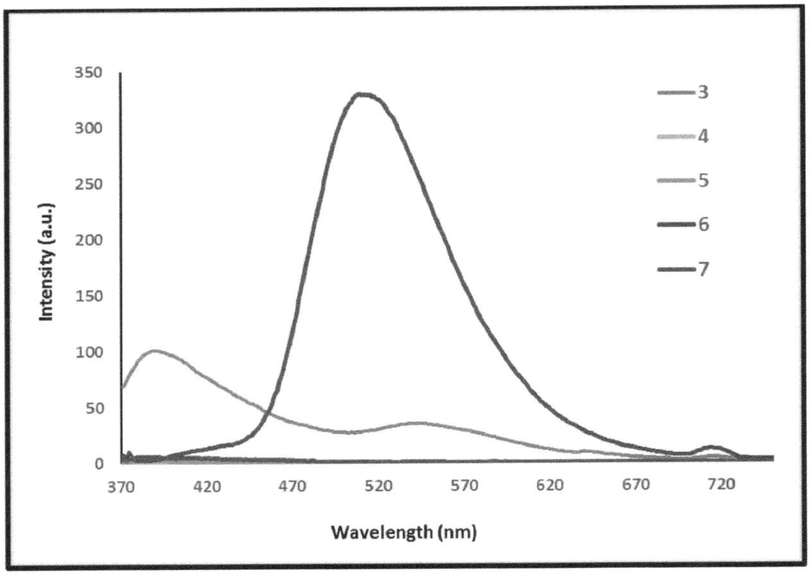

FIGURE 1.3 UV-Vis absorption spectra of L2 and Ni(II) (4), Cu(II) (5), Zn(II) (6) and Co(II) (7) complexes [8].

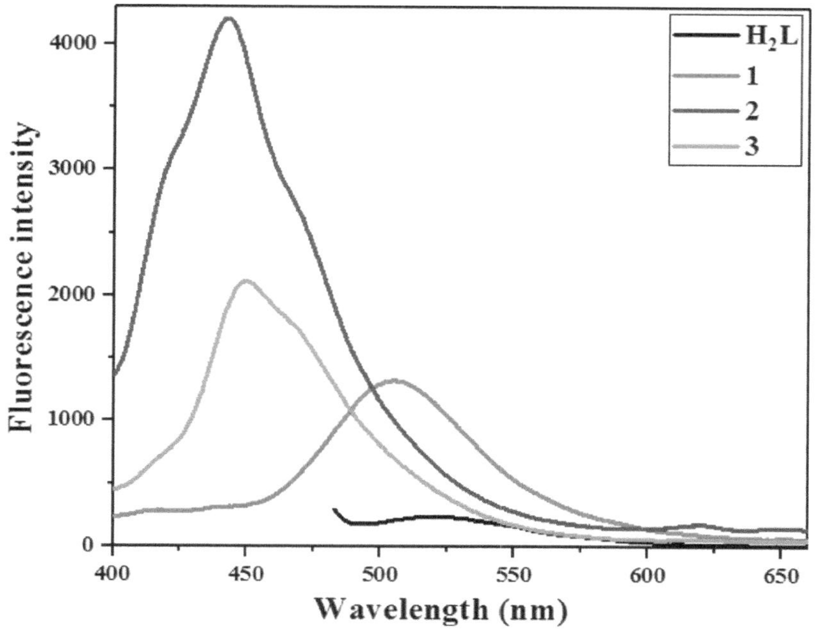

FIGURE 1.4 The fluorescence spectra of the ligand H_2L and its metal complexes 1–3 [9].

Dong et al. synthesized (Scheme 1.4) and studied the fluorescence data and spectra of the SB ligand H_2L and its metal complexes (Complex 1, 2, and 3), in at room temperature (Figure 1.4). All three metal complexes display a wide band from 405 nm to 540 nm with peaks at 504, 442, and 449 nm, respectively, which corresponds to the transition between ligands [14]. Compared to the unbound ligand H_2L, the fluorescence spectras of metal Complexes 1–3 exhibited a strong blue shift and amplification, with a larger degree in heteronuclear complexes than in mononuclear complexes. The blue shifts in the emission spectra are attributable to the electron deficient metal ions attached to H_2L ligand, which increases the energy gap of the molecular orbitals. The development of complexes that greatly improve the stiffness of the ligand to reduce the energy decadence caused by thermal vibrations is responsible for the higher intensities of the emission peaks [15] [9].

Ayoub et al. synthesized (Scheme 1.5) and analyzed SB ligand L1 ligand and its metal complexes fluorescence characteristics in CH_3CN at room temperature (Figure 1.5). A highest emission peak was visible in the ligand's excitation spectra at 397 nm when stimulated at 320 nm. SB system often exhibited fluorescence caused by intra-ligand π-π^* transitions. Fluorescence spectral peaks at 379, 396, 397, and 362 nm were visible in the excitation spectra of the Ni(II), Co(II), Cu(II), and Mn(II) complexes when stimulated at 316, 310, 330, and 312 nm. Metal ions have been seen to either enhance or quench molecules that include nitrogen in their fluorescence emission [10].

Mohamed et al. prepared ligand of 2, 4-diaminopyridine and isatin and its metal chelates (Scheme 1.6) and studied their spectra at 298 K. A band at 247 nm in the electronic spectra of the SB ligand can be assigned to conjugation. Another band, designated for the aromatic ring transition, is seen at 216 nm. These bands, corresponding to the azomethine group present in metal chelates, are displaced from 208 to 241 nm. A prominent band that indicates the n-π^* transition corresponding to the azomethine group also occurred at 337 nm in (Ligand). Complexes exhibit the n-π^* transition between 337 and 361 nm. Both the $\pi \rightarrow \pi^*$ and $n \rightarrow \pi^*$ transitions exhibit a substantial bathochromic shift in respect to SB. Transitions shift as a result of the chelate formation, showing the interaction of metal ion and the SB ligand [11].

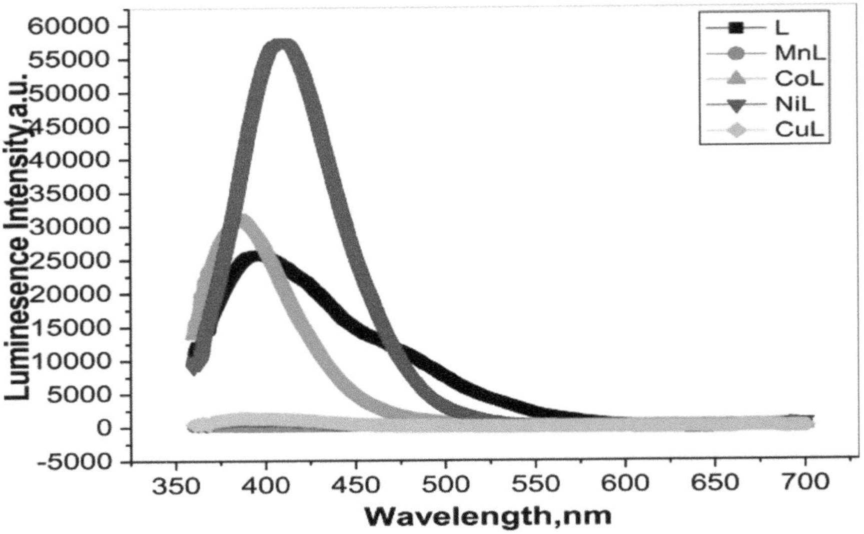

FIGURE 1.5 Fluorescence emission spectra of Ligand (L1); $[Mn(L1)Cl_2].4H_2O$; $[Ni(L1)Cl_2].6H_2O$ and $[Cu(L1)Cl_2].2H_2O$).

1.4 THERMAL PROPERTIES

Andiappan et al. studied the thermal properties of SB metal complexes of anthracene-9-carbaldehyde and pyridine-3,4-diamine including some rare earth alkali metals (Pr, Yb, and Er). The thermal stability of SB and its metal complexes was investigated using method of thermogravimetry (TG) and differential thermogravimetry (DTG), as shown in Figure 1.6a and b, respectively. SB indicated two phases of weight loss as a result of the degradation of anthracene-9-carbaldehyde and 3,4-diaminopyridine molecules [17] whereas, the complexes exhibited approximate 2–3% weight loss between the temperature range 35 and 110°C due the release of water molecules. Following this, pyridine and anthracene carbaldehyde breakdown, azomethine nitrogen and metal ion bond distortion and free amino pyridine losses all resulted in three weight loss processes [18]. SB-Yb and SB-Er complexes had approximate weight reduction between temperature range 350 and 500°C as a result of the dissolution of the anthracene 9-carbaldehyde, but the SB-Pr complex showed just a minor weight reduction between 360 and 500°C. The SB-Pr complex displayed less weight reduction and more residue. The DTG curves for SB and Pr, Er, and Yb metal complex, all samples display high exothermic decomposition peaks, with SB having an exothermic decomposition peak at 457°C. The DTG curves of metal complexes displayed two strong peaks at 447°C and 510°C, which corresponded to the decompositions of anthracene-9-carbaldehyde and condensed pyridine units, respectively [7].

Ayoub et al. performed thermal analysis of complexes of SB ligand N, N'-bis (1-(thiophen-2-yl) ethylidene) pyridine-2, 6-diamine. Complex $[Mn(C_{17}H_{15}N_3S_2)Cl_2].4H_2O$ thermogravimetric analysis comprised four breakdown phases. The initial transition between 44–113°C resulted in an estimated mass reduction of 4.27. The second transition between 113–205°C resulted in approximate mass reduction of 6.587. The third stage between 204–426°C revealed approximate mass reduction of 17.930. The final phase between 427–1000 °C exhibited approximate mass reduction of 41.112. $[Co(C_{17}H_{15}N_3S_2)(Cl)_2].6H_2O$ thermogravimetric analysis comprised three breakdown phases. The initial step between 47–204°C resulted in an approximate mass reduction of 7.113. The second

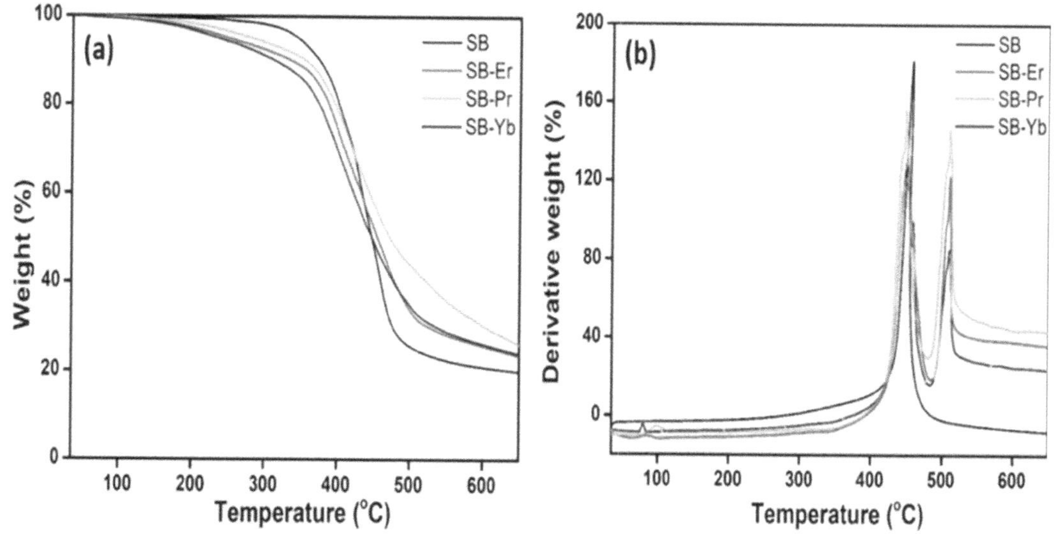

FIGURE 1.6 Thermal characteristics of SB (a) thermogravimetric (TG) curves, (b) differential TG (DTG) curves [7].

stage between 204–281°C revealed an approximate mass reduction of 3.575. The final transition in the range of 282–1000°C resulted in an approximate mass reduction of 50.705. $[Ni(C_{17}H_{15}N_3S_2)(Cl)_2]6H_2O$ thermogravimetric analysis comprises three breakdown phases. The initial step between 43–346°C revealed an approximate mass reduction of 14,768. The second step take places between 346–533°C revealed an approximate mass reduction of 25.271. The final transition in the range of 533–1000°C resulted in an approximate mass reduction of 52.222.

The $[Cu(C_{17}H_{15}N_3S_2)(Cl)_2].2H_2O$ thermogravimetric analysis requires three breakdown phases. The initial transition between 44–144°C revealed an approximate mass reduction of 7.371. The estimated mass loss for the second step takes place from 145–310°C was 16.879. The final step takes place in the range of 311–1000 °C resulted in an estimated mass reduction of 45.971 [10].

Ali et al. studied the thermal properties of 2-[(pyridin-2-ylmethylidene)amino]-6-aminopyridine (L3), 2-[(2-furylmethylene)]phenylenediamine(L4) and its Mn(II), Pd(II) metal chelates. In Complex $[MnL3\ (OH_2)_2Cl_2].2H_2O$, the temperature range of 45–721°C exhibited four stages of breakdown. The first two stages of disintegration, which occurred between 45–246°C, resulted in an estimated 22.50% mass loss. In the temperature range of 246–457°C, the second breakdown stage resulted in a 17.42% mass loss. The final stage of decomposition corresponded to a 45.90% mass loss. The thermal breakdown of $[PdL3\ Cl_2]$ was the two-step process. The initial phase between 47–344°C, observed percentage of mass loss was 18.41%. The second stage, which occured between 344–650°C, the observed percentage of mass loss at this stage was 52.96%. Four steps were required to break down $[MnL4\ (OH_2)_2Cl_2].2H_2O)$. The first two decompositions took place between 45–370°C. The mass loss measured was 29.91%. Between 370–483°C, a mass loss of 8.96% occurred during the third stage. The final stage of decomposition occurred in the temperature range of 483–999°C, accompanied by a 41.71% mass loss. $[PdL_2Cl_2]$ underwent three stages of breakdown between 47–850°C. The initial stage of breakdown occurred between 47–380°C with a net mass reduction of 14.10%. The second stage occurs between 380–650°C with a mass reduction of 24.73%. The third step occurs between 650–850°C with a mass loss of 27.99% [19].

TABLE 1.1
Thermogravimetric Data of Complexes

Complex	STEPS	Temperature Range (°C)	DTG Peak	TG Weight Loss % Calculated	TG Weight Loss % FOUND	Inference	Ref
$[Mn(C_{17}H_{15}N_3S_2)(Cl)_2].4H_2O$	1	44-113	82	3.449	4.277	Loss of one lattice water	[10]
	2	113-205	154	7.144	6.587	Loss of two lattice water	
	3	113-205	309	17.845	17.930	Loss of $1/2Cl_2$, $2CH_3$, H_2O	
	4	427-1000	635	40.973	41.112	Loss of, 2SH, 2C ¼ N, $2H_2$, $1/2Cl_2$	
$[Co(C_{17}H_{15}N_3S_2)(Cl)_2].6H_2O$	1	47-204	175	6.407	7.113	Loss of two lattice water	
	2	204-281	258	3.423	3.575	Loss of one lattice water	
	3	281-1000	377	50.601	50.705	Loss of $3H_2O$, 2SH, 2C ¼ N, Cl_2, CH_3	
$[Ni(C_{17}H_{15}N_3S_2)(Cl)_2].6H_2O$	1	43-346	284	14.241	14.768	Loss of 4.5 lattice water	
	2	346-533	445	25.795	25.271	Loss of 1.5 lattice water, Cl_2, $1/2C_2$, CH_3	
	3	533-1000	648	52.425	52.222	Loss of C_4H_3S, C_5H_3N, C]N	
$[Cu(C_{17}H_{15}N_3S_2)(Cl)_2].2H_2O$	1	44-144	88	7.280	7.371	Loss of two lattice water	
	2	145-310	240	16.575	16.879	Loss of Cl_2, $1/2C_2$	
	3	311-1000	378	45.752	45.971	Loss of 2C ¼ N, $2CH_3$, C_4H_3S, $1/2C_2$	
$[MnL^1 (OH_2)_2Cl_2].2H_2O$	1,2	45–246		22.48	22.5	Loss of $4H_2O$ and NH_3	[19]
	3	246–457		17.89	17.42	Loss of Cl_2	
	4	457–721		45.74	46.3	Loss of HCN, 2CN, 3HC:CH and 2C	
$[PdL^1 Cl_2]$	1	47–344		18.88	18.41	Loss of Cl_2	
	2	344–650		52.79	52.96	Loss of L^1	
$[MnL^2 (OH_2)_2Cl_2].2H_2O$	1,2	45–370		30.21	29.91	Loss of $4H_2O$, NH_3 and HCN	
	3	370–483		9.2	8.96	Loss of $1/2Cl_2$	
	4	483–999		42.07	41.71	Loss of $1/2Cl_2$, 3HC:CH and 4C	
$[PdL^2Cl_2]$	1	47–380		14.43	14.1	Loss of $1/2Cl_2$ and NH_3	
	2	380–650		24.34	24.73	Loss of $1/2Cl_2$, HCN and HC:CH	
	3	650–850		27.52	27.99	Loss of 2HC:CH and 4C	

1.5 CONCLUSION

Since SB and their metal complexes have wide range of medical applications, including as antiviral, antiparasitic, antibacterial, antifungal, antidiabetic, and anticancer medicines, they are among the most important classes of chemical compounds. They are one of the most extensively used organic compounds in coordination chemistry. SB and its metal complexes are simple to synthesize utilizing condensation reactions. TG-DTG analysis was carried out by different authors for determining the thermal stabilities of these complexes which demonstrated that SB metal complexes showed improved thermal stabilities due to incorporating metal or metal salts. Fluorescence spectra and UV-Vis spectra showed the coordination interaction between SB ligand and metal. SB and their metal complexes both can be designated as photoactive material as indicated from its optical properties.

BIBLIOGRAPHY

[1] D. Osypiuk, B. Cristóvão, and L. Mazur, "New heteronuclear complexes of PdII–LnIII–PdII with Schiff base ligand: Synthesis, crystal structures and chemical properties," *J. Mol. Struct.*, vol. 1261 , 2022, doi: 10.1016/j.molstruc.2022.132924.

[2] E. Raczuk, B. Dmochowska, J. Samaszko-Fiertek, and J. Madaj, "Different Schiff Bases—Structure, Importance and Classification," *Molecules*, vol. 27, no. 3, 2022, doi: 10.3390/molecules27030787.

[3] M. S. More, P. G. Joshi, Y. K. Mishra, and P. K. Khanna, "Metal complexes driven from Schiff bases and semicarbazones for biomedical and allied applications: a review," *Mater. Today Chem.*, vol. 14, p. 100195, 2019, doi: 10.1016/j.mtchem.2019.100195.

[4] L. H. Abdel-Rahman, A. M. Abu-Dief, R. M. El-Khatib, S. M. Abdel-Fatah, and A. A. Seleem, "New Cd(II), Mn(II) and Ag(I) Schiff Base Complexes: Synthesis, Characterization, DNA Binding and Antimicrobial Activity," *Int. J. Nanomater. Chem.*, vol. 2, no. 3, pp. 83–91, 2016, doi: 10.18576/ijnc/020303.

[5] A. M. Abu-Dief and I. M. A. Mohamed, "A review on versatile applications of transition metal complexes incorporating Schiff bases," *Beni-Suef Univ. J. Basic Appl. Sci.*, vol. 4, no. 2, pp. 119–133, 2015, doi: 10.1016/j.bjbas.2015.05.004.

[6] S. Afrin Dalia, M. Saddam Hossain, M. Nuruzzaman Khan,CM Zakaria,F. Afsan,M. Kudrat-E-Zahan andM. Mohsin Ali, "A short review on chemistry of Schiff base metal complexes and their catalytic application Evaluation of groundwater quality View project Schiff base View project A short review on chemistry of Schiff base metal complexes and their catalytic application," *Int. J. Chem. Stud.*, vol. 6, no. 3, 2018, [Online]. Available: www.researchgate.net/publication/325845095.

[7] K. Andiappan, A. Sanmugam, E. Deivanayagam, K. Karuppasamy, H. S. Kim, and D. Vikraman, "Schiff base rare earth metal complexes: Studies on functional, optical and thermal properties and assessment of antibacterial activity," *Int. J. Biol. Macromol.*, vol. 124, pp. 403–410, 2019, doi: 10.1016/j.ijbiomac.2018.11.251.

[8] G. K. Karaoğlan, "Synthesis of new Schiff base and its Ni(II), Cu(II), Zn(II) and Co(II) complexes; photophysical, fluorescence quenching and thermal studies," *J. Mol. Struct.*, vol. 1256, 2022, doi: 10.1016/j.molstruc.2022.132534.

[9] J. Dong, T. Wan, K. Li, X. Kong, Q. Shen, and H. Wu, "Mononuclear Dy(III)/heteropolynuclear Zn(II)–Dy(III) Schiff base complexes: Synthesis, structures, fluorescence and antioxidant properties," *J. Mol. Struct.*, vol. 1264, no. 3, p. 133340, 2022, doi: 10.1016/j.molstruc.2022.133340.

[10] M. A. Ayoub, E. H. Abd-Elnasser, M. A. Ahmed, and M. G. Rizk, "Some new metal(II) complexes based on bis-Schiff base ligand derived from 2-acetylethiophine and 2,6-diaminopyridine: Syntheses, structural investigation, thermal, fluorescence and catalytic activity studies," *J. Mol. Struct.*, vol. 1163, pp. 379–387, 2018, doi: 10.1016/j.molstruc.2018.03.006.

[11] G. G. Mohamed, M. M. A. Omar, B. S. Moustafa, H. F. AbdEl-Halim, and N. A. Farag, "Spectroscopic investigation, thermal, molecular structure, antimicrobial and anticancer activity with modelling studies of some metal complexes derived from isatin Schiff base ligand," *Inorg. Chem. Commun.*, vol. 141, no. June, p. 109606, 2022, doi: 10.1016/j.inoche.2022.109606.

[12] J. Joseph, K. Nagashri, and G. A. B. Rani, "Synthesis, characterization and antimicrobial activities of copper complexes derived from 4-aminoantipyrine derivatives," *J. Saudi Chem. Soc.*, vol. 17, no. 3, pp. 285–294, 2013, doi: 10.1016/j.jscs.2011.04.007.

[13] R. Antony, S. Theodore David Manickam, K. Saravanan, K. Karuppasamy, and S. Balakumar, "Synthesis, spectroscopic and catalytic studies of Cu(II), Co(II) and Ni(II) complexes immobilized on Schiff base modified chitosan," *J. Mol. Struct.*, vol. 1050, no. Ii, pp. 53–60, 2013, doi: 10.1016/j.molstruc.2013.07.006.

[14] Y. Qu, Y. Wu, C. Wang, K. Zhao, and H. Wu, "A new 1, 8-naphthalimide-based fluorescent 'turn-off' sensor for detecting Cu 2 + and sensing mechanisms," 2020, doi: 10.1177/1747519819886540.

[15] G. Z. Deng, N. Xin, Y. N. Han, and Y. Q. Sun, "Hydrothermal Synthesis, Crystal Structures, and Fluorescence Properties of Ni (II)– Ln (III) Complexes (Ln = Sm, Pr, Eu) 1," vol. 44, no. 5, pp. 365–371, 2018, doi: 10.1134/S1070328418050019.

[16] O. A. El-Gammal, D. A. Saad, and A. F. Al-Hossainy, "Synthesis, spectral characterization, optical properties and X-ray structural studies of S centrosymmetric N2S2 or N2S2O2 donor Schiff base ligand and its binuclear transition metal complexes," *J. Mol. Struct.*, vol. 1244, p. 130974, 2021, doi: 10.1016/j.molstruc.2021.130974.

[17] J. E. Santos, E. R. Dockal, and É. T. G. Cavalheiro, "Thermal Behavior Of Schiff Bases From Chitosan," *J. Therm. Anal. Calorim.*, vol. 79, no. 1388, pp. 243–248, 2005.

[18] K. D. Trimukhe and A. J. Varma, "Metal complexes of crosslinked chitosans: Correlations between metal ion complexation values and thermal properties," vol. 75, pp. 63–70, 2009, doi: 10.1016/j.carbpol.2008.06.011.

[19] O. A. M. Ali, "Characterization, thermal and fluorescence study of Mn(II) and Pd(II) Schiff base complexes," *J. Therm. Anal. Calorim.*, vol. 128, no. 3, pp. 1579–1590, 2017, doi: 10.1007/s10973-016-6055-9.

2 Therapeutic Approaches of Schiff Base Metal Complexes in the Treatment of Diabetes Mellitus

Amreeta Preetam[1], *Jyoti Singh*[2], *Sangeeta Talwar*[2], *and Sunny Manohar*[2*]
[1] Department of Applied Sciences, Bharati Vidyapeeth's College of Engineering, New Delhi-110063, India
[2] Department of Chemistry, Deen Dayal Upadhyaya College, University of Delhi, New Delhi-110078, India
*Corresponding author
Email: sunnymanohar@ddu.du.ac.in

CONTENTS

2.1 INTRODUCTION

2.1.1 About Diabetes Mellitus

Diabetes Mellitus is a global health problem. The number of people affected with diabetes is about 422 million and about 1.5 million people yearly lose their life due to diabetes all over the world.[1] Both the numbers have been increasing steadily in the past few decades and are continuing to do so. Diabetes is a metabolic disorder in which the sugar levels of the blood are increased. Pancreatic β-cells produce and release insulin hormone in response to blood sugar levels and insulin is responsible for the movement of blood sugar to various cells of the body. The body of a person suffering from diabetes either does not make enough insulin or if it does make, it is not used effectively which leads to diabetes. It is of two types: Type 1 and Type 2 diabetes. Type 1 diabetes is an autoimmune disease where the body cells start destroying the β-cells of the pancreas where insulin is formed. As

DOI: 10.1201/9781003291459-2

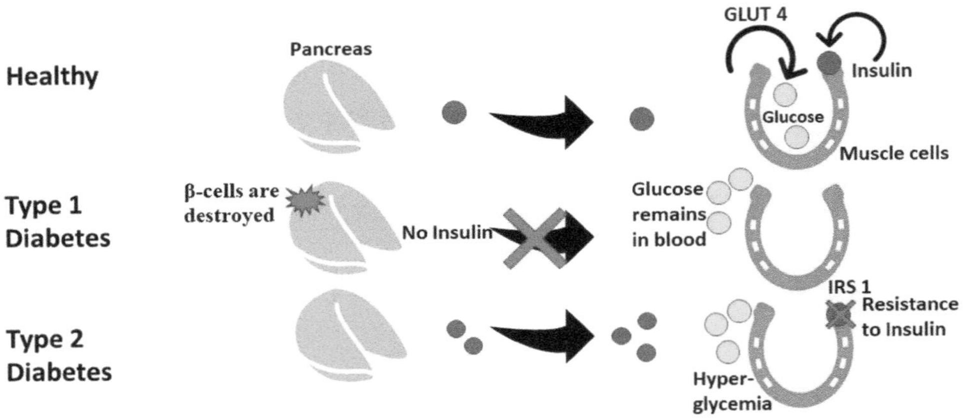

FIGURE 2.1 Type 1 and Type 2 Diabetes. Insulin signals causes the glucose transporter (GLUT 4) to move to the membrane so that glucose uptake occurs inside the cell. In insulin resistance state (IRS 1), there is a failure of insulin signals to promote this uptake. Uncontrolled diabetes leads to hyperglycaemia.

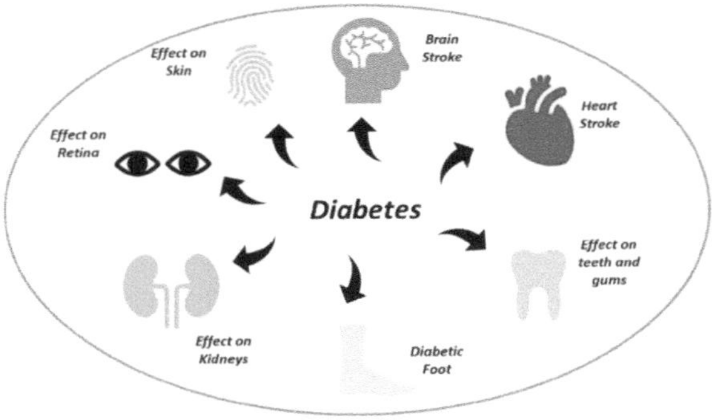

FIGURE 2.2 Harmful effect of diabetes on body organs.

a result, insulin is not produced in the body and the blood sugar levels increase. The reasons for this type of diabetes are not known exactly but can be attributed to genetic factors to some extent. Type 2 diabetes occurs when the insulin produced in the body is not able to perform its role,that is, the body becomes resistant to insulin and results in increased levels of blood sugar (Figure 2.1). This type of diabetes results from the combination of genetic factors and lifestyle. Obesity can increase the risk of Type 2 diabetes.[2,3] High blood sugar levels destroy the tissues all over the body and diabetes if not treated can cause severe damage to the eyes, kidneys and heart (Figure 2.2). It can lead to vision loss, hearing loss, heart stroke, neuropathy, nephropathy, depression and dementia. Diabetic patients complains of increased hunger and thirst. The reason is excessive urination along with excess glucose in the urine due to increased catabolic mobilization of proteins and fats. Other symptoms include weight loss, tiredness and sores that do not heal easily.[2,3]

A large number of enzyme changes occur in diabetes mellitus. There is reduced secretion of trypsin, chymotrypsin and cholinesterase whereas the secretion of phosphohexose isomerase, aminotransferases and dehydrogenases is increased. The activities of glucokinase, phosphofructokinase and pyruvate kinase decrease in diabetes while β-glucuronidase, N-acetyl-β-glucosaminidase, acid phosphatase, alkaline phosphatase, trehalase, amylase, and glucosidase show increased activity.[4,5] Therapeutic agents are developed based on the studies involving enzyme inhibition activities. For example, α-amylase and α-glucosidase enzymes promotes the breakdown of carbohydrates to glucose through hydrolysis. Therefore, α-amylase/α-glucosidase amylase inhibition assay is performed for various therapeutic agents so that their viability as effective medication for the treatment of diabetes can be tested.[6] Standard drug acarbose is generally used as a control while determining the IC_{50} values in these inhibition assays. Alloxan is a common drug which induces a condition similar to diabetes and is administered to the test animals for studying the enzyme changes related to diabetes *in vivo*.[7]

People with diabetes have high sugar levels in their blood, and it can cause damage to various body organs. Therefore, the medication is aimed to lower blood sugar levels and to prevent or delay the complications arising out of the effect of blood sugar on the body's organs. The treatment involves medicines which lower blood pressure, those reducing the risks of blood clots and also the ones lowering cholesterol. Antidiabetics are prescribed only if the blood sugar levels are not lowered by exercise or changing the diet and lifestyle. These can be given in the form of insulin hormones, insulin-mimetic agents or oral hypoglycemic agents (Table 2.1 and 2.2). Insulin

TABLE 2.1
Types of Antidiabetics, Their Mode of Action and Side Effects

Name of the Drug	Structure	Class of the Drug	Mode Of Action	Side Effects
Metformin	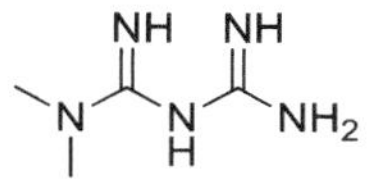	Biguanides	Reduces the blood sugar levels	Acidosis, nausea, breathing problems
Pioglitazone	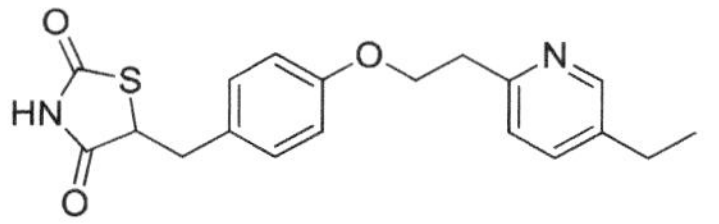	Thiazolidinediones	Reduces the blood sugar levels	Headaches, muscle aches, throat inflammation, swelling
Glimepiride	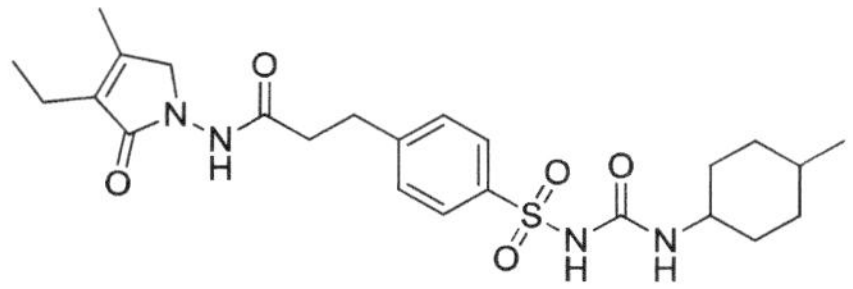	Sulfonylureas	Stimulate the production of insulin	Hypoglycemia, weight gain, allergic reactions in some cases
Repaglinide	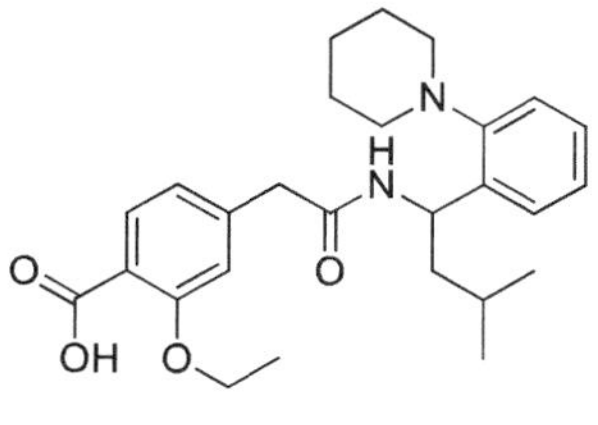	Glinides	Stimulate the production of insulin	Mild or moderate hypoglycemia, weight gain
Sitagliptin	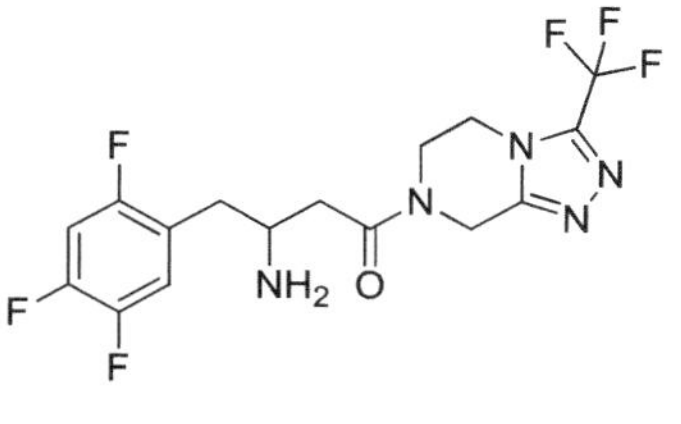	Gliptins	Stimulate the production of insulin	Headaches, gastrointestinal problems

(*continued*)

TABLE 2.1 (Continued)
Types of Antidiabetics, Their Mode of Action and Side Effects

Name of the Drug	Structure	Class of the Drug	Mode Of Action	Side Effects
Dapaglifozin	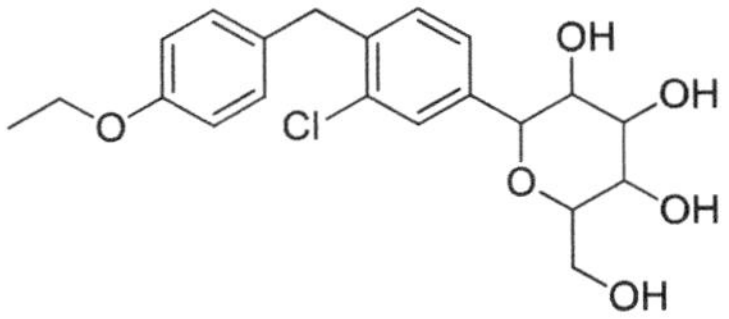	Gliflozins	Causes more sugar excreted in urine	Acidosis, dehydration, kidney problems
Acarbose	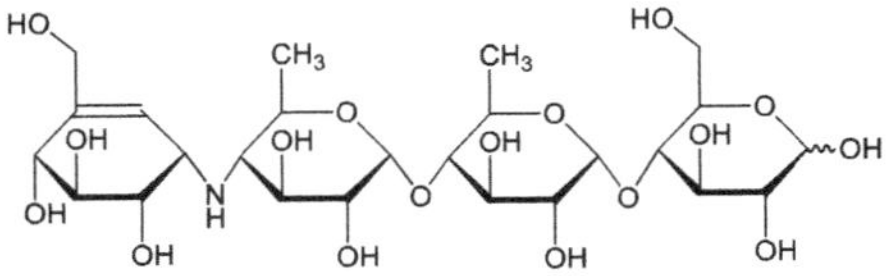	α-Gucosidase inhibitors	Slowing or delaying the digestion of carbohydrates	Allergic skin reaction, yellowing of the whites of eyes and skin, diarrhea, constipation

TABLE 2.2
Types of Insulin Hormone and Commercial Names of the Insulin Pen

Type of Insulin	Insulin Pen
Short acting Insulin	Humulin, Novolin
Rapid acting Insulin	Novolog, Flex pen
Intermediate acting Insulin	Humulin N, Novolin N
Long-acting Insulin	Lantus, Levemir
Combination of Insulin	Novolog Mix 70/30

hormones need to be injected into the body through an insulin pen which is a kind of injection that can be injected inside the fatty tissue beneath the skin through a small and thin needle (Table 2.2). Insulin-mimetic agents, for example, selenium and several arylalkylamine vanadium compounds, mimic insulin hormone in performing its functions and can be administered to control and treat diabetes.[8,9]

Although the above-mentioned antidiabetics are there in the market and are prescribed and taken frequently for the control and management of diabetes, their efficacy to maintain a normoglycemia with no or minimum side effects remains a major challenge. Also, as diabetes leads to a progression of secondary ailments and complications to a variable extent in different persons, personalization of drugs based on several etiological factors including age, gender, lipid profile and co-morbidities is necessary. The development of new targets, new drugs and newer technologies are hence the need of the hour and scientists should work together toward the development of potent and robust antidiabetic drugs to control the escalation of diabetes.

2.1.2 About Schiff Bases

A Schiff base, named after the scientist Hugo Schiff, is a substituted imine (or azomethine) class of compounds with the general formula $R_1R_2C{=}NR_3$.[10,11] Herein, the carbonyl group (>C=O-) of the aldehyde or ketone is replaced by the >C=N- moiety. These can be easily prepared by the

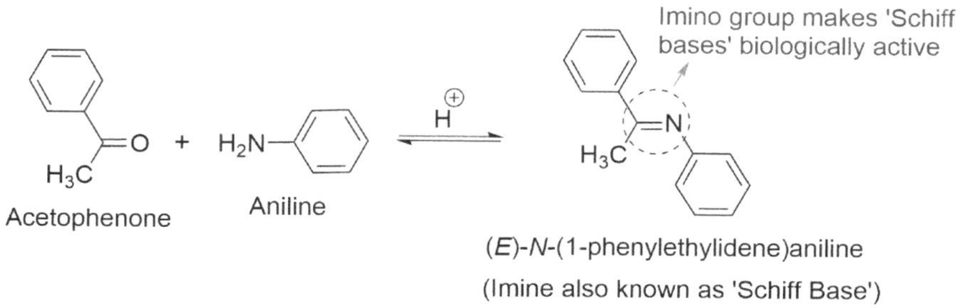

FIGURE 2.3 A Schiff base formed by the condensation of acetophenone and aniline in the presence of acid catalyst.

condensation reaction between various aldehydes/ketones with amino compounds resulting in imines known as Schiff bases (Figure 2.3). Schiff bases are used as an intermediate and catalyst in various organic transformations. It has been proved in the past that because of the presence of the imino group in their chemical structure, they are biologically active,[14–17] and have been discovered to exhibit a wide spectrum of activities including anti-inflammatory,[18] analgesic,[19] antiproliferative,[20] antioxidant,[21] antimicrobial,[22]anticancer[23,24] and antidepressant[25] properties.

Schiff bases are designed and developed as potential antidiabetic agents by several research groups in the last decade.[26–30] Due to their relative ease of synthesis, wide applicability of starting materials and metal-coordination ability, they are widely employed as a versatile ligand in modern coordination chemistry for the synthesis of metal complexes of material and pharmaceutical importance. It has been observed that the resultant Schiff base metal complexes were generally found to have improved antidiabetic activity as discussed in the next section.

2.2 THE ROLE OF SCHIFF BASE METAL COMPLEXES IN THE TREATMENT OF DIABETES MELLITUS

Metal ions play an essential role in maintaining the function of the human body. Deficiency of the metal ions in the body may lead to many kinds of complications/diseases including anemia (iron/cobalt deficiency), growth retardation (zinc deficiency), heart diseases in babies (copper deficiency), disturbance in carbohydrate/lipid metabolism (chromium deficiency) and infertility (manganese deficiency). Chemists, biologists and pharmacists across the world are putting their resources and energy into exploring newer and more effective ways to overcome these diseases for the healthy survival of mankind. Scientists discovered that the interplay between organic ligands and metal ions can play a vital role in exploring novel complexes of biological interest.[31–34] Several studies have proved that metal-ligand interaction works'hand in glove' in bringing synergistic effects to increase the *in vitro* and *in vivo* biological activity. In this quest, a number of research groups in the last few decades indicated that the transition metal complexes produced by organic ligands can impart insulin-mimetic action as well as inhibit the enzymes responsible for diabetes mellitus.[35–38] For the synthesis of the transition metal complexes exhibiting antidiabetic properties, the ligands based on Schiff base remain the ligand of choice. The rationale behind the selection being the fact that Schiff bases are found to show antidiabetic activities and hence their conjunction with a metal ion can lead to synergism in improving efficacy. Although numerous metal ions (Cr, Cu, Ni, Co, Cu, Zn, V etc.) have been utilized in the preparation of these Schiff base metal complexes, zinc and vanadium metal ions, because of their favorable activities in antidiabetic drug development program,[39–51] are found to be most commonly used.

It should be noted that various Schiff base metal complexes with antidiabetic activity have been explored in the last decades, the discussion of which will make this chapter very lengthy, so in this chapter the authors particularly focus on demonstrating the synthesis and antidiabetic activity of Schiff base metal complexes that have been reported in recent years.

2.3 ANTIDIABETIC ACTIVITY OF SOME RECENTLY SYNTHESIZED ZINC-BASED SCHIFF BASE COMPLEXES

In 2016, Miyazaki et al discovered novel Zn (II) Schiff base complexes ([N-sβ-Zn]; **1**, [N-bsE-Zn]; **2**, [N-bsP-Zn]; **3**, and [$Zn_2\{(DMN)_2Cl_2\}$]; **4**) and investigated their α-glucosidase inhibitory activity.[52] The synthesis of the complexes (**1–4**) was achieved by first preparing the desired Schiff base ligands ([N-sβ], [N-bsE], [N-bsP] and DMN) and then reacting the methanolic solution of the ligands with appropriate zinc salts (Scheme 2.1). The prepared complexes (**1–4**) along with their starting ligands were tested *in vitro* against the α-glucosidases obtained from yeast and rat small intestine.

The results obtained in this assay is shown in Table 2.3. The *in vitro* study revealed the Zn (II) Schiff base complexes (**1–4**) to have an enhanced α-glucosidase inhibitory activity than their corresponding Schiff base ligands. Complex (**4**), which showed highest activity against rat intestinal α-glucosidase, along with its corresponding ligand was further subjected to *in vivo* activity through oral maltose and glucose tolerance test on mouse models. Complex (**4**) showed the best activity in the maltose tolerance test with significantly lower postprandial blood glucose levels. In the glucose tolerance test, however, the ligand (DMN) showed the best anti-hyperglycemic activity. Careful examination of the experiments illustrated two things: (a) it may be anticipated that the α-glucosidase inhibitory activity of these complexes is due to the synergism between metal and ligands and, (b) switching from one ligand to another ligand in the Zn (II) complex can lead to an enhanced synergistic effect between the metal ions and the ligand for imparting better α-glucosidase inhibitory activity.

Balan et al evaluated N_2O_2 Schiff base Zn complex (ZnL;**5**) for their α-amylase and α-glucosidase inhibitory activity in an attempt to explore their potential as antidiabetic drugs.[53]

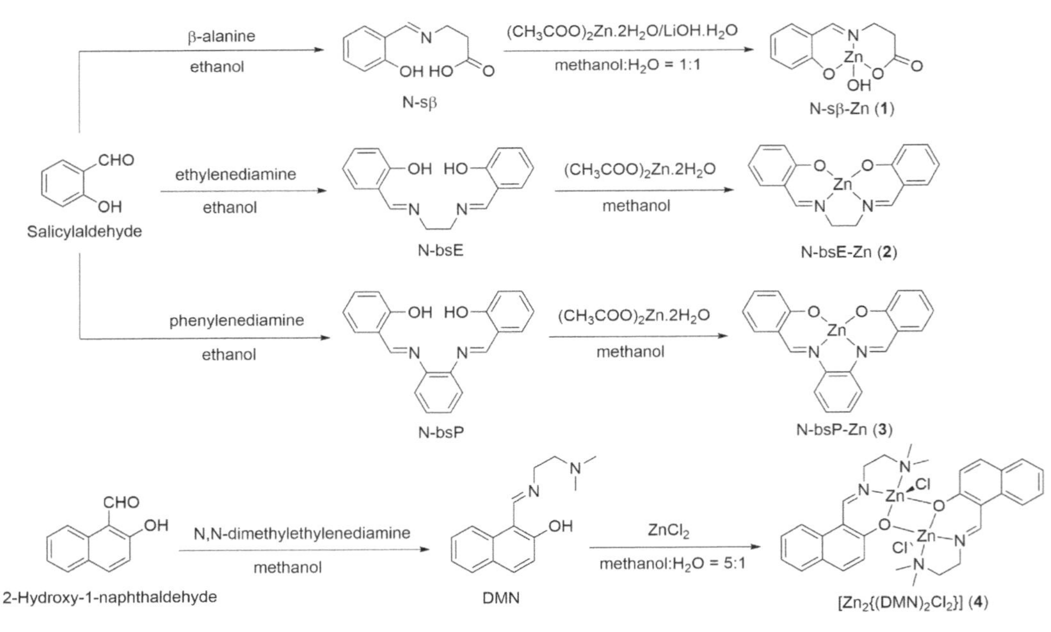

SCHEME 2.1 Synthesis of Zn (II) Schiff base complexes (**1–4**).

TABLE 2.3
***In vitro* α-Glucosidase Inhibitory Activity of Complexes (1–4) and Their Corresponding Ligands**

Compound	*In vitro* α-Glucosidase Inhibitory Activity	
	Against Yeast α-Glucosidase (IC_{50})	Against Rat Intestinal α-Glucosidase (IC_{20})
[N-sβ-Zn]; **1**	2.89 ± 0.91 μM	126 ± 39.1 μM
[N-bsE-Zn]; **2**	3.10 ± 0.74 μM	189 ± 57.3 μM
[N-bsP-Zn]; **3**	16.1 ± 2.11 μM	NA
$[Zn_2\{(DMN)_2Cl_2\}]$; **4**	4.06 ± 0.45 μM	86.0 ± 26.4 μM
[N-sβ]	NA	NA
[N-bsE]	NA	NA
[N-bsP]	NA	NA
DMN	NA	NA

NA = Not Active

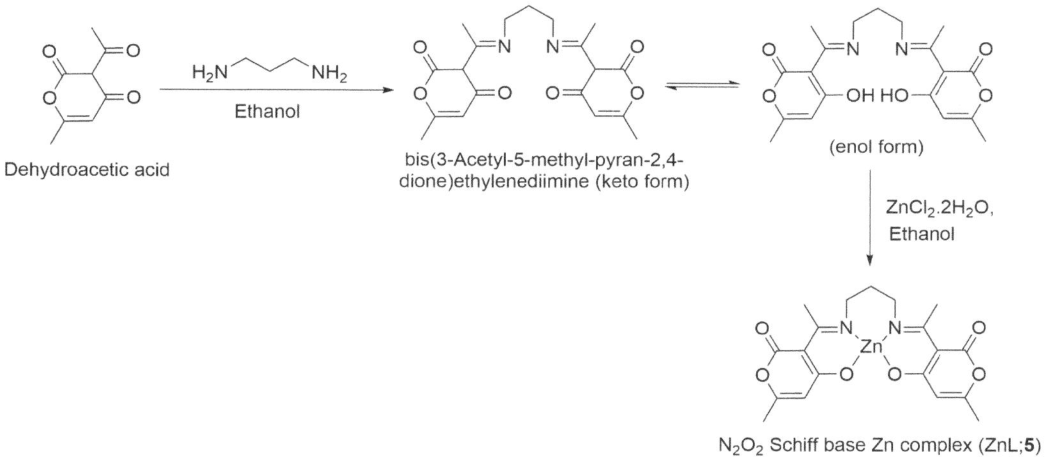

SCHEME 2.2 Synthesis of N_2O_2 Schiff base Zn complex (ZnL;**5**).

The synthesis of the ZnL (**5**) was achieved by a two-step procedure using Schiff base ligand, bis(3-acetyl-5-methyl-pyran-2,4-dione)ethylenediimine (Scheme 2.2). In their investigation, the synthesized ZnL (**5**) showed IC_{50} values of 0.18 and 0.23 mg against α-amylase and α-glucosidase enzymes respectively. The research group further investigated the mechanism of inhibition of ZnL (**5**) by carrying out enzyme kinetic studies using Dixon and Lineweaver-Burk (LB) plots. The results indicated the complex to be an efficient, mixed, non-competitive and non-competitive inhibitor of α-amylase (having inhibitory constant (K_i) value = 77.8 μg and Michaelis constant (K_m) value = 0.35) and α-glucosidase (having inhibitory constant (K_i) value = 31.6 μg and Michaelis constant (K_m) value = 1.4) respectively. These *in vitro* analysis provided a good scientific background for the use of Zn based complexes as potential antidiabetic agents.

In year 2018, Koothapan and co-workers evaluated the antidiabetic properties of a novel zinc-metformin-3-hydroxyflavone complex against T2DM.[54] The synthesis was based on their previous findings, in which they tested several oraganozinc complexes[55–57] along with the

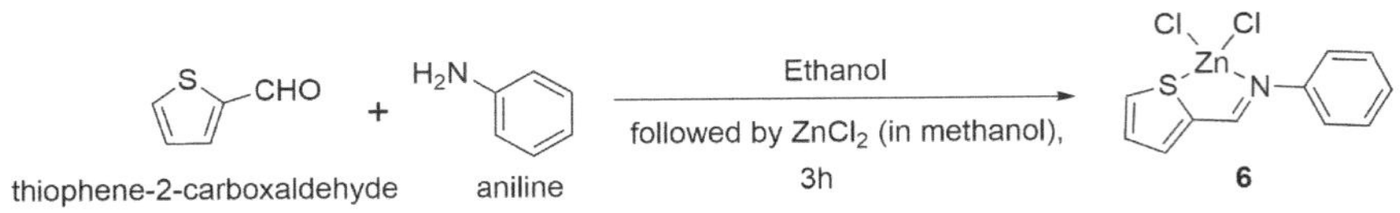

SCHEME 2.3 Synthesis of water soluble Zn (II) complex containing *(E)*-N-(thiophen-2-ylmethylene)aniline pharmacophore (**6**).

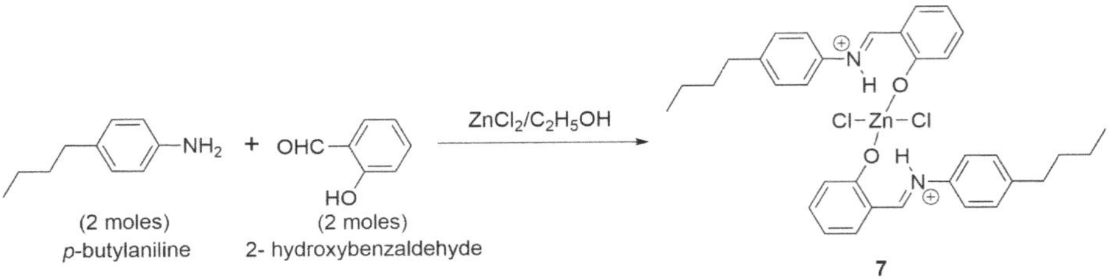

SCHEME 2.4 Synthesis of mono(bis(2-(4-butylphenyl)imino)methyl)phenoxy)zinc(II) dichloride complex (7).

metformin-3-hydroxyflavone complex,[58,59] and found them to have enhanced antidiabetic efficacy. In the present study, the zinc-metformin-3-hydroxyflavone complex termed as zinc-mixed ligand complex was tested for its hypoglycemic efficacy through a series of assays including glucose tolerance test, insulin resistance assessment and insulin sensitivity check. The data from these experiments revealed the complex to have considerable antidiabetic activities at a lower concentration than metformin-3-hydroxyflavone complex alone. The findings clearly illustrated the synergistic effect of zinc in increasing antidiabetic activity.

A series of water soluble Zn (II) complexes containing *(E)*-N-(thiophen-2-ylmethylene)anilines pharmacophore and having either tetrahedral or octahedral geometry were also investigated for their antidiabetic activity.[60,61] The most active complex (**6**) among the screened complexes showed an α-glucosidase inhibitory activity of IC_{50} = 24.45 μg/ml (compared to IC_{50} = 15.55 μg/ml exhibited by the reference drug acarbose). The synthesis of the complex (**6**) took place by a one pot reaction between thiophene-2-carboxaldehyde, aniline and $ZnCl_2$ as shown in Scheme 2.3.

In a separate research work, Sathishkumar et al. investigated *in vitro* α-glucosidase inhibitory activity of mononuclear, mono(bis(2-(4-butylphenyl)imino)methyl)phenoxy) zinc(II) dichloride complex (**7**) using the kinetic end point assay methodology.[62] The desired complex (**7**) was prepared by reacting *p*-butylaniline, 2- hydroxybenzaldehyde and $ZnCl_2$ in 2:2:1 ratio using ethanol as the solvent (Scheme 2.4). In comparison to the reference drug acarbose (having IC_{50} = 0.54 μM), the synthesized complex (**7**) showed an IC_{50} value of 3.04 μM. The researchers also performed docking experiments of the complex (**7**) with the targeted human intestinal α-glucosidase enzyme to further understand the mechanism of interaction between the zinc complex and the enzyme. The zinc complex (**7**) interacted well with amino acid residues of the α-glucosidase enzyme via hydrogen and ionic bonding and was found to be well located in the binding site of the enzyme.

Later in the year 2020, an interesting piece of work was carried by Manimohan et al where the researchers designed a water soluble biopolymer Schiff base ligand (3b-OCMCS) and then prepared Co(II), Ni(II) and Zn(II) complexes with it.[63] The *in-situ* synthesis of the ligand (3b-OCMCS) took place by first making a Schiff base by the reaction of 4-methoxy benzohydrazide and dibenzoyl methane. The Schiff base precursor was then added to a uniform mixture of deacetylated chitosan, acetic acid and methanol followed by the treatment with a methanolic solution of mono-chloroacetic

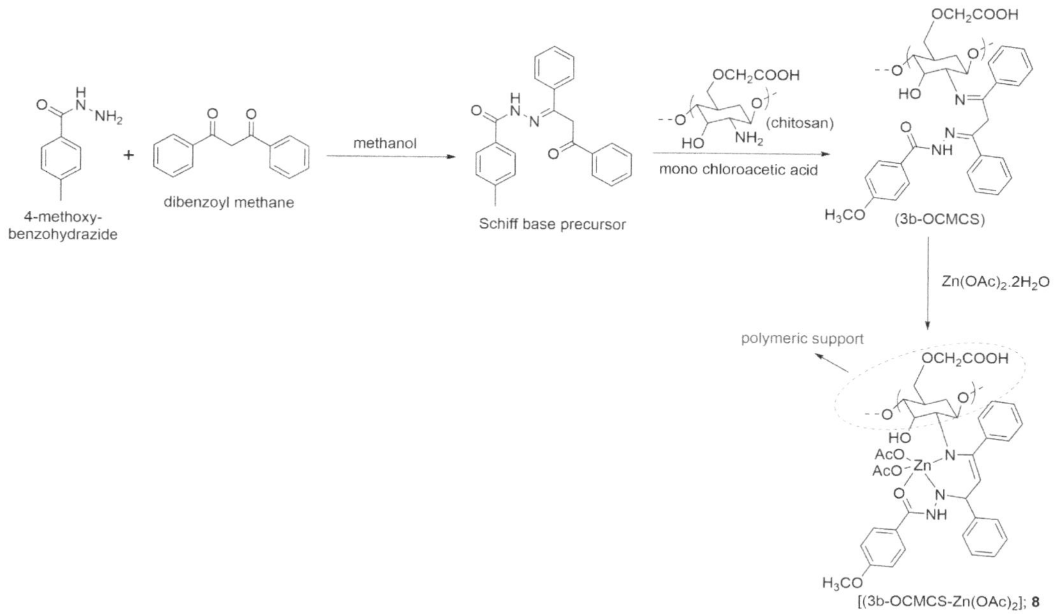

SCHEME 2.5 Synthesis of biopolymer Schiff base incorporated Zn complex (**8**).

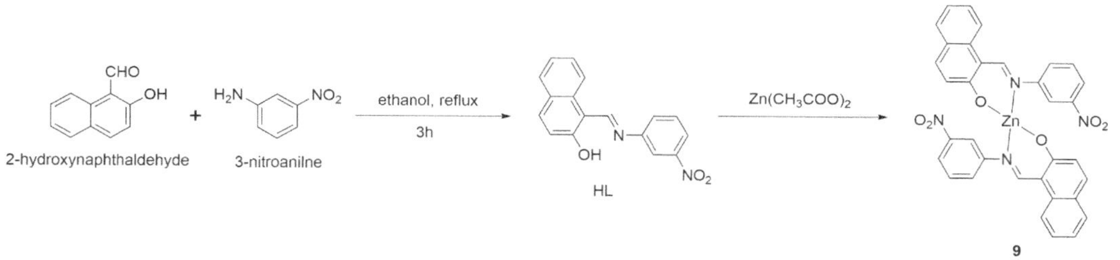

SCHEME 2.6 Synthesis of Zn (II) complex (**9**).

acid. The prepared methoxybenzhydrazide grafted biopolymer Schiff base ligand was then treated with metal acetate salt to get the respective metal complexes (Scheme 2.5). The α-amylase inhibitory activity in the antidiabetic assay revealed the prepared Zn complex [3b-OCMS-Zn(OAc)$_2$;**8**] to possess better or comparable antidiabetic characteristics (with respect to standard drug acarbose), at all concentrations, with a maximum inhibition of 86.99%. This is also found to be better in comparison to its corresponding ligand (3b-OCMCS) (inhibition of 82.46%). The synthesized complex (**8**) also exhibited promising antibacterial and antifungal activity in the same work.

In vivo biological screening of a novel Schiff base Zn complex (**9**) was done by Rauf et al. to evaluate its antidiabetic efficacy.[64] The synthesis of the complex (**9**) took place by first reacting 2-hydroxynaphthaldehyde and 3-nitroanilne to get the required Schiff base ligand 1-((3-nitrophenylimino)methyl)naphthalen-2-olate (HL) (Scheme 2.6). The ligand was then subjected to complexation with Zn (II) acetate salt to get the required zinc complex (**9**). The zinc complex (**9**) didn't exhibit any enhanced antidiabetic activity in their experiments when compared to its corresponding ligand (HL), showing that there are exceptions to the general trend observed in most of the cases where complexation with the metal ions generally enhanced antidiabetic efficacy.

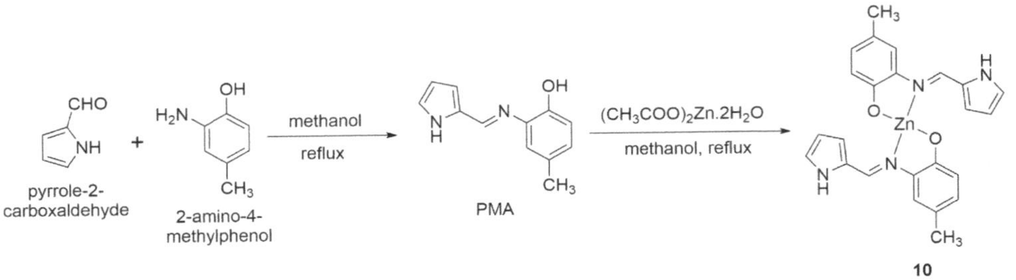

SCHEME 2.7 Synthesis of Zn complex (**10**) derived from a heterocyclic Schiff base (PMA).

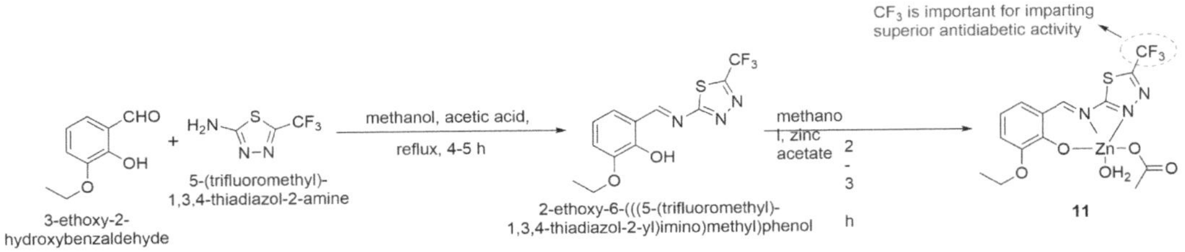

SCHEME 2.8 Synthesis of zinc (II) complex of thiadiazole based Schiff base ligand (**11**).

α-Amylase inhibitory activity of a novel zinc (II) complex (**10**) along with other metal [Ni(II) and Cu(II)] complexes derived from a heterocyclic Schiff base was reported by Shanty and co-workers using the DNS method.[65] The zinc complex (**10**) synthesized through a two-step reaction (Scheme 2.7) using pyrrole-2-carboxaldehyde and 2-amino-4-methylphenol as the starting material was found to enhance α-amylase inhibitory activity.

In 2022, Deswal et al. synthesized a new series of coordination complexes of Co(II), Ni(II), Cu(II) and Zn(II) transition metals containing thiadiazole based Schiff base ligands.[66] The complexes were further tested for their potential antidiabetic activity α-amylase and α-glucosidase enzymes. It was found that the Zn complex (**11**) prepared by the reaction between 2-ethoxy-6-(((5-(trifluoromethyl)-1,3,4-thiadiazol-2-yl)imino)methyl) phenol ligand and zinc acetate in methanol demonstrated excellent inhibitory activity with an IC_{50} value of 1.33 μmol/mL and 0.60 μmol/mL for α- amylase and α-glucosidase respectively (in comparison, reference drug acarbose displayed IC_{50} value of 1.28 μmol/mL and 0.58 μmol/mL for α- amylase and α-glucosidase respectively) (Scheme 2.8). The inhibitory activity of the metal complex (**11**) was also found to be almost two-fold higher than its parent ligand, which further supported the idea of getting improved efficacy by the synergistic combination of metal and ligand. SAR (structure activity relationship) studies revealed that the electron withdrawing CF_3 group of the thiadiazole ring is crucial for imparting superior antidiabetic activity in the complex. The complex (**11**) was further subjected to *in-silico* experiments to examine its drug likeness and the mode of inhibition pathway. The study of ADME properties revealed it to have excellent drug-like properties (the bioavailability value is 0.55), while the molecular docking studies showed it to have good binding affinity (chiefly through hydrogen bond and metallic bond interactions) in the active site of the enzyme. Overall, the computational data was found to be in good agreement with the experimental results.

Some metal complexes attached with a new Schiff based ligand 2-(((3Z)-1,5-dimethyl-2-phenyl-4-((thiophen-2-ylmethylene)amino)-1,2-dihydro-3H-pyrazol-3-ylidene)amino)pyri dine-3-ol have recently been reported by Sudha et al. as potential antidiabetic agent.[67] In their study, the Zn complex (**12**) was prepared by a two-step procedure. The first step involved the synthesis of the ligand

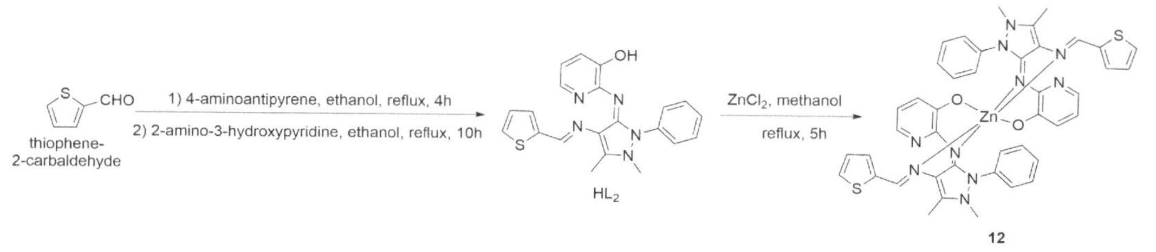

SCHEME 2.9 Synthesis of zinc (II) complex (**12**).

(HL_2) by reacting thiophene-2-carbaldehyde and 4-aminoantipyrine together to get the Schiff base intermediate, followed by the addition of 2-amino-3-hydroxypyridine (Scheme 2.9). The prepared ligand (HL_2) was then treated with $ZnCl_2$ in hot methanolic solution to obtain the required Zn complex (**12**). The Zn complex (**12**) showed good inhibitory activity against pancreatic α-amylase when tested *in vitro*. The activity was found to be similar to that of the reference drug, acarbose. The Zn complex (**12**) also interacted well with the targeted human pancreatic α-amylase in molecular modeling studies with a good docking score.

2.4 ANTIDIABETIC ACTIVITY OF SOME RECENTLY SYNTHESIZED VANADIUM-BASED SCHIFF BASE COMPLEXES

Several studies in the past indicated increased inflammation, particularly in obese persons as the cause of occurrence of Type 2 diabetes mellitus (T2DM). The inflammatory mediators cause the cells to become non-respondent to insulin signaling which in-turn creates an insulin resistance (IR) state. In the presence of this IR state, the cellular glucose uptake can't be enabled as the glucose transporter 4 (GLUT4) translocation from the cytoplasm to the plasma membrane does not take place. Combination therapeutic approaches in the past have been explored by scientists to overcome this reduced insulin sensitivity along with inflammation. In combination therapy, usually two different medicines were given to the patient to overcome the disease; however they carry a potential risk of undesirable side effects owing to the occurrence of incompatible pharmacokinetics and increased toxicity due to higher dosages. In order to overcome these shortcomings of combination therapy, Ki et al. thought to synthesize a hybrid type of compounds possessing pharmacophores having both insulin-mimetic and anti-inflammatory properties in the same molecule.[68] Through a literature survey, the group noticed that insulin-mimitic activity was possessed by vanadium oxo complexes while tryptamine imparted an additional anti-inflammatory effect. Keeping this in mind, Ki et al. prepared bifunctional activity imparting vanadyl-Schiff base complex, bis(pyridoxylidinetryptamine) vanadium (IV) (VOTP; **13**), having tryptamine pharmacophore, and further carried out significant research to analyse their potential as an antidiabetic drug. The one pot synthesis of the complex (**13**) was done using the Schiff base ligand (TPL) as shown in Scheme 2.10. The complex (**13**) was then treated with inflammation induced IR state in HEK-293 cell-based assay using total internal reflection fluorescence (TIRF) microscopy technique to check their insulin signaling restoration and GLUT4 translocation ability. The *in vitro* biological activity result showed VOTP (**13**) to have significant activity even at a nano-molar dosages range and they can be applied to keep normal glucose homeostatis.

Jia et al. in 2017 synthesized a new dioxidovanadium (V) complex (**14**) by the complexation reaction between oxovanadium sulphate and Schiff base N'-(pyridin-2-ylmethylene)picolinohydrazide (HPPH) (Scheme 2.11).[69] The complex (**14**) was further investigated as a selective inhibitor of a signaling enzyme known as protein tyrosine phosphatase 1B (PTP1B). PTP1B is involved in the negative regulation of signaling pathways mediated through leptin and insulin receptors and is regarded

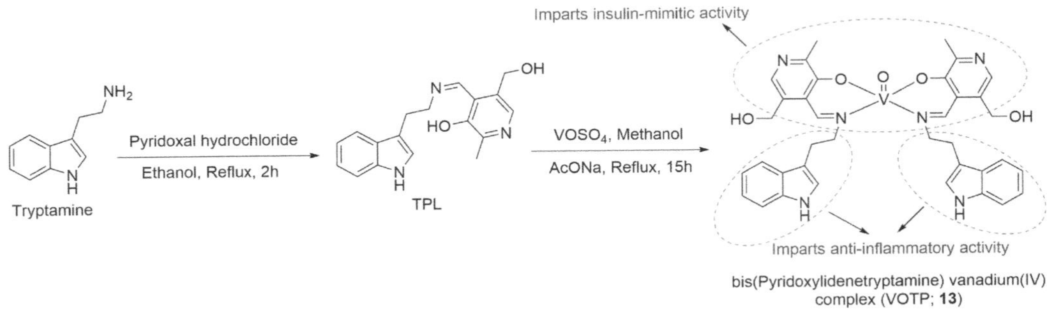

SCHEME 2.10 Synthesis of vanadyl-Schiff base complex (VOTP; **13**).

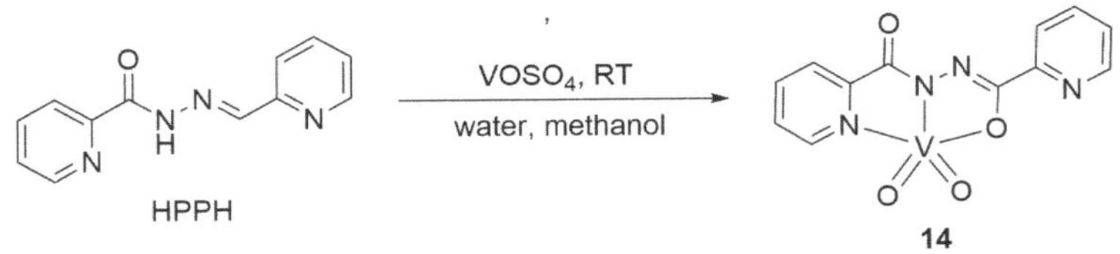

14

SCHEME 2.11 Synthesis of dioxidovanadium (V) complex (**14**).

TABLE 2.4
The Effect of Complex 15, Complex 16 and Metformin on Survival % and Blood Glucose Levels in the *In Vivo* Experiment (Results Are Indicated as Mean ± SEM)

Groups	Vehicle	Diabetic	Meformin (20 mg/kg)	Meformin (40 mg/kg)	Complex 15 (20 mg/kg)	Complex 15 (40 mg/kg)	Complex16 (20 mg/kg)	Complex16 (40 mg/kg)
Survival %	100	83.33	100	100	83.33	83.33	100	100
Blood glucose/ mg dL^{-1}	79.99 ± 6.66	401.67 ± 35.87	212.3 ± 34.95	188.5 ± 1.8	163.4 ± 17.9	144.25 ± 47.67	165.5 ± 13.64	101.2 ± 8.08

as one of the well-validated targets for the treatment of Type 2 diabetes and obesity. The outcome of the experiments carried out by the research group was extremely promising. The synthesized vanadium complex selectively inhibited PTP1B with good potency (IC_{50} = 0.13 µm) *in vitro*. The complex (**14**) was also discovered to be a good insulin-mimetic agent as it effectively increased the phosphorylation of the PTP1B substrates. The complex (**14**) showed decreased cytotoxicity in the MTT assays as well. It is envisioned that additional modifications in the organic moieties of the complex (**3**) may lead to even more effective and safe antidiabetic agent.

A series of oxovanadium (IV) complexes prepared by using metformin Schiff bases were explored to have insulin enhancing characteristics.[70] Different metformin Schiff bases in the study were prepared by combining metformin with various aromatic aldehydes through a template reaction. In the *in vivo* insulin enhancing studies, two of the complexes (**15** and **16**) synthesized by using Schiff bases HL^1 and H_2L^4, respectively (Scheme 2.12), exhibited considerable blood glucose lowering effect through extrapancreatic mechanisms (Table 2.4). The activity was found to be superior in comparison to when only metformin is administered to mice. The complexes were also indicated to be safe as the treated mice showed a survival rate of 83.3–100%.

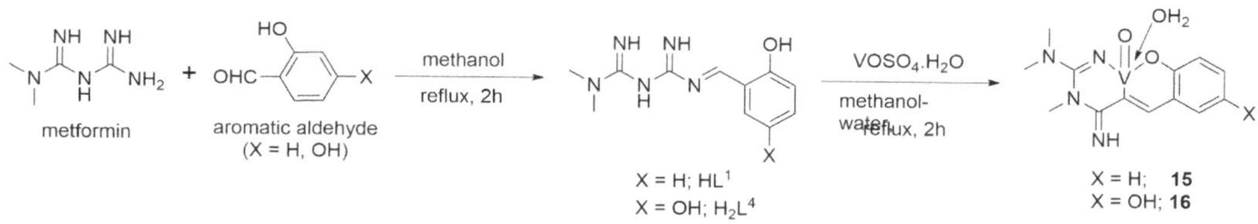

SCHEME 2.12 Synthesis of oxovanadium (IV) complex using metformin Schiff base (**15** and **16**).

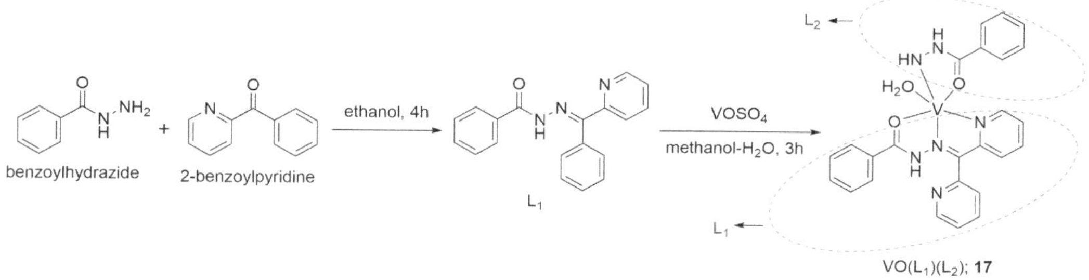

SCHEME 2.13 Synthesis of oxovanadium (IV) complex having mixed ligands (**17**).

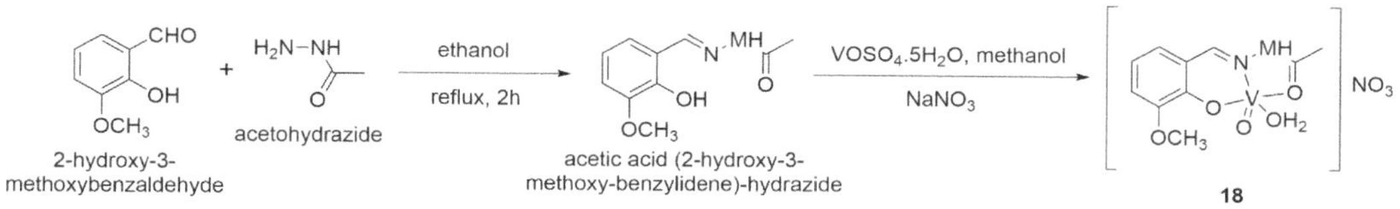

SCHEME 2.14 Synthesis of NNO donor Schiff base vanadium (IV) complex (**18**).

Oxovanadium (IV) complex, having mixed ligands of type [VO(L_1)(L_2)] (**17**), was later synthesized by Patel et al.[71] The synthesis of this complex was found to be quite interesting. First, the tridentate Schiff base ligand L_1 (N'-[(Z)-phenyl(pyridin-2-yl)methylidene]benzohydrazide) is prepared by refluxing benzoylhydrazide and 2-benzoylpyridine in ethanol for 4h. The ligand L_1 is then treated with $VOSO_4$ (in 2:1 ratio) using aquamethanol in aerobic conditions (Scheme 2.13). This leads to the formation of a new type of mixed ligand containing oxovanadium (IV) complex (**17**), where it is seen that metal assisted hydrolysis of ligand L_1 *in situ* leads to the formation of a new bidentate ligand L_2 (benzohydrazide). The complex, when evaluated for its *in vitro* α-glucosidase inhibitory activity, displayed moderate % inhibition of 14.75 μM (in comparison, the standard drug acarbose showed % inhibition of 18.59 μM).

The antidiabetic efficacy of vanadium complexes is further exploited by Patel et al.[72] The research group synthesized few NNO donor Schiff base vanadium (IV/V) complexes and checked their α-amylase and α-glucosidase inhibition activity. The complex (**18**), prepared by the reaction of vanadyl sulphate monohydrate with the Schiff base ligand, acetic acid (2-hydroxy-3-methoxy-benzylidene)-hydrazide (Scheme 2.14), exhibited maximum potency *in vitro* with an inhibitory activity of IC_{50} = 4.16 μg/ml against α-glucosidase and IC_{50} = 122.28 μg/ml against α-amylase. The activities were found to be significantly higher than its corresponding ligand. The complex (**18**) also displayed concentration dependent activities in the assays. In the molecular docking studies against α-glucosidase enzyme, the complex (**18**) formed five interactions involving one pi-alkyl interaction

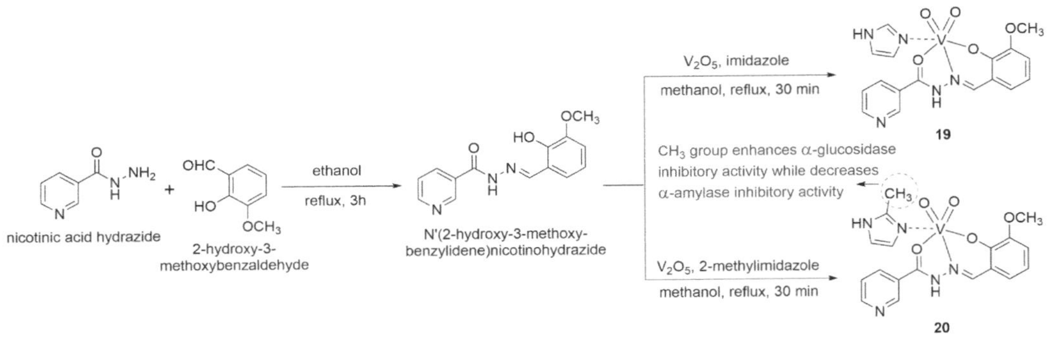

SCHEME 2.15 Synthesis of new dioxidovanadium (V) complexes (**19** and **20**).

and four hydrogen bonds with the different amino acid residue of the enzyme. A favorable binding energy of -4.950 Kcal/mol was observed with the protein receptor of the enzyme.

The same research group later synthesized new dioxidovanadium (V) complexes (**19** and **20**) coordinated with nicotinic acid and imidazole ligands.[73] The tridentate ONO Schiff base ligand, N'(2-hydroxy-3-methoxy-benzylidene)nicotinohydrazidewas synthesized by the reaction of starting materials nicotinic acid hydrazide and 2-hydroxy-3-methoxybenzaldehyde. The ligand was further treated with V_2O_5 and imidazole/2-methylimidazole to get the final complex (**19** and **20**) (Scheme 2.15). In the α-glucosidase inhibition activity assay, complexes **19** and **20** showed an inhibition with IC_{50} value 153.03 and 32.54 μg/ml respectively. The activity of complex **20** was found to be comparable to the control acarbose used in the α-glucosidase inhibition assay. On the other hand, in the α-amylase inhibition activity, complex **19** was found to be more active with an IC_{50} value of 23.66 μg/ml. This shows that even a slight change of H to CH_3 group in the complex can bring a marked effect on activity. The activity of complex **20** was found to be comparable to the control acarbose used in the assay. The findings of this study were proved to be consistent with the earlier reported vanadium complexes.

Patel et al., in progressing their work, carried out the synthesis of a series of anionic dioxidovanadium (V) complexes integrated with a hydrazone-based Schiff base ligand.[74] The synthesis of the ligand took place by combining isonicotinic acid hydrazide and o-vanillin in ethanolic solution. The most active complexes (**21** and **22**) were synthesized by reacting the isolated ligand with V_2O_5 in the presence of 2-ethyl-imidazole and 2-methyl-benzimidazole respectively (Scheme 2.16). Complex **21** showed maximum inhibitiory activity against α-amylase enzyme with an IC_{50} value 82.99 μg/ml while complex **22** displayed maximum inihibitory activity against α-glucosidase enzyme with an IC_{50} value 53.88 μg/ml. It was observed that both the complexes (**21** and **22**) exhibited increased efficacy on increasing their concentrations.

Abd-Elaziz et al., in a quest to find out anti-hyperglycemic agents against Type 2 diabetes mellitus, synthesized few vanadium (III/IV) based metalopharmaceutical complexes.[75] The most active complexes (**23** and **24**) of the series were synthesized in a two-step procedure as shown in Scheme 2.17. Complex **23** was prepared by reacting a mixture of dimethyl tartrate with hydrazine hydrate, followed by the addition of salicylaldehyde to get the desired ligand (L^1). The ligand (L^1) was then subjected to reaction with vanadium pentoxide in ethanolic solution to attain complex **23**. For complex **24**, the preparation of ligand (L^2) took place by reacting dimethyl tartrate with o-phenylenediamine together for some time, followed by the addition of 3-(hydroxyimino)pentane-2,4-dione. In the *in vivo* studies, it was seen that the blood glucose level significantly dropped when the streptozotocin-induced diabetic rats were treated with oral doses (30 and 100 μM) of complex (**23** and **24**).

An extensive study was done by Szklarzewicz et al. in exploring the antidiabetic activity of Schiff base vanadium complex. In 2020, the research group synthesized a series of 17 vanadium (III-V)

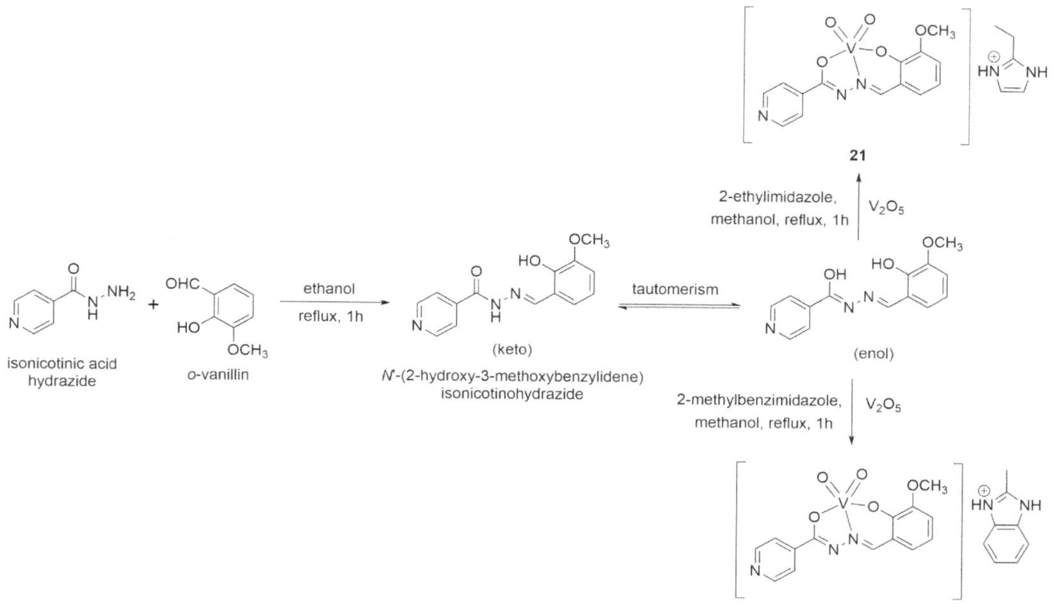

SCHEME 2.16 Synthesis of anionic dioxidovanadium (V) complexes (**21** and **22**).

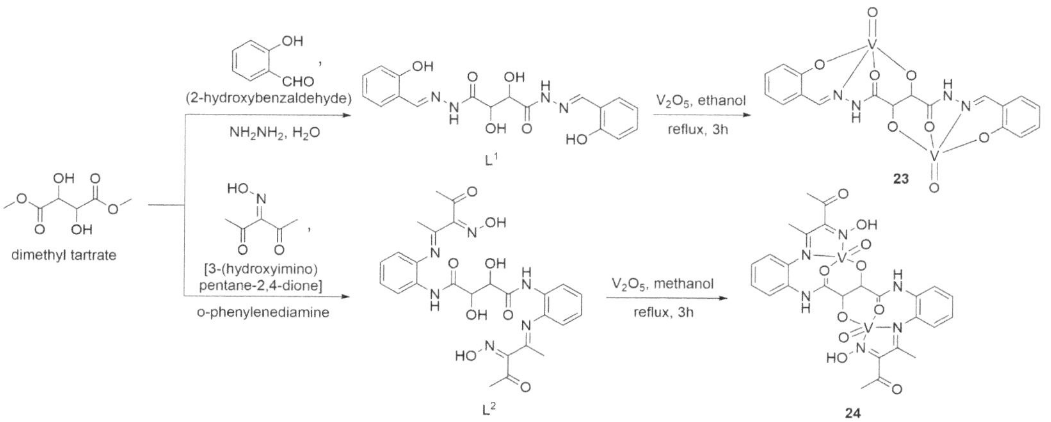

SCHEME 2.17 Synthesis of vanadium (III/IV) based metalopharmaceutical complexes (**23** and **24**).

complex from hydrazido-hydrazone Schiff bases and determined their biological activity for the inhibition of human phosphatases.[76] The synthesis of two of the active complexes (**25** and **26**) took place in anaerobic conditions using argon and is shown in Scheme 2.18. Complex **25** was synthesized by first refluxing a mixture of 5-bromosalicylaldehyde and 4-nitrobenzhydrazide together in ethanol for about 15 minutes, followed by the addition of [V(acac)$_3$] and further refluxing for 30 minutes. Complex **26** was prepared by a similar procedure, first refluxing 5-bromosalicylaldehyde and 4-chlorobenzhydrazide in ethanol for 15 minutes. [VO(acac)$_2$] was added to the reaction mixture, followed by addition of 1,10-phenanthroline therein. The final complex (**26**) was obtained after refluxing the reaction mixture for an additional 20 minutes. It has been found that, in the case of all

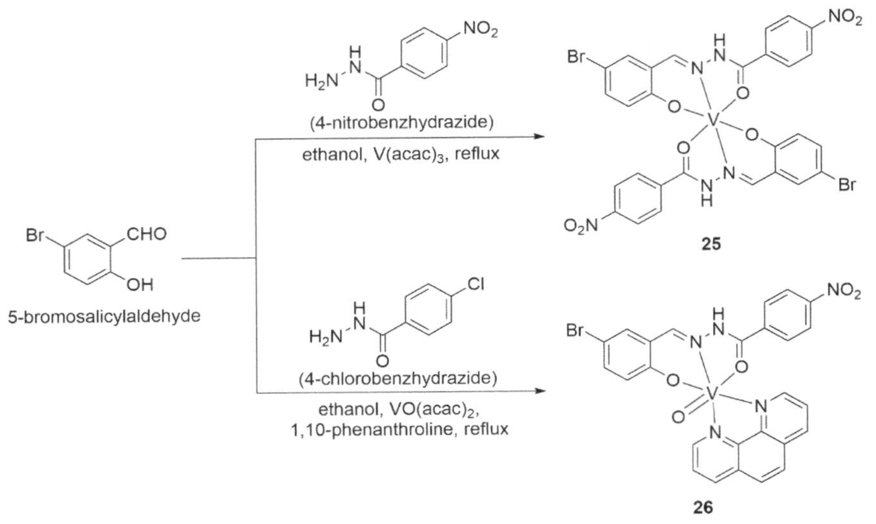

SCHEME 2.18 Synthesis of Schiff base vanadium complexes (**25** and **26**).

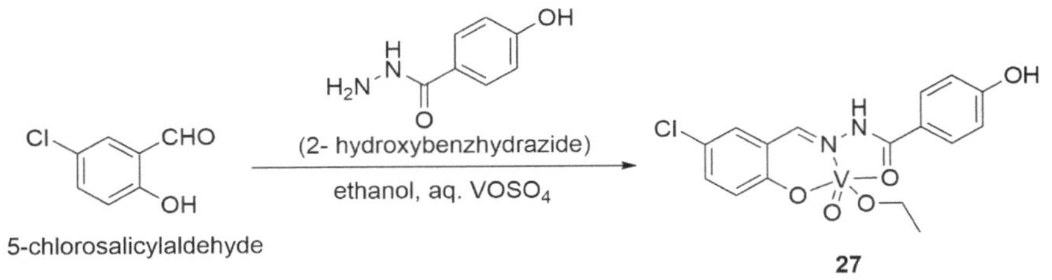

SCHEME 2.19 Synthesis of a new vanadium complex having ONO Schiff base ligand (**27**).

tyrosine phosphatases tested, complexes **25** and **26** had much higher IC_{50} values than IC_{50} for reference, $VOSO_4$ and bis(maltolato)oxovanadium (IV) (BMOV).

Further, the research group synthesized a series of new vanadium complexes having tridentate ONO Schiff base ligands and checked their human tyrosine phosphatases inhibitory activity.[77] The most active complex (**27**) of the series was synthesized by reacting 5-chlorosalicylaldehyde and 4-hydroxybenzhydrazide anaerobically in ethanol under reflux conditions to get the corresponding Schiff base ligand (Scheme 2.19). Aqueous $VOSO_4$ was then added to the reaction mixture followed by refluxing for 1 h to yield the required complex (**27**). The non-cytotoxic complex **27** displayed excellent tyrosine phosphatases inhibitory activity with IC_{50} values lower than reference BMOV.

The insulin-mimetic behavior of several vanadium (IV and V) complexes with the similar kind of ONO Schiff base ligand as described earlier (Schemes 2.17 and 2.18) were further studied by the same research group.[78] During the synthesis of these complexes, 1,10-phenanthroline was used as a co-ligand. In the biological assay, the complexes demonstrated a wide range of tyrosine inhibition activity, with some showing similar or superior activity compared to $VOSO_4$ and reference compound BMOV. The researchers discovered that inhibiting tyrosine phosphatases is not a crucial factor in determining the biological activity of the complexes they studied. It was also visualized that the activity of the complexes might be connected to their stability in the investigated biological system. Szklarzewicz et al. later studied several oxidovanadium (IV) and (V) complexes having tridentate salicylhydrazone Schiff base ligand.[79] This time they took different types of aldehydes

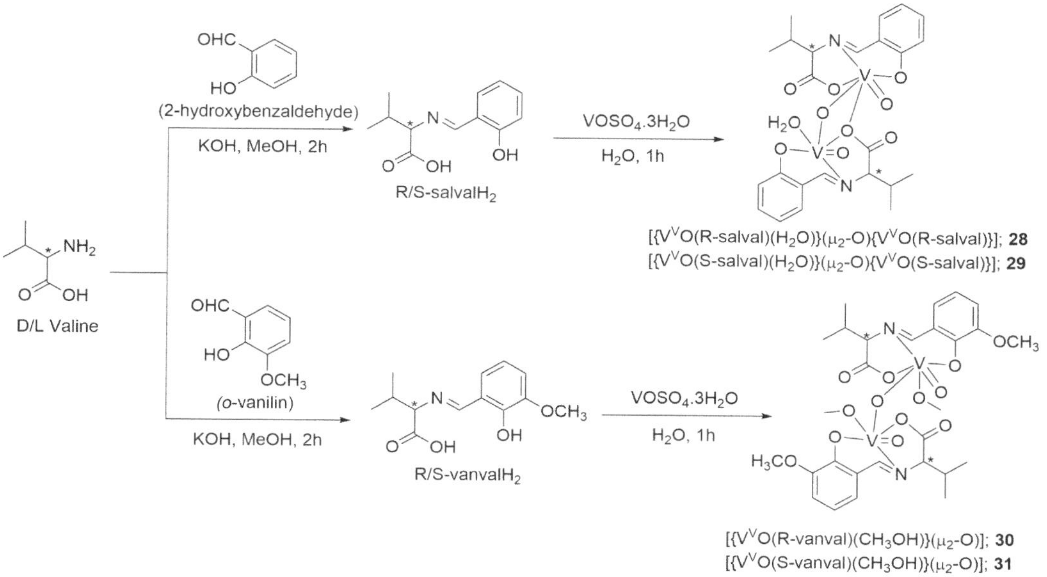

SCHEME 2.20 Synthesis of oxidovanadium (V) complexes coordinated by L/D-valine Schiff bases (**28–31**).

along with 1,10-phenanthroline (phen)/2,2'-bipyridine (bpy) as co-ligands for the synthesis of the complexes. Two of the active complexes $[VO(L_1)(bpy)]\cdot 0.5EtOH\cdot 0.5H_2O$ (L_1 = N'-(2-hydroxy-5-nitrobenzylidene)benzohydrazide) having flexible bpy ligand showed superior activity over BMOV, while $[VO(L_1)(phen)]\cdot 1.5H_2O$ (L_1 = N'-(2-hydroxy-5-nitrobenzylidene)benzo hydrazide) having relatively rigid ligand, phen displayed activity similar to BMOV. The results clearly supported the fact that changing the ligand type has a noteworthy impact on biological activity.

The *in vitro* insulin-mimetic activity of a family of binuclear oxidovanadium (V) complexes (**28–31**) coordinated by L/D-valine Schiff bases were reported by Turtoiet al.[80] The synthetic route for the preparation of these complexes (**28–31**) is given in Scheme 2.20. The synthesized complex (**28–31**) reduced α-amylase and total intracellular protein tyrosine phosphatase activity in the biological assay. In the human hepatoma (HepG2) cell line, the complexes improved phosphorylation of the insulin receptor. The complexes were also found to possess either no cytotoxicity or very low cytotoxicity when checked against HepG2 cells.

2.5 ANTIDIABETIC ACTIVITY OF SOME RECENTLY SYNTHESIZED SCHIFF BASE COMPLEXES BASED ON COPPER/NICKEL/COBALT/CHROMIUM/RUTHENIUM

Although zinc and vanadium-based Schiff base complexes are exploited more frequently by the researchers in order to find out improved antidiabetic agents, substantial research is also carried out to evaluate antidiabetic activity for other metal (such as copper, nickel, cobalt, chromium and ruthenium) based Schiff base complexes. Significant antidiabetic activities have been reported for the copper based Schiff base complex $[Cu_2(L)_2(H_2O)].H_2O$ {(L) = N-salicylidene-β-alanine(2-)} (**32**).[81] The synthesis of Cu complex **32** took place by adopting a one pot synthetic procedure. An aqueous solution of copper (II) acetate is first added to the aqueous solution of β-alanine. This was followed by the addition of a solution of salicylaldehyde in ethanol and further heating the reaction mixture at about 60°C for 1 h to achieve the desired complex **32** (Scheme 2.21). *In vivo* studies on pancreatic β-cells were performed to determine the antidiabetic activity of the complex **32**. The oxidative stress

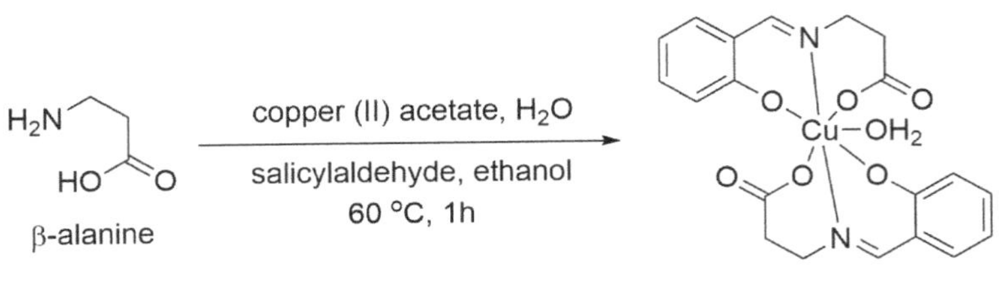

SCHEME 2.21 Synthesis of Cu based Schiff base complex (**32**).

model of alloxan-induced diabetes mellitus was followed, where the complex **32** showed positive results depicting its nature as a good preventing agent against the stress induced complications of the disease.

Schiff base copper (II) complexes, [CuL^1(tmen)] and [Cu$_2$L$_2^2$(tmen)] have been synthesized by Lakshmi et al. and tested for their *in vitro* antidiabetic activity. The Schiff base ligand, N-(salicylidene)-L-valine (H$_2$L^1) for the preparation of complex [CuL^1(tmen)] was synthesized by adding an aqueous solution of L-valine and KOH to a solution of salicylaldehyde in ethanol and further heating the resulting mixture at 60°C for 1 h. In a similar manner, synthesis of Schiff base ligand N-(3,5-dichlorosalicylidene)-L-valine (H$_2$L^2) for the preparation of complex [Cu$_2$L$_2^2$(tmen)] took place where instead of salicylaldehyde, 3,5-dichlorosalicylaldehyde was used as the starting material. The synthesized ligands (H$_2$L^1 and H$_2$L^2) were then made to react with hydrated copper (II) acetate and N,N,N′,N′-tetramethylethylene-1,2-diamine (tmen) for the synthesis of respective Schiff base metal complexes [CuL^1(tmen)] and [Cu$_2$L$_2^2$(tmen)]. The IC$_{50}$ values for complex [CuL^1(tmen)] and [Cu$_2$L$_2^2$(tmen)] were reported to be 941.20 μg/mL and 389.01 μg/mL for α-amylase inhibition assay and 919.02 μg/mL and 350.02 μg/mL for α-glucosidase inhibition assay, respectively. These IC$_{50}$ values were much higher than the standard drug acarbose (48.33 for α-amylase inhibition assay and 42.19 for α-glucosidase inhibition assay) and hence the use of these complexes as antidiabetic agents was stressed.[82]

Karthik et al. reported the synthesis and antidiabetic activity evaluation of a Cu (II) Schiff base complex (**33**).[83] The synthesis of the ligand, 2,6-dimethoxy-4-((quinolin-3-ylimino)methyl) phenol for the preparation of complex (**33**) took place by refluxing the methanolic solutions of 4-hydroxy-3,5-dimethoxybenzaldehyde and 3-aminoquinoline for 3 h. The desired Cu complex (**33**) was obtained by reacting methanolic solution of its divalent metal salt copper (II) acetate with the prepared ligand (Scheme 2.22). During the formation of complex, the metal to ligand ratio was kept at 1:2. In the α-amylase inhibition assay, the complex (**33**) possessed reasonable antidiabetic activity with IC$_{50}$ value 12.85 μg/mL.

The synthesis of Cu(II), Co(II) and Ni(II) complexes of a Schiff base ligand was reported by Sakthivel et al.[84] The ligand [2,2'-((1E,1'E)-((4-nitro-1,2-phenylene)bis(azanylylidene)) bis(methanylylidene))diphenol] was prepared by condensing 4-nitro-o-phenylenediamine and 2-hydroxybenzaldehyde in methanol as shown in Scheme 2.23. The ligand was then stirred with the corresponding metal chlorides in ethanol at room temperature to get the corresponding Cu (II), Co (II) and Ni (II) complexes (**34–36**). All the complexes were tested for antidiabetic activity using the α-amylase inhibition assay. The copper complex (**36**) showed better activity as compared to the other two metal complexes (**34–35**). The molecular docking results showed that there is a strong bonding interaction of the Schiff base metal complexes with the active site of the enzyme which resulted in reasonable activities.

The Cu (II) complex of the Schiff base, 2-ethoxy-6-(((5-(trifluoromethyl)-1,3,4-thiadiazol-2-yl) imino)methyl) phenol has also been synthesized and evaluated for their α-amylase and α-glucosidase inhibitory activity.[66] The Zn (II) complex of the ligand is discussed earlier in this chapter under section 3 (Scheme 8). The complex was synthesized by adding dropwise a methanolic solution

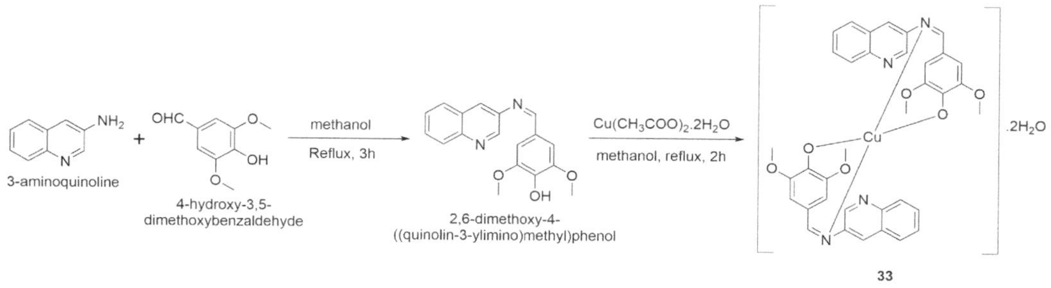

SCHEME 2.22 Synthesis of Cu based Schiff base complex (**33**).

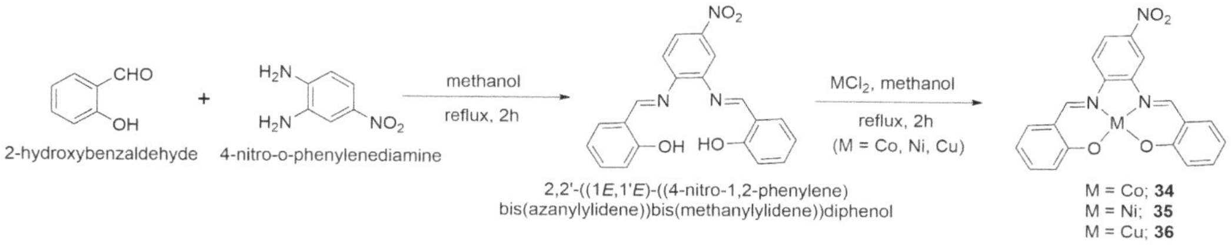

SCHEME 2.23 Synthesis of Co, Ni and Cu based Schiff base complex (**34–36**).

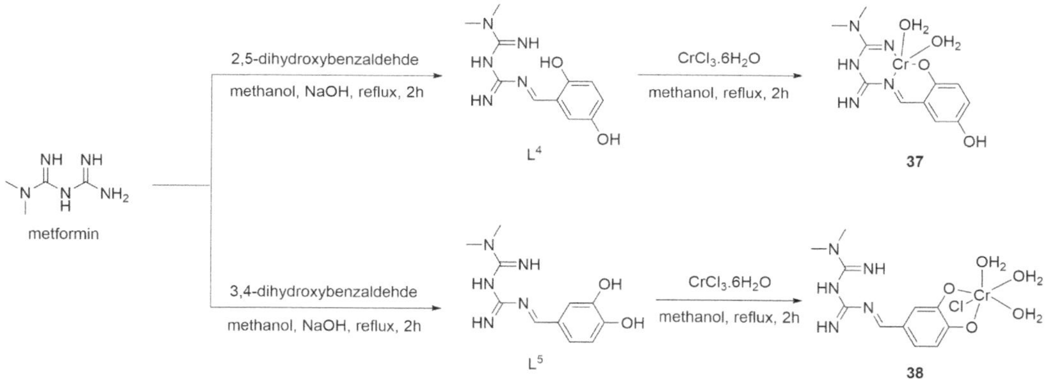

SCHEME 2.24 Synthesis of Cr based Schiff base complex (**37–38**).

of 3-ethoxy-2-hydroxybenzaldehyde to a solution of 5-(trifluoromethyl)-1,3,4-thiadiazole-2-amine, then refluxing the resulting solution for 4–5 h in acidic medium to get the corresponding Schiff base ligand. The Cu (II) complex of the Schiff base was finally obtained by stirring a 1:1 molar mixture solution of the Cu (II) acetate and the Schiff base ligand in methanol for 23 hours at room temperature. IC_{50} value of the complex was found to be 1.41 μmol/mL for α-amylase and 0.62 μmol/mL for α-glucosidase which was close to the standard drug, acarbose, used in the assay.

Cr(III) complexes (**37** and **38**) with Schiff base ligands (L^4 and L^5) made by combining metformin with 2,5-dihydroxybenzaldehyde and 3,4-dihydroxybenzaldehyde produced significant decrease in the blood glucose level when tested *in vivo*.[85] The ligand synthesis (L^4 and L^5) involved the dropwise addition of a methanolic solution of 2,5-dihydroxybenzaldehyde/3,4-dihydroxybenzaldehyde to a methanolic solution of metformin and further refluxing the resulting solution for 2 hours at a pH 10 (Scheme 2.24). The Schiff base ligands (L^4 and L^5) were further treated with $CrCl_3.6H_2O$ to get the

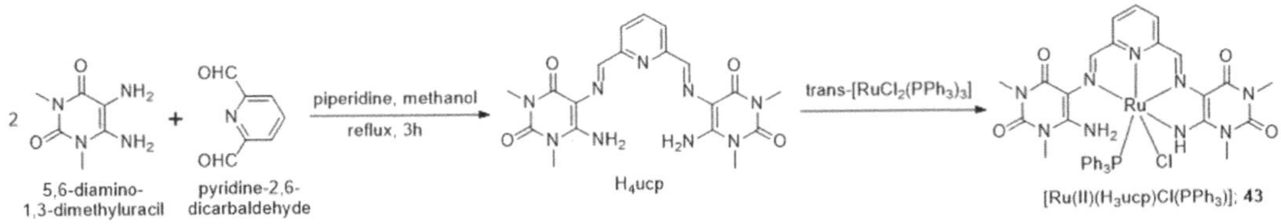

SCHEME 2.25 Synthesis of ruthenium based Schiff base complex (**43**).

TABLE 2.5
% α-Amylase Inhibition Reported for the Chromium Complexes

Compound	% α-Amylase Inhibition
Acarbose	95.95
$[CrL^1Cl_3(H_2O)]$; **39**	80.50
$[Cr(L^1)_2Cl_2]Cl$; **40**	90.89
$[CrL^2Cl_3(H_2O)]$; **41**	70.65
$[Cr(L^2)_2Cl_2]Cl$; **42**	78.83

respective Cr complexes (**37** and **38**). Both the complexes (**37** and **38**) showed the blood glucose lowering effect up to 65.74% and 66.76% (upon a dose of 40 mg/kg of the complex) in the *in vivo* studies conducted with alloxan-induced diabetes mellitus in mice.

Shukla et al. prepared chromium complexes (**39–42**) of Schiff base ligands, (Z)-N'-benzylideneisonicotinohydrazide (L^1) and (E)-N'-(4-chlorobenzylidene) isonicotinohydrazide (L^2) using chromium (III) chloride and further checked its antidiabetic properties.[86] The ligands were prepared by the reaction of isonicotinohydrazide with benzaldehyde for L^1 and with *p*-chlorobenzaldehyde for L^2 using methanol as solvent. The complexes (**39–42**) exhibited reasonable percentage α-amylase inhibition when compared with the standard drug acarbose (Table 2.5). The inhibition is attributed to the synergistic combination of the ligand to metal resulting in better absorption on the surface of the α-amylase enzyme.

Another Schiff base ligand, 2,6-bis-((6-amino-1,3- dimethyluracilimino)methylene) pyridine (H_4ucp) was prepared by refluxing 5,6-diamino-1,3-dimethyluracil and pyridine-2,6-dicarbaldehyde in methanol for 3 hours (Scheme 2.25).[87] The reaction was catalyzed by piperidine. Later, a novel ruthenium (II) complex, [Ru(II)(H_3ucp)Cl(PPh_3)] (**43**), was synthesized by the reaction of trans-[$RuCl_2(PPh_3)_3$] with the ligand (H_4ucp) and further evaluated for its antidiabetic efficacy.[88] The complex (**43**) was found to delay the onset of Type 2 diabetes in prediabetic rats. It restored insulin sensitivity thus regulating the blood glucose levels. The complex when administered to the animals also resulted in decreased HbA1c concentration. The compound was also shown to decrease the risk of cardiovascular diseases[89] and hepatic complications[90] in a prediabetic condition.

2.5 CONCLUDING REMARKS

In summary, the authors in this chapter presented an overview about diabetes mellitus, Schiff bases and the development of Schiff base metal complexes as potential antidiabetic agents. Transition metal complexes bring an opportunity to conjugate different metal ions to pharmaceutically relevant organic ligands, in order to increase biological response. The selected papers illustrated how Schiff base ligands when coordinated to metal ions such as Zn, V, Cu, Ni, Co, Cr and Ru were found to enhance antidiabetic efficacy most of the time. For the successful development of a potent and safe next generation antidiabetic drug, these complexes need to be further evaluated in detail for their mode of action studies, structure activity relationship, molecular target interaction and cytotoxicity parameter.

REFERENCES

1. World Health Organization, WHO, www.who.int/health-topics/diabetes#tab=tab_1 (Accessed: 11-Feb-2022).
2. Singh, N., Kesherwani, R., Tiwari, A. K. & Patel, D. K. 2016. A review on diabetes mellitus. *The Pharma Innovation Journal* 5:36–40.
3. Deshmukh, C. D. & Jain, A. 2015. Diabetes mellitus: A review. *International Journal of Pure & Applied Bioscience* 3:224–230.
4. Belfiore, F., Vecchio, L. L. & Napoli, E. 1973. Serum enzymes in diabetes mellitus. *Clinical Chemistry* 19:447–452. https://doi.org/10.1093/clinchem/19.5.447
5. Forman, D. T. & Wiringa, K. 1973. Enzyme changes in diabetes mellitus. *Annals of Clinical Laboratory Science* 3:374–385.
6. Silavwe, H. N., Villa-Rodriguez, J. A., Ifie, I., Holmes, M., Aydin, E., Jensen, J. M. & Williamson, G. 2015. Inhibition of human α-amylase by dietary polyphenols. *Journal of Functional Foods* 19:723–732. https://doi.org/10.1016/j.jff.2015.10.003
7. Vanco, J., O. Svajlenova, O., Racanska, E., Muselik, J. & Valentova, J. 2004. Antiradical activity of different copper(II) Schiff base complexes and their effect on alloxan-induced diabetes. *Journal of Trace Elements in Medicine and Biology* 18:155–161. doi: 10.1016/j.jtemb.2004.07.003
8. Stapleton, S. R. 2000. Selenium: An insulin-mimetic. *Cellular and Molecular Life Sciences* 57:1874–1879. doi: 10.1007/PL00000669
9. Zorzano, A., Palacin, M., Marti, L. & Garcia-Vicente, S. 2009. Arylalkylamine vanadium salts as new antidiabetic compounds. *Journal of Inorganic Biochemistry* 103:559–566. doi: 10.1016/j.jinorgbio.2009.01.015
10. Qin, W., Long, S., Panunzio, M. & Biondi, S. 2013. Schiff bases: A short survey on an evergreen chemistry tool. *Molecules* 18:12264–12289. doi: 10.3390/molecules181012264
11. Raczuk, E., Dmochowska, B., Samaszko-Fiertek, J. & Madaj, J. 2022. Different Schiff bases–Structure, importance and classification. *Molecules* 27:787. doi: 10.3390/molecules27030787
12. Dhar, D. N. & Taploo, C. L. 1982. Schiff-bases and their applications. *Journal of Scientific & Industrial Research* 41:501–506.
13. Sen, S., De, B. & Easwari, T. S. 2014. Synthesized 2-substituted-3-phenylthiazolidine-4-ones as potent antioxidants and antidiabetic agents. *Tropical Journal of Pharmaceutical Research* 13:1445–1454. doi: 10.4314/tjpr.v13i9.10
14. Kajal, A., Bala, S., Kamboj, S., Sharma, N. & Saini, V. 2013. Schiff bases: A versatile pharmacophore. *Journal of Catalysts* 2013. http://dx.doi.org/10.1155/2013/893512
15. Hameed, A., Al-Rashida, M., Uroos, M., Abid Ali, S. & Khan, K. M. 2017. Schiff bases in medicinal chemistry: A patent review (2010–2015). *Expert Opinion on Therapeutic Patents* 27:63–79. doi: 10.1080/13543776.2017.1252752
16. Hameed, S. A., Varkey, J. & Jayasekhar, P. 2019. Schiff bases and Bicyclic derivatives comprising 1, 3, 4-thiadiazole moiety: A review on their pharmacological activities. *Asian Journal of Pharmaceutical Research* 9:299–306. doi: 10.5958/2231-5691.2019.00047.9
17. Arulmurugan, S., Kavitha, H. P. & Venkatraman, B. R. 2010. Biological activities of Schiff base and its complexes: A review. *Rasayan J Chem.* 3:385–410.
18. Vazzana, I., Terranova, E., Mattioli, F. & Sparatore, F. 2004. Aromatic Schiff bases and 2, 3-disubstituted-1,3-thiazolidin-4-one derivatives as antiinflammatory agents. *Arkivoc* 5:364–374. doi: http://dx.doi.org/10.3998/ark.5550190.0005.531
19. Chinnasamy, R. P., Sundararajan, R. & Govindaraj, S. 2010. Synthesis, characterization, and analgesic activity of novel schiff base of isatin derivatives. *Journal of Advanced Pharmaceutical Technology & Research* 1:342–347. doi: 10.4103/0110-5558.72428
20. Sztanke, K., Maziarka, A., Osinka, A. & Sztanke, M. 2013. An insight into synthetic Schiff bases revealing antiproliferative activities *in vitro*. *Bioorganic &Medicinal Chemistry* 21:3648–3666. doi: 10.1016/j.bmc.2013.04.037

21. Al Zoubi, W., Al-Hamdani, A. A. S. & Kaseem, M. 2016. Synthesis and antioxidant activities of Schiff bases and their complexes: A review. *Applied Organometallic Chemistry* 30:810–817. https://doi.org/10.1002/aoc.3506
22. Da Silva, C. M., da Silva, D. L., Modolo, L. V., Alves, R. B., de Resende, M. A., Martins, C. V. & de Fátima, Â. 2011. Schiff bases: A short review of their antimicrobial activities. *Journal of Advanced Research* 2:1–8. https://doi.org/10.1016/j.jare.2010.05.004
23. Mahal, A., Wu, P., Jiang, Z. H. & Wei, X. 2019. Schiff bases of tetrahydrocurcumin as potential anticancer agents. *ChemistrySelect* 4:366–369. https://doi.org/10.1002/slct.201803159
24. Matela, G. 2020. Schiff bases and complexes: a review on anti-cancer activity.*Anti-Cancer Agents in Medicinal Chemistry (Formerly Current Medicinal Chemistry-Anti-Cancer Agents)* 20:1908–1917. doi: 10.2174/1871520620666200507091207
25. Sadia, M., Khan, J., Naz, R., Zahoor, M., Shah, S. W. A., Ullah, R., Naz, S., Bari, A., Mahmood, H. M., Ali, S. S., Ansari, S. A. & Sohaib, M. 2021. Schiff base ligand L synthesis and its evaluation as anticancer and antidepressant agent. *Journal of King Saud University-Science* 33:101331. https://doi.org/10.1016/j.jksus.2020.101331
26. Chigurupati, S., Palanimuthu, V. R., Kanagaraj, S., Sundaravadivelu, S. & Varadharajula, V. R. 2021. Green synthesis and in silico characterization of 4-Hydroxy-3-methoxybenzaldehyde Schiff bases for insulysin inhibition–a potential lead for type 2 diabetes mellitus. *Journal of Applied Pharmaceutical Science* 11:063–071. doi: 10.7324/japs.2021.110706
27. Afzal, H. R., Khan, N. U. H., Sultana, K., Mobashar, A., Lareb, A., Khan, A., Gull, A., Afzaal, H., Khan, M. T., Rizwan, M. & Imran, M. 2021. Schiff bases of pioglitazone provide better antidiabetic and potent antioxidant effect in a streptozotocin–nicotinamide-induced diabetic rodent model. *ACS Omega* 6:4470–4479. https://doi.org/10.1021/acsomega.0c06064
28. Rahim, F., Zaman, K., Taha, M., Ullah, H., Ghufran, M., Wadood, A., Rehman, W., Uddin, N., Adnan, S., Shahe, A., Sajid, M., Nawaz, F. & Khan, K. M. 2020. Synthesis, *in vitro* alpha-glucosidase inhibitory potential of benzimidazole bearing bis-Schiff bases and their molecular docking study. *Bioorganic Chemistry* 94:103394. https://doi.org/10.1016/j.bioorg.2019.103394
29. Aslam, M., Anis, I., Afza, N., Hussain, A., Iqbal, L., Iqbal, J., Ilyas, Z., Iqbal, S., Chaudhry, A. H. & Niaz, M. 2012. Structure-activity relationship study: Synthesis, characterization and biological investigation of Schiff bases derived from 2-aminophenol and 4-haloacetophenones. *International Journal of Current Pharmaceutical Research* 4:42–46.
30. Manimohan, M., Pugalmani, S. & Sithique, M. A. 2020. Biologically active water soluble novel biopolymer/hydrazide based O-carboxymethyl chitosan Schiff bases: Synthesis and characterisation. *Journal of Inorganic and Organometallic Polymers and Materials* 30:3658–3676. doi: 10.1007/s10904-020-01487-9
31. Thompson, K. H. & Orvig, C. 2006. Metal complexes in medicinal chemistry: New vistas and challenges in drug design. *Dalton Transactions* 6:761–764. doi: 10.1039/b513476e
32. Zhang, C. X. & Lippard, S. J. 2003. New metal complexes as potential therapeutics. *Current Opinion in Chemical Biology* 7:481–489. doi: 10.1016/s1367-5931(03)00081-4
33. Rafique, S., Idrees, M., Nasim, A., Akbar, H. & Athar, A. 2010. Transition metal complexes as potential therapeutic agents. *Biotechnology and Molecular Biology Reviews* 5:38–45.
34. Bakhtiar, R. & Ochiai, E. I. 1999. Pharmacological applications of inorganic complexes. *General Pharmacology: The Vascular System* 32:525–540. https://doi.org/10.1016/S0306-3623(98)00223-7
35. Azam, A., Raza, M. A. & Sumrra, S. H. 2018. Therapeutic application of zinc and vanadium complexes against diabetes mellitus a coronary disease: A review. *Open Chemistry* 16:1153–1165. doi: 10.1515/chem-2018-0118
36. Maanvizhi, S. A. B. A., Boppana, T., Krishnan, C. & Arumugam, G. 2014. Metal complexes in the management of diabetes mellitus: A new therapeutic strategy. *International Journal of Pharmacy and Pharmaceutical Sciences* 6:40–44.
37. Sakurai, H., Yoshikawa, Y. & Yasui, H. 2008. Current state for the development of metallopharmaceutics and antidiabetic metal complexes. *Chemical Society Reviews* 37:2383–2392. https://doi.org/10.1039/B710347F
38. Sakurai, H., Katoh, A. & Yoshikawa, Y. 2006. Chemistry and biochemistry of insulin-mimetic vanadium and zinc complexes: Trial for treatment of diabetes mellitus. *Bulletin of the Chemical Society of Japan* 79:1645–1664. https://doi.org/10.1246/bcsj.79.1645

39. Jansen, J., Karges, W. & Rink, L. 2009. Zinc and diabetes–clinical links and molecular mechanisms. *The Journal of Nutritional Biochemistry* 20:399–417. doi: 10.1016/j.jnutbio.2009.01.009
40. Coulston, L. & Dandona, P. 1980. Insulin-like effect of zinc on adipocytes. *Diabetes* 29:665–667. doi: 10.2337/diab.29.8.665
41. Chabosseau, P. & Rutter, G. A. 2016. Zinc and diabetes. *Archives of Biochemistry and Biophysics* 611:79–85. doi: 10.1016/j.abb.2016.05.022
42. Salgueiro, M. J., Krebs, N., Zubillaga, M. B., Weill, R., Postaire, E., Lysionek, A. E., Caro, R. A., De Paoli, T., Hager, A. & Boccio, J. 2001. Zinc and diabetes mellitus: Is there a need of zinc supplementation in diabetes mellitus patients?*Biological Trace Element Research* 81:215–228. doi: 10.1385/BTER:81:3:215
43. Cruz, K. J. C., de Oliveira, A. R. S. & do NascimentoMarreiro, D. 2015. Antioxidant role of zinc in diabetes mellitus. *World Journal of Diabetes* 6:333–337. doi: 10.4239/wjd.v6.i2.333
44. Poucheret, P., Verma, S., Grynpas, M. D. & McNeill, J. H. 1998. Vanadium and diabetes. *Molecular and Cellular Biochemistry* 188:73–80.
45. Thompson, K. H. & Orvig, C. 2006. Vanadium in diabetes: 100 years from Phase 0 to Phase I. *Journal of Inorganic Biochemistry* 100:1925–1935. https://doi.org/10.1016/j.jinorgbio.2006.08.016
46. Thompson, K. H. 1999. Vanadium and diabetes. *Biofactors* 10:43–51. doi: 10.1002/biof.5520100105
47. Brichard, S. M. & Henquin, J. C. 1995. The role of vanadium in the management of diabetes. *Trends in Pharmacological Sciences* 16:265–270. doi: 10.1016/s0165-6147(00)89043-4
48. Thompson, K. H., Lichter, J., LeBel, C., Scaife, M. C., McNeill, J. H. & Orvig, C. 2009. Vanadium treatment of type 2 diabetes: A view to the future. *Journal of Inorganic Biochemistry* 103:554–558. https://doi.org/10.1016/j.jinorgbio.2008.12.003
49. Shechter, Y., Goldwaser, I., Mironchik, M., Fridkin, M. & Gefel, D. 2003. Historic perspective and recent developments on the insulin-like actions of vanadium; toward developing vanadium-based drugs for diabetes. *Coordination Chemistry Reviews* 237:3–11. https://doi.org/10.1016/S0010-8545(02)00302-8
50. Sakurai, H., Fujisawa, Y., Fujimoto, S., Yasui, H. & Takino, T. 1999. Role of vanadium in treating diabetes. *Journal of Trace Elements in Experimental Medicine* 12:393–401. https://doi.org/10.1002/(SICI)1520-670X(1999)12:4<393::AID-JTRA13>3.0.CO;2-P
51. Domingo, J. L. & Gómez, M. 2016. Vanadium compounds for the treatment of human diabetes mellitus: A scientific curiosity? A review of thirty years of research. *Food and Chemical Toxicology* 95:137–141. doi: 10.1016/j.fct.2016.07.005
52. Miyazaki, R., Yasui, H. & Yoshikawa, Y. 2016. α-Glucosidase inhibition by new Schiff base complexes of Zn (II). *Open Journal of Inorganic Chemistry* 6:114–124. doi: 10.4236/ojic.2016.62007
53. Balan, K., Ratha, P., Prakash, G., Viswanathamurthi, P., Adisakwattana, S. & Palvannan, T. 2017. Evaluation of invitro α-amylase and α-glucosidase inhibitory potential of N2O2 schiff base Zn complex. *Arabian Journal of Chemistry* 10:732–738. https://doi.org/10.1016/j.arabjc.2014.07.002
54. Koothappan, M., Vellai, R. D., Subramanian, I. P. & Subramanian, S. P. 2018. Synthesis of a new zinc-mixed ligand complex and evaluation of its antidiabetic properties in high fat diet: Low dose streptozotocin induced diabetic rats. *Diabetes & Metabolism Journal* 42:244–248. doi: 10.4093/dmj.2018.0002
55. Vijayaraghavan, K., Pillai, S. I. & Subramanian, S. P. 2012. Design, synthesis and characterization of zinc-3 hydroxy flavone, a novel zinc metallo complex for the treatment of experimental diabetes in rats. *European Journal of Pharmacology* 680:122–129. doi: 10.1016/j.ejphar.2012.01.022.
56. Maheswari, J. U., Pillai, S. I. & Subramanian, S. P. 2015. Zinc-silibinin complex: Synthesis, spectral characterization and biochemical evaluation of antidiabetic potential in high fat fed low dose STZ induced type 2 diabetic rats. *Journal of Chemical and Pharmaceutical Research* 7:2051–2064.
57. Gopalakrishnan, V., IyyamPillai, S. & Subramanian, S. P. 2015. Synthesis, spectral characterization, and biochemical evaluation of antidiabetic properties of a new zinc-diosmin complex studied in high fat diet fed-low dose streptozotocin induced experimental type 2 diabetes in rats. *Biochemistry Research International* 2015:1–11. https://doi.org/10.1155/2015/350829
58. Jaiganesh, C., Devi, V. R., Pillai, S. I. & Subramanian, S. 2017. Synthesis, characterization and evaluation of antidiabetic properties of a new metformin-3-hydroxyflavone complex studied in high fat diet fed-low dose streptozotocin induced experimental type 2 diabetes in Wistar rats. *International Journal of Pharma and Bio Sciences* 8:1–15. doi: 10.22376/IJPBS.2017.8.3.P1-15

59. Jaiganesh, C. & Subramanian, S. 2017. Metformin-3-hydroxyflavone, a new schiff base complex modulates the activities of carbohydrate regulatory enzymes in high fat diet fed-low dose streptozotocin induced type 2 diabetes in experimental rats. *Journal of Chemical and Pharmaceutical Research* 9:90–100.
60. Paria, D., Kundu, S., Singh, K. K., Singh, S. S. & Singh, K. S. 2018. Synthesis, characterization and antidiabetic activity of some water soluble zn (ii) complexes with (e)-n-(thiophen-2-ylmethylene) anilines. *Asian Journal of Chemistry* 30:1115–1119. https://doi.org/10.14233/ajchem.2018.21216
61. Kundu, S. & Singh, K. S. 2020. Study of synthesis, spectroscopic characterization and their antidiabetic activity of some water soluble zn (ii) complexes with n, s-donor ligands derived from (e)-n-(thiophen-2-ylmethylene) anilines. *Current Perspectives on Chemical Sciences Vol.5*, ed. W. M. M. Sad, Chapter 2, Page 18–28. B. P. International. https://doi.org/10.9734/bpi/cpcs/v5
62. Sathishkumar, R., Magesh, C. J., Tamilselvan, S., Lavanya, G., Venkatapathy, K., Vimalan, M. & Muthu, S. 2018. Synthesis, crystal structure, spectroscopic and docking studies of mononuclear, mono (bis (2-(4-butylphenyl) imino) methyl) phenoxy) zinc (II) dichloride complex as a promising candidate for α-glucosidase inhibition. *Chemical Data Collections* 17:187–195. doi: 10.1016/j.cdc.2018.08.012
63. Manimohan, M., Pugalmani, S. & Sithique, M. A. 2020. Synthesis, spectral characterisation and biological activities of novel biomaterial/n, n, o donor tridentate co (ii), ni (ii) and zn (ii) complexes of hydrazide based biopolymer schiff base ligand. *Journal of Inorganic and Organometallic Polymers and Materials* 30:4481–4495. doi: 10.1007/s10904-020-01578-7
64. Rauf, A., Shah, A., Munawar, K. S., Ali, S., Tahir, M. N., Javed, M. & Khan, A. M.2020. Synthesis, physicochemical elucidation, biological screening and molecular docking studies of a Schiff base and its metal (II) complexes. *Arabian Journal of Chemistry* 13:1130–1141. https://doi.org/10.1016/j.ara bjc.2017.09.015
65. Shanty, A. A. & Mohanan, P. V. 2021. Synthesis, characterization, DNA binding, antibacterial, antidiabetic, molecular docking and DFT studies of Ni (II), Cu (II) and Zn (II) complexes derived from heterocyclic Schiff base. *Inorganic and Nano-Metal Chemistry*, 1–16. https://doi.org/10.1080/24701 556.2021.1988977
66. Deswal, Y., Asija, S., Dubey, A., Deswal, L., Kumar, D., Jindal, D. K. & Devi, J. 2022. Cobalt (II), nickel (II), copper (II) and zinc (II) complexes of thiadiazole based Schiff base ligands: synthesis, structural characterization, DFT, antidiabetic and molecular docking studies. *Journal of Molecular Structure* 1253:132266. doi: 10.1016/j.molstruc.2021.132266
67. Sudha, A.& Askarali., S. J. 2022. Evaluation of characterization, biological and computational studies of new Schiff base ligand and some metal (II) complexes. *InorganicaChimicaActa*, 120817. https://doi.org/10.1016/j.ica.2022.120817
68. Ki, J., Mukherjee, A., Rangasamy, S., Purushothaman, B. & Song, J. M. 2016. Insulin-mimetic and anti-inflammatory potential of a vanadyl-Schiff base complex for its application against diabetes. *RSC Advances* 6:57530–57539. doi: 10.1039/C6RA11111D
69. Jia, Y., Lu, L., Zhu, M., Yuan, C., Xing, S. & Fu, X. 2017. A dioxidovanadium (V) complex of NNO-donor Schiff base as a selective inhibitor of protein tyrosine phosphatase 1B: Synthesis, characterization, and biological activities. *European Journal of Medicinal Chemistry* 128:287–292. doi: 10.1016/j.ejmech.2017.02.003
70. Mahmoud, M. A., Zaitone, S. A., Ammar, M. A. & Sallam, S. A. 2017. Synthesis, spectral, thermal and insulin-enhancing properties of oxovanadium (IV) complexes of metformin Schiff-bases. *Journal of Thermal Analysis & Calorimetry* 128:957–969. doi: 10.1007/s10973-016-6018-1
71. Patel, R. N. & Singh, Y. P. 2018. Synthesis, structural characterization, DFT studies and in-vitro antidiabetic activity of new mixed ligand oxovanadium (IV) complex with tridentate Schiff base. *Journal of Molecular Structure* 1153:162–169. doi: 10.1016/j.molstruc.2017.10.010
72. Patel, N., Prajapati, A. K., Jadeja, R. N., Patel, R. N., Patel, S. K., Gupta, V. K., Tripathi, I. P.&Dwivedi, N.2019. Model investigations for vanadium-protein interactions: Synthesis, characterization and antidiabetic properties. *InorganicaChimicaActa* 493:20–28. https://doi.org/10.1016/j.ica.2019.04.050
73. Patel, N., Prajapati, A. K., Jadeja, R. N., Patel, R. N., Patel, S. K., Tripathi, I. P., Dwivedi, N., Gupta, V. K. & Butcher, R. J. 2020. Dioxidovanadium (V) complexes of a tridentate ONO Schiff base ligand: Structural characterization, quantum chemical calculations and in-vitro antidiabetic activity. *Polyhedron* 180:114434. doi: 10.1016/j.poly.2020.114434

74. Patel, N., Prajapati, A. K., Jadeja, R. N., Tripathi, I. P. & Dwivedi, N. 2021. Synthesis, characterization and *in vitro* antidiabetic activity of anionic dioxidovanadium (V) complexes. *Journal of the Indian Chemical Society* 98:100047. https://doi.org/10.1016/j.jics.2021.100047
75. AbdElaziz, R. M., El Tabl, A. S., Elmazar, M. M. & AbdElwahed, M. 2020. Metallopharmaceutical complexes based on vanadium as potential anti-hyperglycemic agents. *Egyptian Journal of Chemistry* 63:389–404. doi: 10.21608/EJCHEM.2019.13511.1837
76. Szklarzewicz, J., Jurowska, A., Matoga, D., Kruczała, K., Kazek, G., Mordyl, B., Sapa, J. & Papież, M. 2020. Synthesis, coordination properties and biological activity of vanadium complexes with hydrazone Schiff base ligands. *Polyhedron* 185:114589. https://doi.org/10.1016/j.poly.2020.114589
77. Szklarzewicz, J., Jurowska, A., Hodorowicz, M., Kazek, G., Głuch-Lutwin, M., Sapa, J. & Papież, M. 2021. Tridentate ONO ligands in vanadium (III-V) complexes-synthesis, characterization and biological activity. *Journal of Molecular Structure* 1224:129205. doi: 10.1016/j.molstruc.2020.129205
78. Szklarzewicz, J., Jurowska, A., Hodorowicz, M., Kazek, G., Głuch-Lutwin, M. & Sapa, J. 2021. Ligand role on insulin-mimetic properties of vanadium complexes: Structural and biological studies. *InorganicaChimicaActa* 516:120135. doi: 10.1016/j.ica.2020.120135
79. Szklarzewicz, J., Jurowska, A., Hodorowicz, M., Kazek, G., Mordyl, B., Menaszek, E. & Sapa, J. 2021. Characterization and antidiabetic activity of salicylhydrazone Schiff base vanadium (IV) and (V) complexes. *Transition Metal Chemistry* 46:201–217. doi: 10.1007/s11243-020-00437-1
80. Turtoi, M., Anghelache, M., Patrascu, A. A., Maxim, C., Manduteanu, I., Calin, M. & Popescu, D. L. 2021. Synthesis, characterization, and *in vitro* insulin-mimetic activity evaluation of valineschiff base coordination compounds of oxidovanadium(V). *Biomedicines* 9:562. doi: 10.3390/biomedicines9050562
81. Vanco, J., Marek, J., Travnicek, Z., Racanska, E., Muselik, J. & Svajlenova, O. 2008Synthesis, structural characterization, antiradical and antidiabetic activities of copper(II) and zinc(II) Schiff base complexes derived from salicylaldehyde and b-alanine. *Journal of Inorganic Biochemistry* 102:595–605. doi: 10.1016/j.jinorgbio.2007.10.003
82. Lakshmi, S. S., Geetha, K., Gayathri, M. & Shanmugam, G. 2016. Synthesis, crystal structures, spectroscopic characterization and *in vitro* antidiabetic studies of new Schiff base Copper (II) complexes. *Journal of Chemical Sciences* 128:1095–1102. doi: 10.1007/s12039-016-1099-8
83. Karthik, S., Gomathi, T. & Vedanayaki, S. 2021. Synthesis, characterization, biological activities of Schiff base metal (II) complexes derived from 4-hydroxy-3,5-dimethoxybenzaldehyde and 3-aminoquinoline. *Indian Journal of Chemistry* 60A:1427–1436. http://nopr.niscair.res.in/handle/123456789/58517
84. Sakthivel, R. V., Sankudevan, P., Vennila, P., Venkatesh, G., Kaya, S. & Serdaroglu, G. 2021. Experimental and theoretical analysis of molecular structure, vibrational spectra and biological properties of the new Co(II), Ni(II) and Cu(II) Schiff base metal complexes. *Journal of Molecular Structure* 1233:130097. https://doi.org/10.1016/j.molstruc.2021.130097
85. Mahmoud, M. A., Zaitone, S. A., Ammar, A. M. & Sallam, S. A. 2016. Synthesis, structure and antidiabetic activity of chromium (III) complexes of metformin Schiff-bases. *Journal of Molecular Structure* 1108:60–70. doi: 10.1016/j.molstruc.2015.11.055
86. Shukla, S. N., Gaur, P., Jhariya, S., Chaurasia, B., Vaidya, P., Dehariya, D. & Azam, M. 2018. Synthesis, characterization, *in vitro* anti-diabetic, antibacterial and anticorrosive activity of some Cr(III) complexes of Schiff bases derived from isoniazid. *Chemical Science Transactions* 7:424–444. doi: 10.7598/cst2018.1509
87. Booysen, I. N., Maikoo, S., Akerman, M. P. & Xulu, B. 2014. Novel ruthenium(II) and (III) compounds with multidentate Schiff base chelates bearing biologically significant moieties. *Polyhedron* 79:250–257. doi: 10.1016/j.poly.2014.05.021
88. Mabuza, L. P., Gamede, M. W., Maikoo, S., Booysen, I. N., Nguban, P. S. & Khathi, A. 2018. Effects of a ruthenium Schiff base complex on glucose homeostasis in diet-induced pre-diabetic rats. *Molecules* 23:1721. https://doi.org/10.3390/molecules23071721
89. Mabuza, L. P., Gamede, M. W., Maikoo, S., Booysen, I. N., Nguban, P. S. & Khathi, A. 2019. Cardioprotective effects of a ruthenium (II) Schiff base complex in a diet-induced prediabetic rats. *Diabetes, Metabolic Syndrome and Obesity: Targets and Therapy* 12:217–223. doi: 10.2147/DMSO.S183811
90. Mabuza, L. P., Gamede, M. W., Maikoo, S., Booysen, I. N., Nguban, P. S.&Khathi, A. 2019. Hepatoprotective effects of a ruthenium(II) Schiff base complex in rats with diet-induced prediabetes. *Current Therapeutic Research* 91:66–72. doi: 10.1016/j.curtheres.2019.100570

3 Proliferation Inhibition of Schiff Base Metal Complexes

Rajesh Kumar,[1] Komal Aggarwal,[2] and Khursheed Ahmad[3]*
[1]Department of Chemistry, R.D.S. College, B.R.A. Bihar University, India
[2]Department of Chemistry, Sri Venkateswara College, University of Delhi, India
[3]Department of Chemistry, National Institute of Technology, Patna, India
*Corresponding author
Email: rajeshenzyme@gmail.com

CONTENTS

3.1 INTRODUCTION

Schiff bases are a fascinating and significant class of ligands for coordination compounds containing an imine (-C=N-) or azomethine (-CH=N-) group which is primarily formed by the combination of various aromatic or cyclic carbonyl groups (aldehydes or ketones) and primary amines under specific reaction conditions [1]. Schiff base ligands are easily combined with almost all metal to produce metal complexes with definite geometry such as square planar, octahedral and so on [2]. The lone pair present on nitrogen atom of the azomethine or imine moiety play a significant role to perform biological activities. During the last few years, various metal complexes have been reported as possessing pharmacological and biological applications in anticancer [3], antibacterial [4], antimicrobial [5], antioxidant [6], anti-inflammatory [7], antiviral [8], antitumor [9], and ureases inhibitor situations [10] and occupy a central role in the progress of coordination chemistry [11] as well as acting as a catalyst in several reactions [12]. Apart from these applications, metal complexes have also shown unique importance in fluorescent probes [13], dye industry, agrochemical and fungicidal[14]. Presently, the study of metal complexes has captivated interest specially for the development of novel anticancer drugs.

Cancer is a horrible disease that occurs on abnormal cell growth invasion and it is a very serious health issue all over the world, as well as the second most leading cause of death after heart disease in developed and growing countries in the world [15]. Presently, chemotherapy is the main and alternative method for both metastasized and localized cancer but it has serious side effects [16]. Metal complexes that are synthesized by the reaction of organic Schiff bases are used as potent medicine or diagnostic agents because metal complexes play a unique mechanism on drug action owing to the large range of coordination numbers, kinetic properties and geometries, that is not thinkable with pure organic molecules [17]. Cis platin, which was developed by Rosenberg and his co-workers, is

DOI: 10.1201/9781003291459-3

one of the most popular and best used anticancer drugs in the world. After developing cis platin as an anticancer drug, during the last few years, several Schiff bases coordinate with metal exhibited anticancer activity. Among them, a pyrazolone Schiff base copper complex has drawn attention as a potential drug compound for the treatment of liver cancer, being endowed with a high proliferative inhibitor against hepatocellular carcinoma (HCC) [18]. In addition, Schiff base Mn (II) complex synthesized from the pyridoxal has been utilized for the treatment of breast cancer [19]. In this chapter, we focus on the synthesis and antiproliferative activity of Schiff bases metal complexes during the last few years.

3.2 SYNTHESIS AND ANTICANCER PROPERTIES

Treatment of cancer is one of the most critical challenges in front of a medicinal chemist due to high anticancer drug resistance. Therefore, there is an urgent requirement to produce new anticancer drugs with better activity and low side effects on human health. Subsequently, several studies are ongoing on the class of Schiff base metal complexes to search for better anticancer compounds in comparison to the available drugs on the market. Demir et al. [20] prepared the N, O, and S atoms containing water-soluble Schiff base **3** from easily available starting material compounds **1** and **2** in ethanol solvent under refluxing condition in good yield. Further, copper, nickel and zinc complexes were prepared from the ligand **3** with the corresponding metal acetates in aqueous methanol under reflux at 60 °C and the final products **4–6** were identified by various techniques such as magnetic susceptibility, elemental analysis, and spectroscopic methods (Figure 3.1). In addition, all synthesized

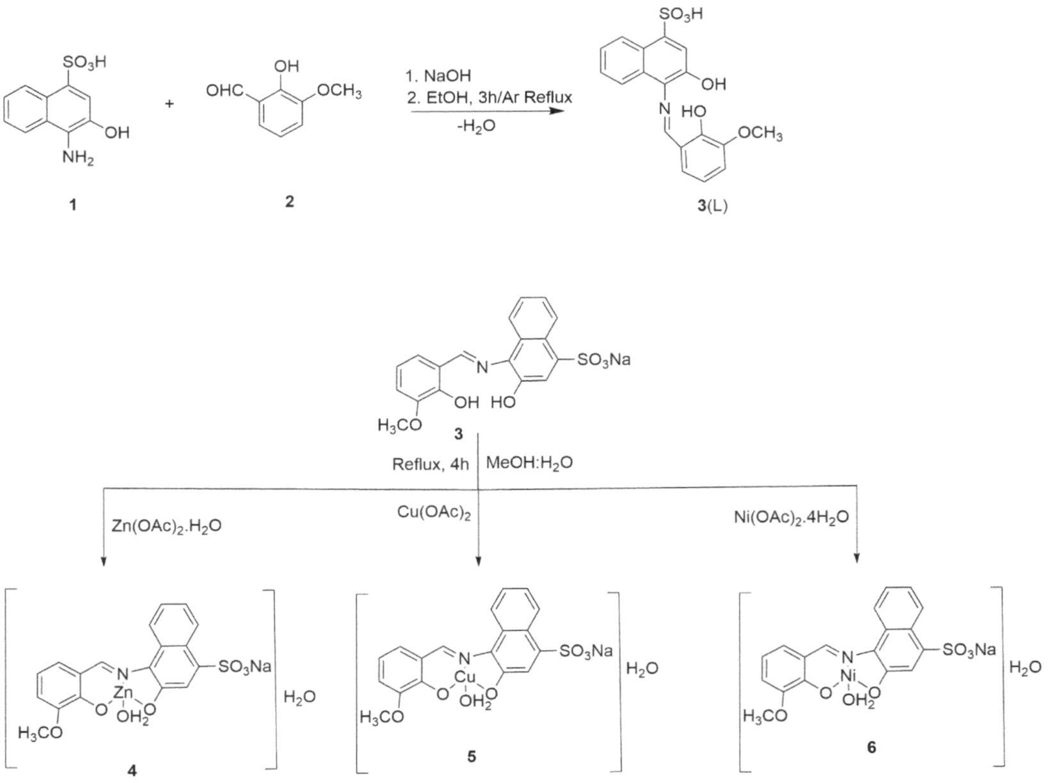

FIGURE 3.1 Synthesis of water-soluble Schiff base and their metal complexes.

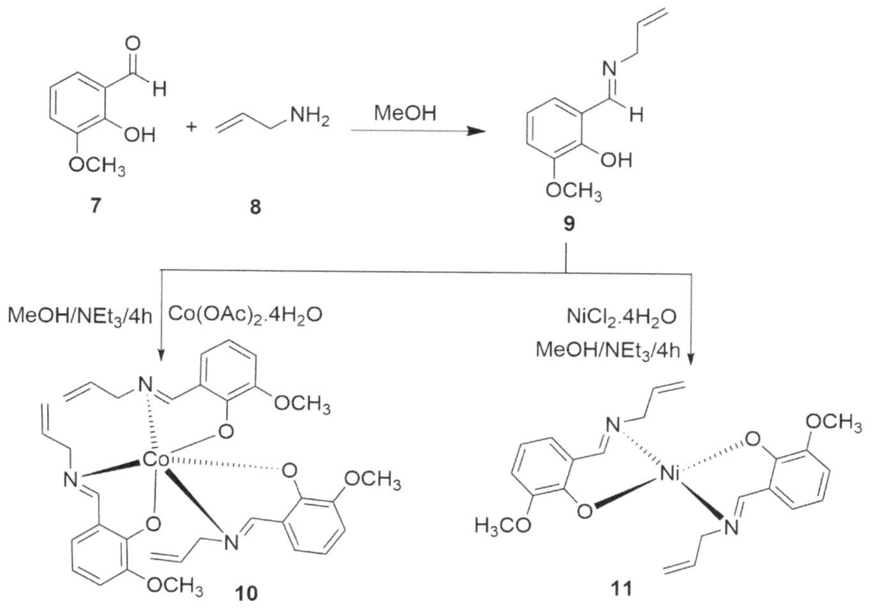

FIGURE 3.2 Synthesis of Co and Ni containing metal complexes.

complexes **4–6** and ligand **3** were evaluated against anticancer activities on A549 lung cancer cell lines. It was found that ligand **3** and Nickel complex **6** did not show any result on A549 cell line, but copper and zinc complexes have good cytotoxic effects on A549 cell line. Complexes **4** and **5** also encouraged apoptosis and restricted cell migration on A549 cells.

The Salehi research group [21] synthesized a novel Schiff base compound **9** from easily available compound substituted benzaldehyde **7** and allylamine **8** in methanol solvent with high yield. Further, novel ligand **9** reacted with $Co(OAc)_2.2H_2O$ and $NiCl_2.4H_2O$ in methanol using triethylamine as a base to produce the corresponding octahedral **10** and square planar **11** complexes, respectively (Figure 3.2). The synthesized ligand **9** and metal complexes **10 &** **11** were characterized by different techniques such as Mass spectra, UV-Vis, IR and X-ray crystallography. After that, all prepared desired complexes and ligand were tested for its antiproliferation properties on human colorectal cancer cells, which demonstrated that novel compounds possessed better potential than the ligand, due to the coordination of metal with various sites of ligands. The synthesized desired compounds showed an anticancer activity with IC 50 value ranges from 0.973 to 102 μg/mL. The largest development inhibitory properties were found on cancer cells at a 01 mg/ml quantity of the novel complexes **10** & **11**. The molecular docking and pharmacophore modeling studies showed different binding sites of the metal complexes that selectively bind to the anticancer target.

Mukherjee et al. [22] synthesized azo-functionalized Schiff base **12** using 2-hydroxy-3-methoxy benzaldehyde and 4-aminoazobenzene in MeOH via reflux. Further, Schiff base **12** reacted with copper acetate in methanol to produce the corresponding metal complex **13** (Figure 3.3). The crystal structure proved that the Cu-complex exists in distorted square planar geometry. Synthesised metal complexes were evaluated for their antiproliferative activity and it was observed that synthesized metal complexes show promising results compared to the breast cancer cell lines, MCF-7.

Rajakkani et al. [23] reported a novel 13-membered tetraaza macrocyclic Schiff base **17** by condensation of compound **14** with two equivalents of 4-aminoantipyrine **15** in ethanol under reflux to produce compound **16** followed by the addition of o-phenylenediamine which gave the desired

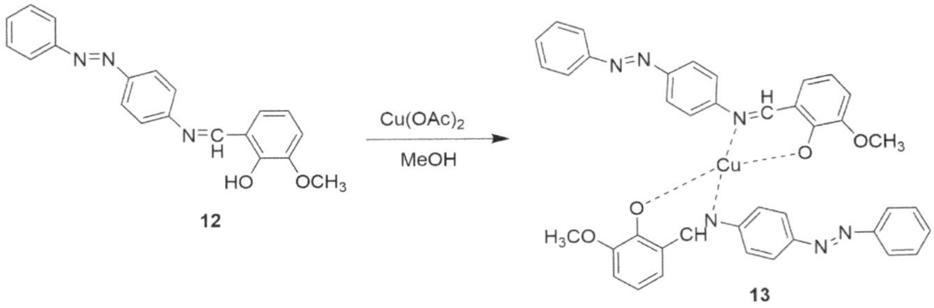

FIGURE 3.3 Synthesis of azo-functionalized Schiff base Cu-complex.

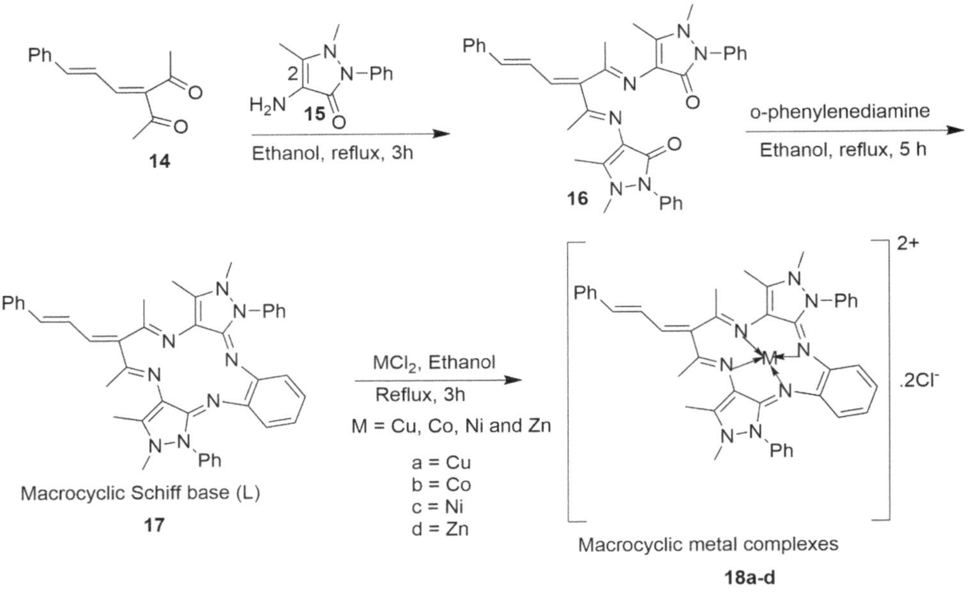

FIGURE 3.4 Synthesis of macrocyclic Schiff base transition metal complexes.

ligand **17** in good yield. Further, ligand **17** has been used for metal complex formation with various transition metals, that is, copper, cobalt and zinc chlorides in ethanol solvent under reflux to produce the corresponding novel complexes **18a–d** in desired quantities (Figure 3.4). The ligand **17** and their complexes **18a–d** have been identified via several methods, that is, nuclear magnetic resonance, Mass spectroscopy, UV-Vis, FTIR, EPR, magnetic moment values and so on. After characterization, it was found that complexes occur in square planar geometry. All synthesized new metal compounds were observed to display better anticancer properties against the three cancerous cell lines HeLa, MCF-7 and Hep-2. As per IC 50 numeric value data, 14, 16 and 12 μM found that Cu (II) complex **18a** was recognized as good cytotoxic agent in contrast to the three cancer cell lines. Overall, a higher anticancer property of copper complex **18a** was found as compared to anticancer drug cis platin.

Kavitha et al. [24] prepared ligand **19** by condensation of pyridoxal and *p*-F-benzaldehyde [25]. Further, ligand **19** has been used for metal complexes **20a–b** and **21** (Figure 3.5) formation with corresponding metal chlorides in ethanol and the desired complexes were identified by several methods such as ^{1}H NMR, FTIR, UV-Vis, powder XRD and so on. The X-ray crystallographic

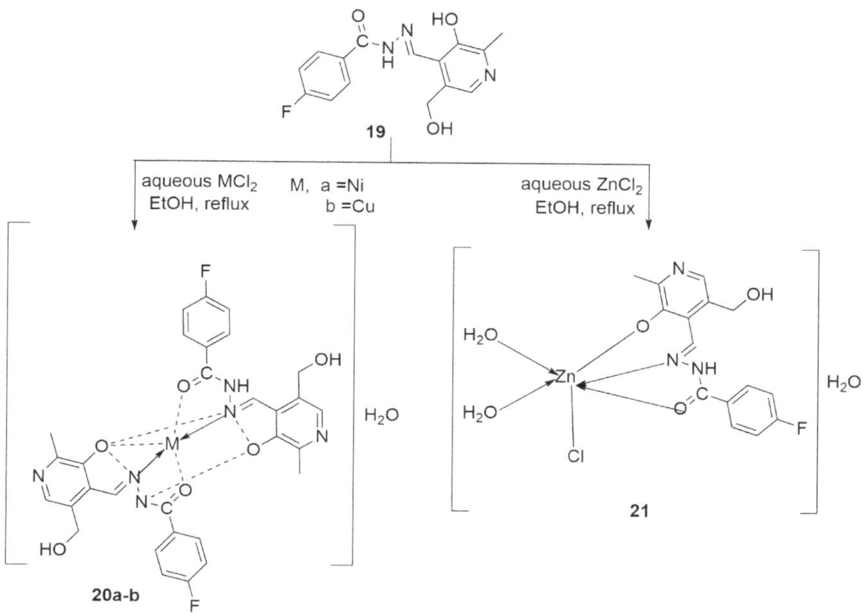

FIGURE 3.5 Synthesis of novel Schiff base containing transition metal complexes.

studies have shown that complexes would be distorted octahedral. The anticancer studies of metal complexes **20a–b** and **21** were carried out among these metal complexes and only metal complex **20b** showed moderate activity against human breast cancer and human cervical cancer and good inhibitory effect on A549 (human lung cancer) cells.

Dhanaraj et al. [26] synthesized new tridentate antipyrine based ligand **28** in three steps: first, compound **22** is treated with **23** in methanol under an acidic environment to produce the corresponding product **24** in good yields. In the second step, compound **25** is treated with hydrazine **26** under the same reaction conditions to afford the corresponding compound **27,** and finally **24** reacted with **27** in methanol under an acidic environment to give the desired Schiff base **28** in high yield (Figure 3.6). Further, ligand **28** reacted with metal acetates to produce the desired metal complexes **29a–d** in good yield. The metal complexes and ligand were identified by several approaches such as Infrared spectroscopy, nuclear magnetic resonance spectroscopy, UV-Vis, molar conductance and so on. The anticancer property of the complexes was checked against liver bilobular cancerous cells (LBir2754) at different concentrations in vitro. It was found that copper complex demonstrated a better inhibitory effect than the ligand and other synthesized compounds in the development of cancer cells. All the other prepared complexes have higher IC50 value in comparison to copper complex on cancer cells. It has been found that ligand and the metal complexes are not very successfully active towards the cancer cells in comparison to anticancer drugs such as estramustine, noscapine and cis platin.

Bao et al. [27] synthesized novel Cu-complex **33** in two steps: first, ethylenediamine **30** reacted with 2′-hydroxyphenylacetone **31** in methanol under reflux to produce the corresponding ligand **32** followed by treatment with copper nitrate to give the desired metal complex **33** in good yield (Figure 3.7). Further, metal complex **33** has been identified by X-ray technique and cytotoxicity of complex **33** has been tested against various cancerous cell lines such as A549, HeLa, LoVo, and ordinary cell lines HUVEC and LO2 by MTT assays, and observed that IC 50 value is somewhat lower than cisplatin. Previously, the anticancer mechanistic pathway demonstrated that the copper complex **33** restricted cell proliferation by hindering the deoxy nucleic acid preparation and worked

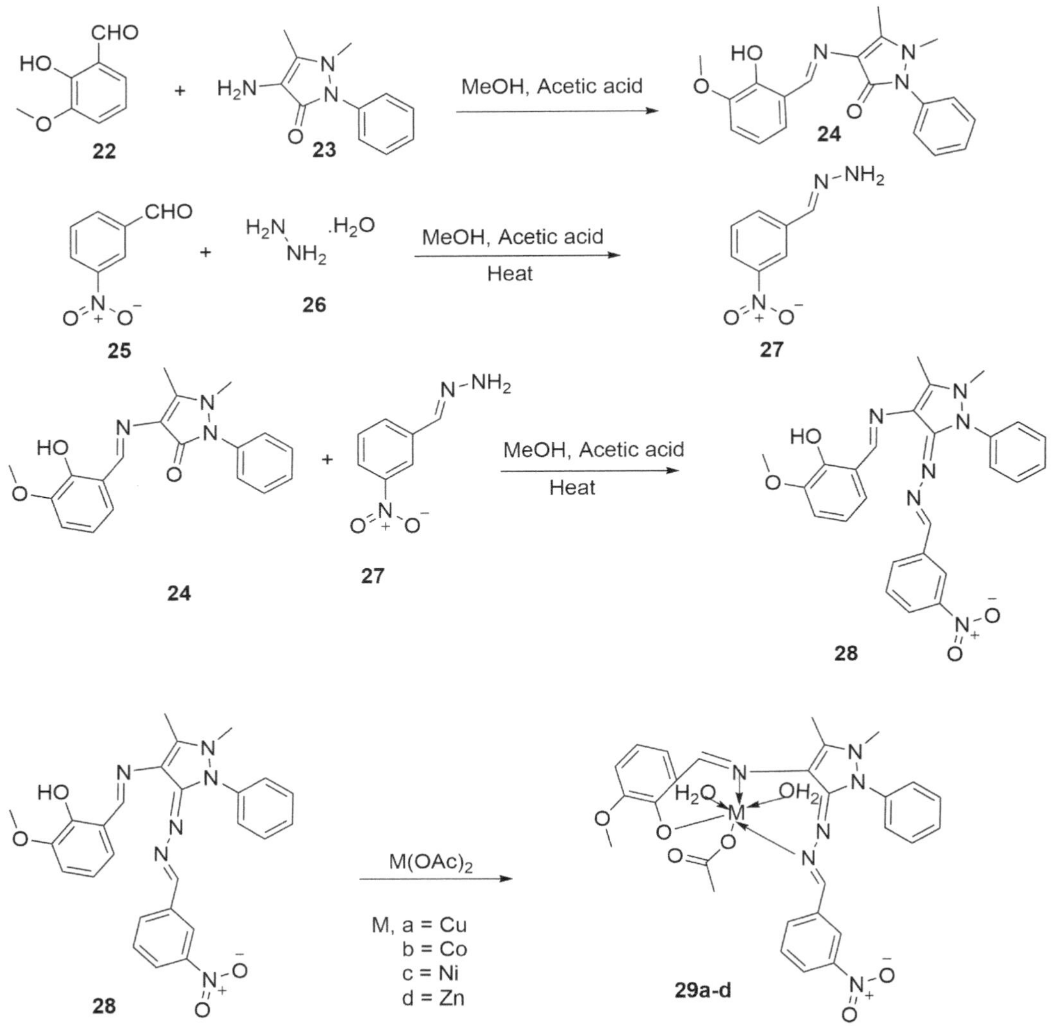

FIGURE 3.6 Synthesis of tridentate antipyrine Schiff base containing metal complexes.

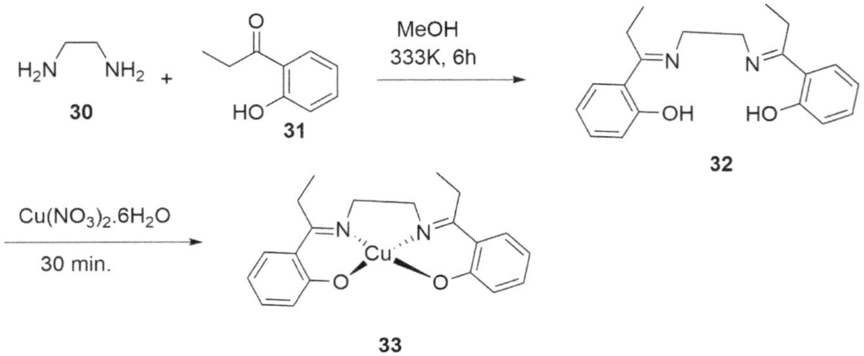

FIGURE 3.7 Synthesis of ethylenediamine containing Cu-complex.

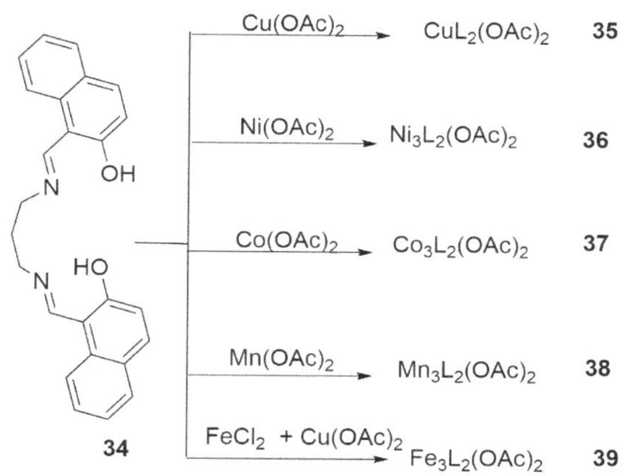

FIGURE 3.8 Synthesis of bis-Schiff base containing transition metal complexes.

on HeLa cells for nuclear division over time. In addition, the complex improved intracellular ROS levels and encouraged apoptosis in a dose reliant on the pathway. Apart from this, the complex does not work as a proteasome inhibitor.

Shi et al. [28] synthesized bis-Schiff base ligand **34** by known procedure [29]. Further, ligand **34** has been used for the synthesis of different complexes **35–39** with suitable transition metal acetates in good yields (Figure 3.8). Previously, these complexes have been screened for antitumor activity by seven human cancer cell lines by the MTT assay in vitro. It demonstrated that complexes **35**, **36** and **39** exhibited better antitumor properties than ligand **34** and metal complexes **37–38**, and even than cis platin. It was observed that copper compound **35** was the best inhibitory property on tumor cells and IC 50 data was less than 0.5 μM for human bladder cancerous cell line T-24. At this quantity, compound **35** demonstrated toxicity to the ordinary cell HL-7702 as exposed by cytometry.

Aslan et al. [30] synthesized ligand **45** in two steps: first, compound **40** reacted with **41** in benzene solvent to produce the corresponding **42**, then **42** treated with 2-hydroxy-1-naphthaldehyde (**44**) produced the final ligand **45** in high yield (Figure 3.9). Further, a series of metal complexes were synthesized with ligand **45** and all metal complexes and ligand were identified by various techniques such as magnetic susceptibility, NMR (^{1}H and ^{13}C), IR, and so on. In addition to, the anticancer activities of metal complexes **46–48** were evaluated against A549 cell line. All metal complexes exhibited anticancer activity on A549 cell line. The Cu (II) complex **47a** exhibited the lowest IC 50 value, therefore it would be established for more studies as an active antitumor candidate to try to avoid lung cancer.

Mbugua et al. [31] synthesized pyrrole-based ligands **53–55** and their corresponding complexes **56–60** with platinum and palladium metal. The Schiff base ligands **53–55** were prepared by reaction of compound **49** either with benzyl amine **50**, compound **51**, or compound **53** via refluxing in MeOH. Further, metal complexes **56–60** were synthesized via the reaction of corresponding transition metal compounds with a solution of Schiff bases in dichloromethane at normal temperature in good yields (Figure 3.10). The complexes **56–60** were identified by various techniques, that is, FTIR, microanalysis, NMR and UV–Vis. Further, all synthesized complexes **56–60** were evaluated against several human cancerous such as Caco-2, HeLa, HepG2, PC-3 and MCF-7 and non-cancerous cell lines (MCF-12A) via MTT and Apopercentage assays, respectively. For calf thymus DNA complex **60** showed extraordinary DNA-binding affinity. Complex **59** reduced the cell possibility of all cell lines, in which five cancerous cell lines by greater than 80%. Compound **60** exhibited better selectivity with no cytotoxicity towards the breast cell line but decreased the capability of five cancerous cell lines, in which one breast cancer cell line was greater than 60%.

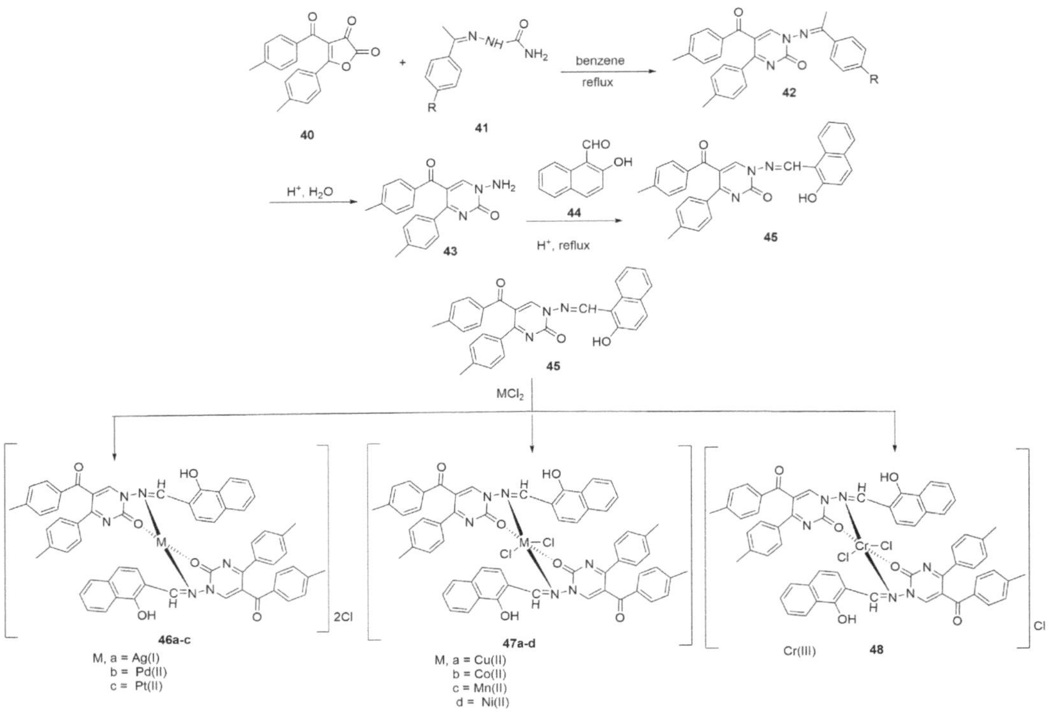

FIGURE 3.9 Synthesis of Schiff base containing novel transition metal complexes.

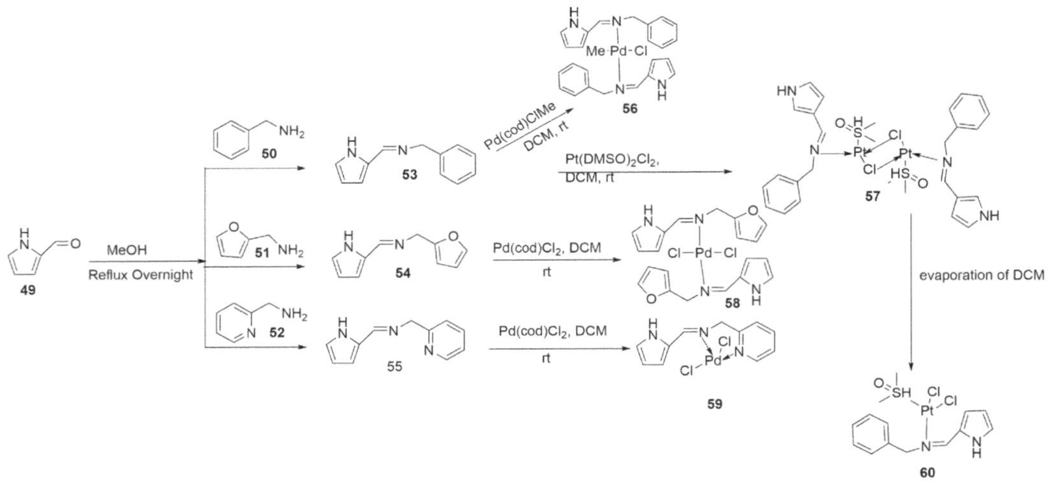

FIGURE 3.10 Synthesis of pyrrole-based transition metal complexes.

Das et al. [32] synthesized ligands **63** and **65** from easily available compounds **61**, **62** and **64** [33]. Further, these ligands **63** and **65** have been used for novel Zn based metal complexes **66–69** formation *via* various zinc salts (Figure 3.11). These Schiff bases and chelate compounds were characterized via several spectroscopic and other approaches. In addition, anticancer properties of azido metal compounds **66** and **67** have been evaluated against breast cancer cell line MCF 7 and

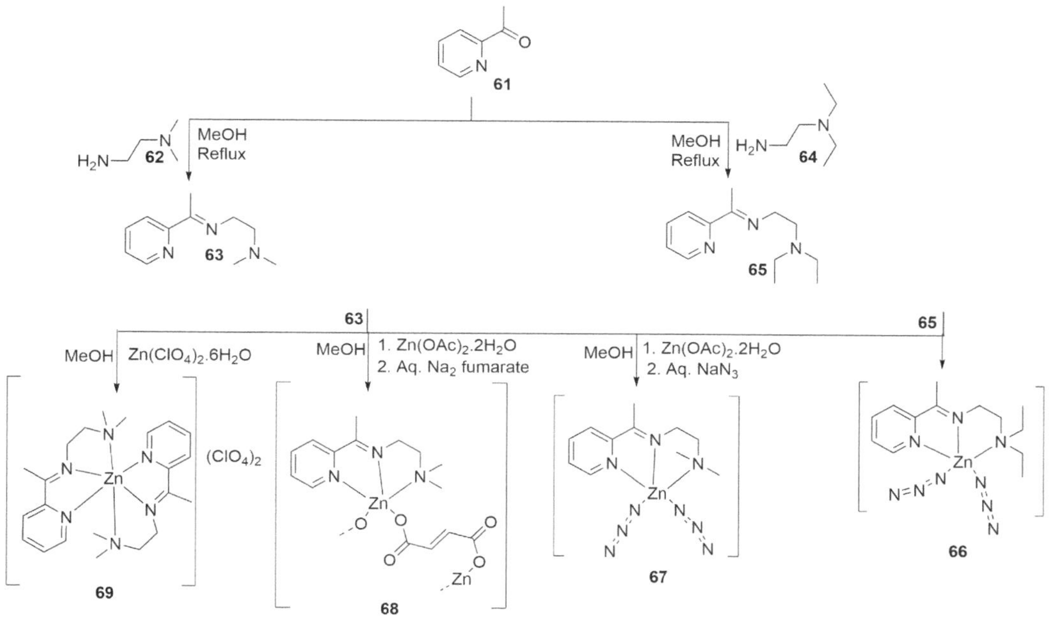

FIGURE 3.11 Synthesis of pyridine containing Zn complexes.

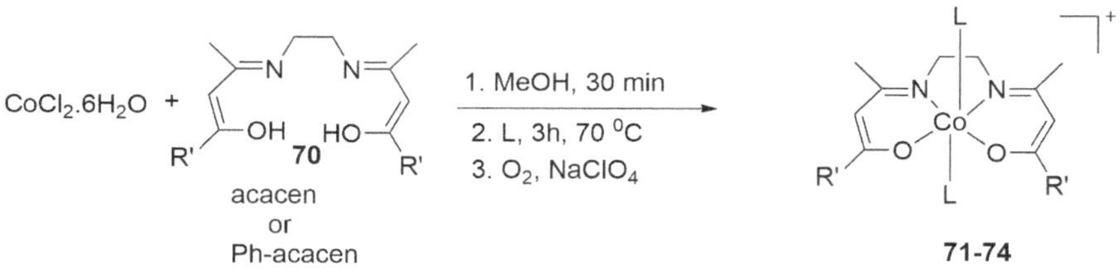

Complex 71: R' = CH_3, L = HA Complex **72**: R' = CH_3, L = DA

Complex **73**: R' = C_6H_5, L = HA Complex **74**: R' = C_6H_5, L = DA

HA =Heptylamine, DA = Dodecylamine

FIGURE 3.12 Synthesis of cobalt complexes via various ligands.

these complexes were found to have outstanding activity towards progressing the inhibition of the human breast cancer cell line and these results show substantial biological importance.

Kumar et al. [34] synthesized four cobalt (III) complexes **71–74** from ligand **70** with $CoCl_2.6H_2O$ in methanol at 70 °C in good yields (Figure 3.12). Further, these metal complexes were identified via various approaches such as mass spectroscopy, infrared spectroscopy and nuclear magnetic resonance. In addition, these complexes interacted with CT-DNA using different spectroscopic measurements and data observed that compounds keeping "Ph-acacen" ligand exhibits best binding capacity for deoxy nucleic acid in comparison to "acacen" ligand -holding compounds. Apart from this, the cytotoxicity potential of all the Schiff base complexes was evaluated against A-549 and VERO cells in vitro. The IC 50 data show that all Schiff base metal complexes have potential as anticancer drug cisplatin, against A-549 cells.

Gastric cancer is a multifactorial disorder that exhibits architectural heterogeneity and is cytological in comparison to other cancers, therefore making it diagnostically interesting to cure. Cis platin is

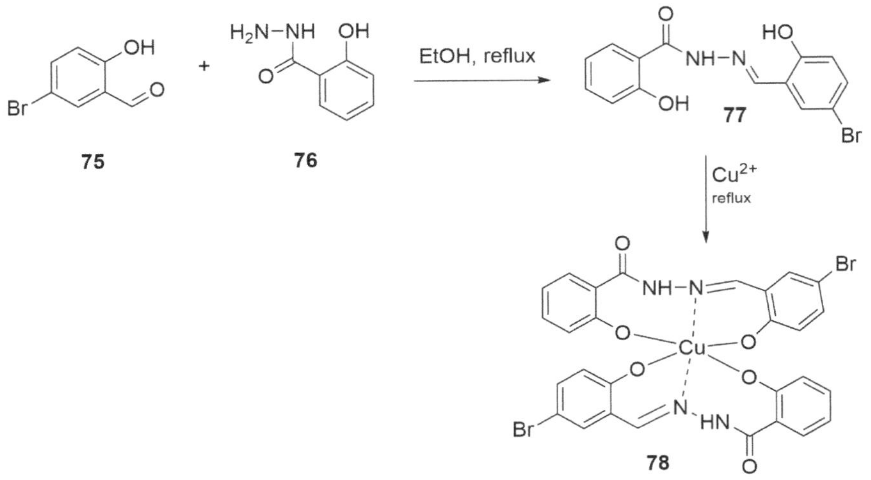

FIGURE 3.13 Synthesis of salicylic hydrazide containing Cu-complex.

commonly utilized for gastric cancer diagnoses but it has lots of side effects and has developed resistance. Liu et al. [35] synthesized Schiff base ligand **77** from commercially available compound **75** and salicylic hydrazide **76** in anhydrous ethanol under reflux for 7 h. Further, synthesized ligand **77** has been used for copper mediated Schiff base complex formation in anhydrous ethanol with copper sulphate under reflux (Figure 3.13). In addition, the synthesized new Schiff base copper complex has been evaluated for antitumor activity on two gastric cancer cell lines BGC-823 and SGC-7901. It was found that copper complex potentially constrains the time and dose dependent way of proliferation of the gastric cancer cells line. The IC 50 of copper complex in BGC-823 and SGC-7901 cells is 01 μM that is much less than the IC 50 value of the anticancer drug cisplatin.

Andiappan et al. [36] synthesized Schiff base ligand **81** using the condensation of compounds **79** and **80** in basic medium (Figure 3.14) and characterized by spectroscopic techniques. Further, this ligand **81** has been used for a novel series of lanthanide-based metal complexes **82a–c** and characterized by various Spectro techniques. The novel prepared metal complexes **82a–c** were evaluated for their cytotoxicity activity versus cervical anticancer and VERO, human breast cancer, cell lines. The Pr and Er complexes showed anticancer activity against VERO, MCF7, and HeLa cancer cells in cytotoxicity studies.

Arafath et al. [37] synthesized novel ligand **85** from easily available compound **83** and **84** in methanol under acidic conditions and subsequently this ligand **85** has been used for metal complexes **86–88** formation with various transition metal under specific conditions (Figure 3.15). The synthesized all Schiff bases and complexes have been identified by various techniques. Further, synthesized chelate compounds were tested for their anticancer property against various human cancer cell lines, that is, cervical, colon and breast cancer. In addition, the synthesized complexes were evaluated for their cytotoxicity property on a standard human cell line. It was observed that among the evaluated complexes, compound **86** showed excellence in halting proliferation of the colon and cervical cancer cells. Complex **87** exhibited cytotoxicity against breast cancer cell line. While ligand exhibited inhibitory effect against colon cancer. The metal chelates **88** (Figure 3.16) demonstrated moderate activity against breast and cervical cancer cells, but it demonstrated irrelevant cytotoxicity against HeLa cells.

Xiao et al. [38] synthesized two novel metal complexes **91** and **92** *via* two flexible Schiff bases, **89** and **90** using solvothermal conditions. Further, X-rays showed that both complexes **91** and **92** exist in distorted octahedral geometry. In addition, the antitumor property of compounds **91** and **92**

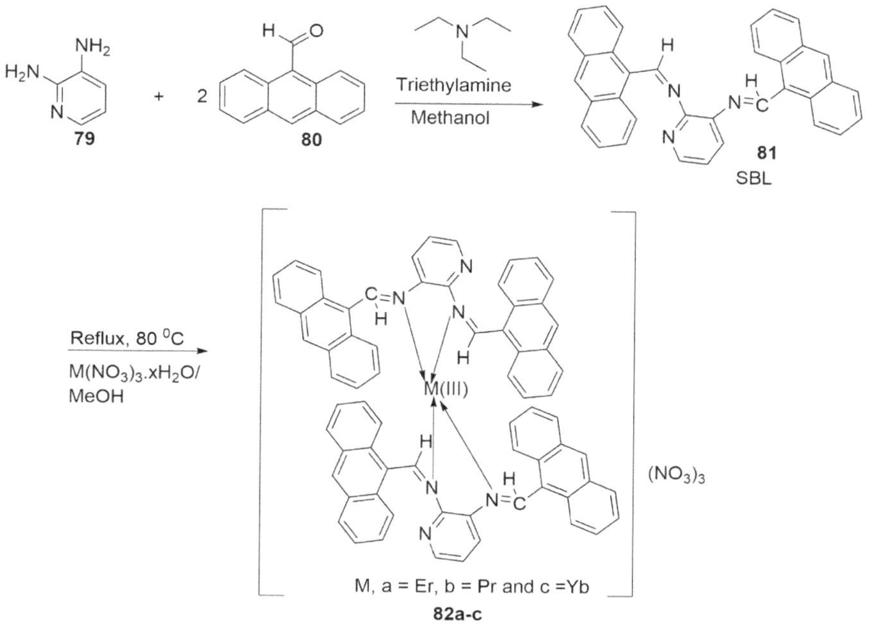

FIGURE 3.14 Synthesis of anthracene containing Schiff base lanthanide complexes.

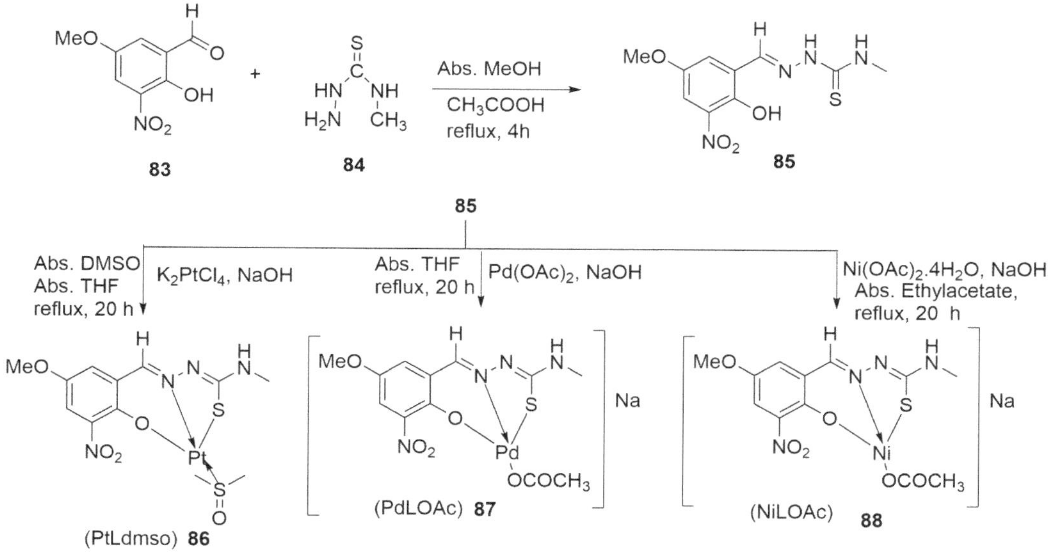

FIGURE 3.15 Synthesis of Ni, Pd, and Pt complexes with novel Schiff base.

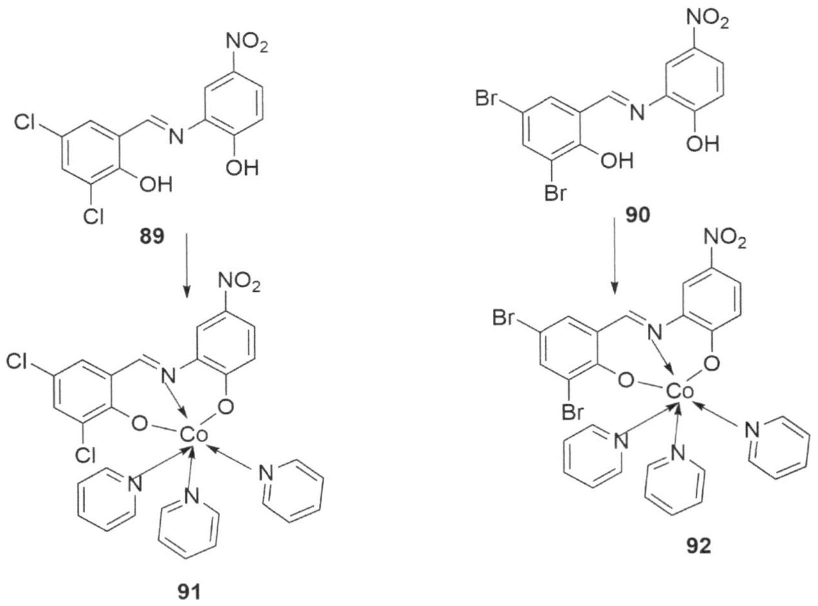

FIGURE 3.16 Synthesis of Pyridine containing Schiff base cobalt complexes.

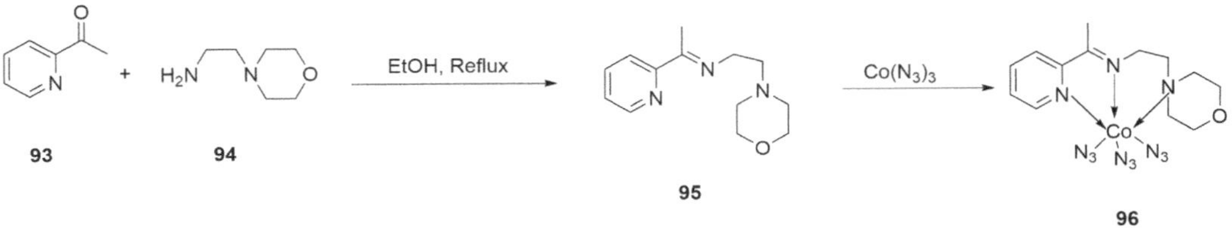

FIGURE 3.17 Synthesis of novel morpholine based Schiff base cobalt complex.

and ligands **89**, and **90** were tested in vitro, against three human skin cancer cell lines, SK-MEL-30, A-431 and HT-144. It was observed that compounds **91** and **92** demonstrated substantial progress inhibition activity on the tumor cell lines compared to ligands **89** and **90**.

Gwaram et al. [39] synthesized cobalt complex **96** from ligand **95**, ligand **95** was synthesized from compound **93** and **94** under reflux in high yield (Figure 17). Crystal data of complex **96** demonstrated distorted octahedral geometry around the metal center with the Schiff base ligand containing N,N,N chelating motif. Complex **96** has been characterized and tested for in vitro anticancer activities. The new complex **96** exhibited excellent antiproliferative property against breast cancer cells MCF-7.

Sathiyanarayanan et al. [40] synthesized Schiff base ligand **99** *via* condensation of compounds **97** and **98** in EtOH. Further, lanthanide complexes **100a–c** were synthesized in moderate to high yields (Figure 18). The cell viability of compound **100a** exhibited apoptosis at IC 50 value of 34.2 mg mL in comparison to compounds **100b** and **100c**. Therefore, desired complexes were synthesized as possibilities for the generation of new anticancer drugs.

Moubeen et al. [41] synthesized a series of mononuclear octahedral Ru (III) complexes **104a–c** in two steps: first, compound **101** treated with **102** in methanol at 80 °C to afford the corresponding Schiff base ligands **103a–c**. In the second step, ligands **103a–c** treated with ruthenium chloride in ethanol to produce the desired complexes **104a–c** in good yields (Figure 19 **and** 20). All ligands

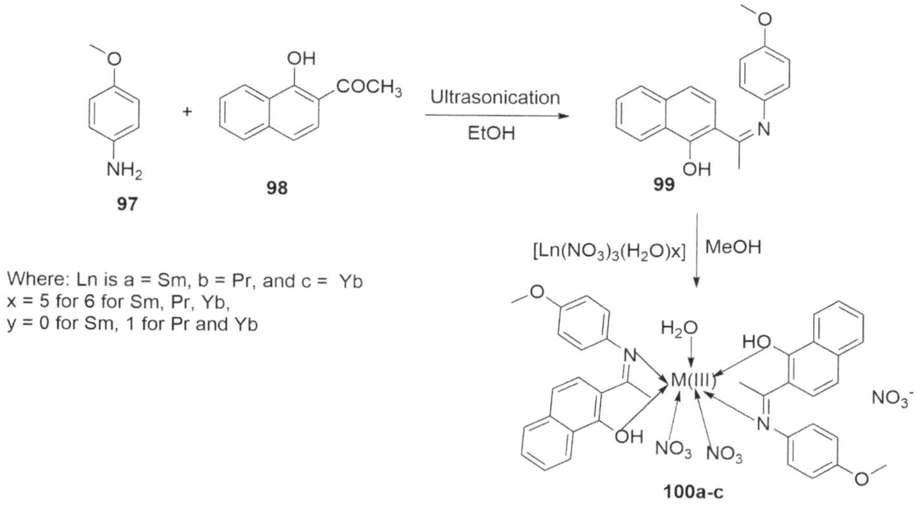

FIGURE 3.18 Synthesis of novel Schiff base containing lanthanide complexes.

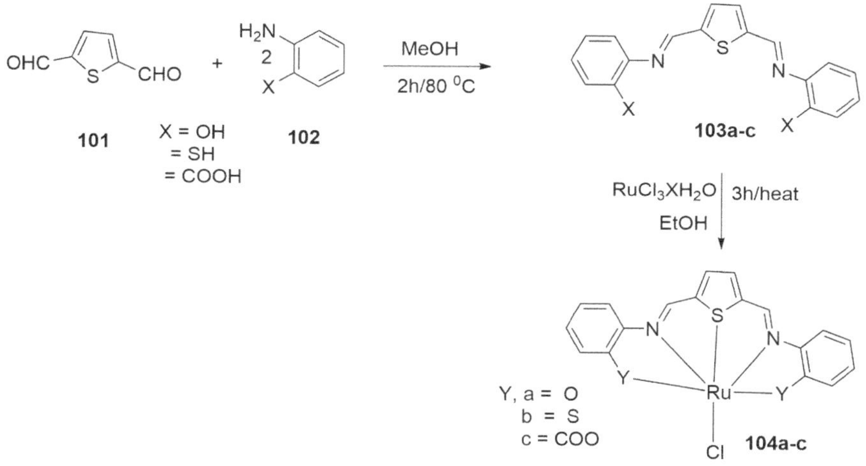

FIGURE 3.19 Synthesis of Ru containing octahedral complex.

and complexes were characterized by various physico-chemical techniques. Further, the anticancer property of the compounds was evaluated against various cancer cell lines such as MCF-7, HeLa and their metal complexes demonstrated strong inhibition against HeLa and MCF-7 cell lines. Ruthenium compound **104c** demonstrated the best antiproliferation activity, permitting its efficacy as a chemotherapeutic candidate for cancer diagnostics.

We know that Platinum-based metal compounds are one of the most fruitful chemotherapeutic agents because they are in substantial demand in cancer treatment despite several side effects. Presently, Ru compounds emerged as an efficient and alternative route due to their promising properties against platinum-resistant cancer. Acharya et al. [42] synthesized Pt and Ru complexes **112–116** with Schiff bases **109–111**. The metal complexes were identified by various spectroscopic methods. Synthesized metal chelates have been evaluated for their anticancer properties and it was demonstrated that compounds containing N, O-coordinating atoms exhibited better than N, N

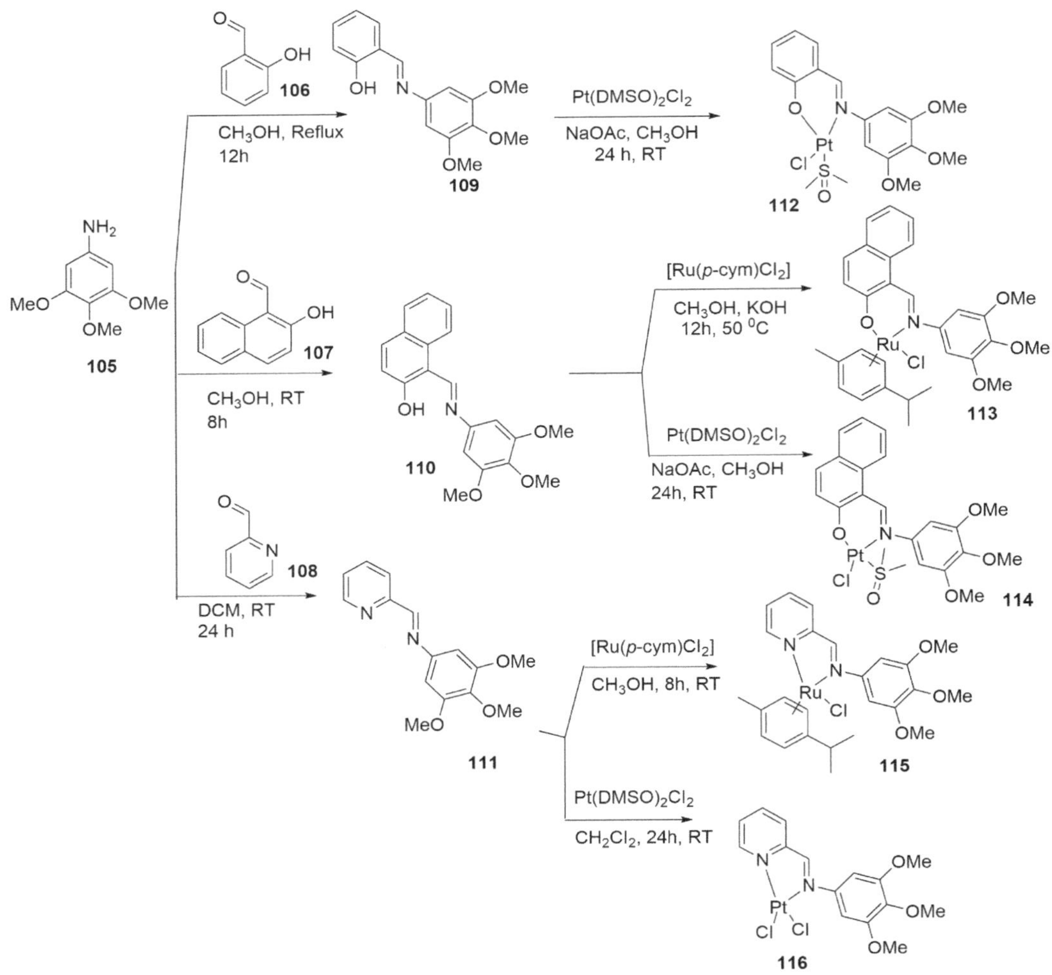

FIGURE 3.20 Synthesis of Pt and Ru complexes via various Schiff bases.

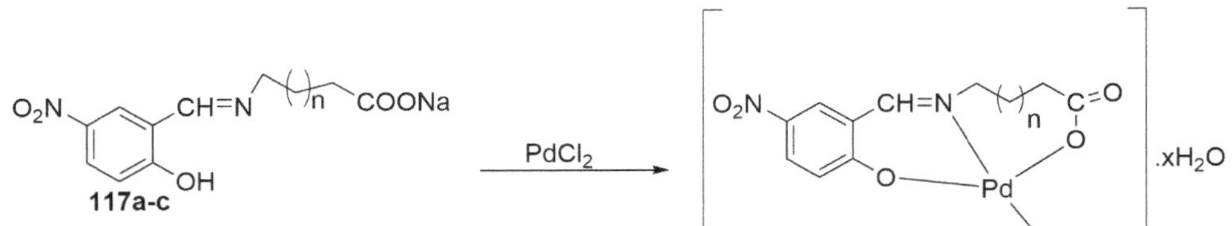

n	x	Ligand	Complex
1	0	117a	118a
2	0	117b	118b
3	1	117c	118c

FIGURE 3.21 Synthesis of novel Schiff base containing transition metal Pd complex.

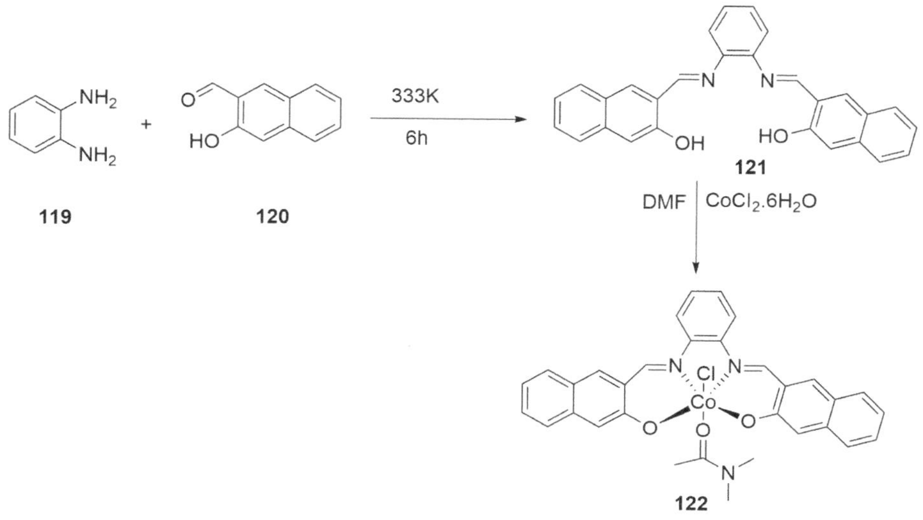

FIGURE 3.22 Synthesis of o-phenylenediamine containing cobalt complex.

coordinating atoms in vitro against various cancer cells, *viz.*, Hep G2, MIA PaCa-2 and MDA-MB-231. The cytotoxicity data recommend that Pt compounds had more potential than the equivalent ruthenium compounds in vitro, and the most cytotoxic compound **114** is 10 to 15 times more poisonous than the available anticancer drugs oxaliplatin and cis platin contrary to MDA-MB-231 cells.

Ozdemir et al. [43] synthesized a series of monosodium salts of Schiff bases **117a–c** [44]. Further, these Schiff bases treated with palladium (II) chloride produced the neutral palladium (II) complexes with good yields (Figure 21). It was found as a green powder, and was stable at 25°C and soluble in DMSO and DMF solvent. All the three complexes **118a–c** were identified by various techniques such as spectroscopic, thermal, conductivity and magnetic susceptibility. In addition, these complexes were evaluated against HeLa and MCF-7 cell lines and a standard HEK-293 cell line. The metal chelates **118a** and **118b** demonstrated a mild antitumor property contrary to HeLa cell lines, but **118c** had better potential than the normal medicine, doxorubicin. All synthesized palladium compounds exhibited the greater super cytotoxicity than doxorubicin contrary to MCF-7 cancer cell lines.

Liao et al. [45] synthesized a novel cobalt (III) complex **122** from ligand **121** and identified by X-ray crystallography (Figure 3.22). Further, the cytotoxicity of cobalt compound **122** was tested against various cancerous cell lines such as LoVo, HeLa, A549 and ordinary cell line LO2 using MTT assays. An anticancer mechanistic route confirmed that the compound **122** repressed cell proliferation via hindering deoxy nucleic acid preparation and worked on HeLa cells.

The Buldurun Research Group [46] synthesized Schiff base ligand **125** from easily available compounds **123** and **124** in methanol. Further, ligand **125** has been used for metal complexes formation **126–127** with the corresponding $CoCl_2.6H_2O$ and $[RuCl_2$ (*p*-cymene)$]_2$ in methanol under reflux for 4 h (Figure 3.23). The structural features were identified from their spectral techniques and magnetic susceptibility measurements and it was found that Ruthenium and Cobalt compounds demonstrated six coordinated octahedral geometry. The anticancer properties of Schiff base, Ruthenium and Cobalt compounds have been tested on the human colon cancer cell line in vitro and biocompatibility properties have been calculated in the fibroblast cell line via the MTT assay. Apart from this, the efficacy of electrochemotherapy on cytotoxic properties of complexes on Caco-2

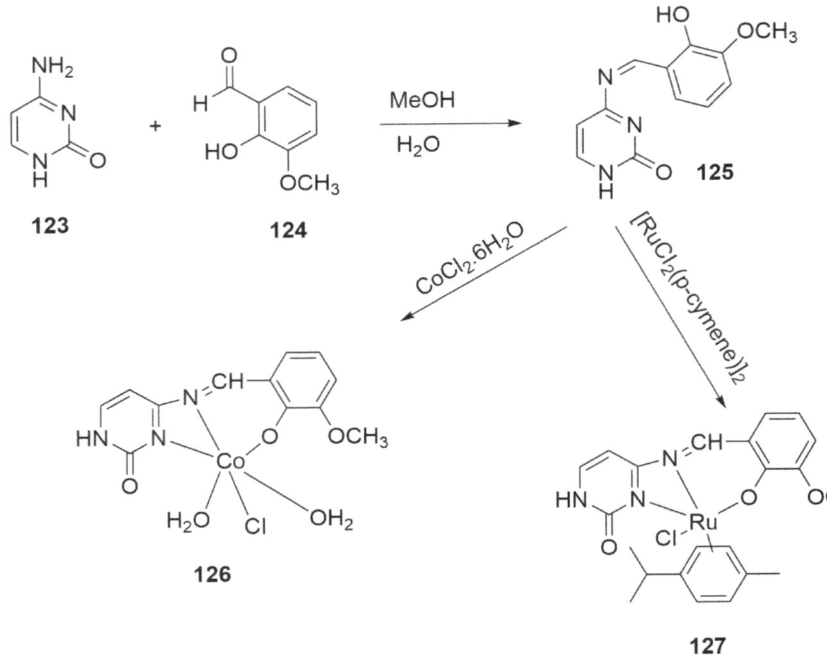

FIGURE 3.23 Synthesis of novel Schiff base containing Cobalt and Ruthenium complexes.

cancer cell line was examined. It was found that Ruthenium and Cobalt complexes exhibited substantial anticancer properties in the colon cancer but the Schiff base had not exhibited important anticancer properties.

Alyar et al. [47] synthesized new Schiff bases **130** and **133** from sulfisoxazole **129**/sulfamethoxazole **132** and substituted salicylaldehyde **128** in ethanol under reflux in good yields and further, these ligands have been used for palladium and copper complex formation. Ligands **130** and **133** were dissolved in ethanol, then appropriate metal chlorides (K_2PdCl_4 and $CuCl_2.2H_2O$) and NaOH solution were added and the resulting solutions were stirred on a hot plate for two days to obtain the complexes in good yields (Figure 3.24). All the synthesized compounds were identified by spectroscopic techniques, conductivity measurements and magnetic susceptibility. The anticancer properties of metal chelates have been tested against human cancer cell lines such as Huh7, MCF7 and HCT116 carcinoma cells and no metal chelates demonstrated potential in human liver and colon cells. Compounds **130**, **133** and their Cu (II) complexes **131b** and **134b** demonstrated hopeful cytotoxic property contrary to all cell lines. IC 50 standards for breast cancer cells are 40 μM for compounds **130**, **133** and their Cu (II) complexes **131b** and **134b** respectively, and Pd (II) complexes **131a** and **134a** exhibited no inhibition effect.

3.3 CONCLUSION

Metal complexes have drawn the attention of researchers in recent years due to their utility in biological processes and applications in the development of new therapeutic candidate such as anti-inflammatory, antibacterial, antioxidant, antifungal, and anticancer. Primarily, this chapter focused on recent research studies carried out using the Schiff bases metal complexes as anticancer agents. It has been found that some metal complexes exhibited fascinating anticancer properties against

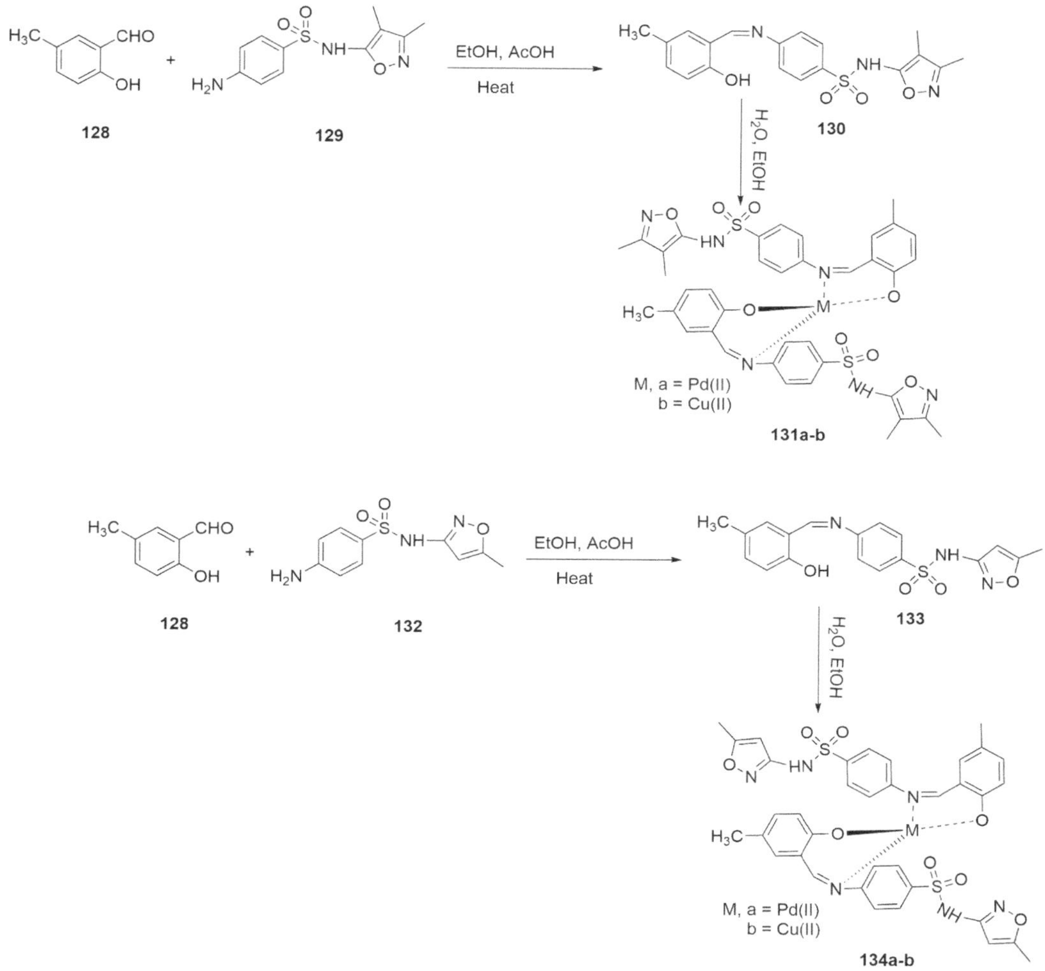

FIGURE 3.24 Synthesis of Pd and Cu complexes *via* sulfisoxazole/sulfamethoxazole based Schiff base.

a wide range of cancer cell lines (MCF7, HepG2, HCT-116, HeLa, etc.), even higher than the approved standard drugs. Mostly anticancer properties have been evaluated *via* well-known and validated practicality assays, revealing IC50 values equal or lower than the clinical used drugs, primarily the cisplatin. In several cases, a better cytotoxic property has been found, the viability of the normal cells being not or slightly affected.

REFERENCES

1 Schiff, H. 1864, Mittheilungen aus dem universitätslaboratorium in Pisa: Eine neue reihe organischer basen. *Justus Liebigs Ann. Chem.* 131:118–119.
2 Vigato, P.A.; Tamburini, S. 2004, The challenge of cyclic and acyclic schiff bases and related derivatives. *Coord. Chem. Rev.* 248:1717–2128.
3 Abdel-Rahman, L.H.; Abu-Dief, A.M.; Shehata, M.R.; Atlam, F.M.; Hassan Abdel-Mawgoud, A.A. 2019, Some new Ag(I), VO(II) and Pd (II) chelates incorporating tridentate imine ligand: Design,

synthesis, structure elucidation, density functional theory calculations for DNA interaction, antimicrobial and anticancer activities and molecular docking studies. *Appl. Organomet. Chem.* 4699–4722.

4 Abu-Dief, M.; Nassr, L.A.E. 2015, Tailoring, physicochemical characterization, antibacterial and DNA binding mode studies of Cu (II) Schiff bases amino acid bioactive agents incorporating 5-bromo-2-hydroxybenzaldehyde. *J. Iran. Chem. Soc.* 12:943–955.

5 Horozi´c, E.; Suljagi´c, J.; Suljkanovic, M. 2019, Synthesis, characterization, antioxidant and antimicrobial activity of Copper (II) complex with Schiff base derived from 2,2-dihydroxyindane-1, 3-dione and Tryptophan. *Am. J. Org. Chem.* 9:9–13.

6 Ejidike, I. 2018, Cu (II) Complexes of 4-[(1E)-N-{2-[(Z)-Benzylidene-amino]ethyl}ethanimidoyl] benzene-1,3-diol Schiff base: Synthesis, spectroscopic, in-vitro antioxidant, antifungal and antibacterial studies. *Molecules* 23: 1581–1598.

7 Al Zoubi, W. 2013, Biological activities of Schiff bases and their complexes: A review of recent works. *Int. J. Org. Chem.* 3:73–95.

8 Sriram, D.; Yogeeswari, P.; Myneedu, N.S.; Saraswat, V. 2006, Microwave assisted synthesis and their evaluation of anti-HIV activities. *Bioorg Med Chem Lett.* 16:2127–2129.

9 Eltayeb, N.E.; Lasri, J.; Soliman, S.M.; Mavromatis, C.; Hajjar, D.; Elsilk, S.E.; Babgi, B.A.; Hussien, M.A. 2020, Crystal structure, DFT, antimicrobial, anticancer and molecular docking of (4E)-4-((aryl) methyleneamino)-1,2-dihydro-2,3-dimethyl-1-phenylpyrazol-5-one. *J. Mol. Struct.* 1213:128185.

10 de Fátima, Â.; de P. Pereira, C.; Olímpio, C.R.S.D.G.; de Freitas Oliveira, B.G.; Franco, L.L.; da Silva, P.H.C. 2018, Schiff bases and their metal complexes as urease inhibitors-A brief review. *J. Adv. Res.* 13:113–126.

11 (a) Shah, S.S.; Shah, D.; Khan, I.; Ahmad, S.; Ali, U.; ur Rahman, A. 2020, Synthesis and antioxidant activities of schiff bases and their complexes: An updated review. *Biointerf. Res. Appl. Chem.* 10:6936–6963.
(b) Ibrahim, M.M.; Fathy, A.M.; Al-Harbi, S.A.; Sallam, S.A.; Al-Juaid, S.; Ramadan, A.E.M.M. 2021, Palladium (II) based imines; synthesis, characterization, X-ray structural analysis; DFT and catalytic hydrogenation study. *J. Organometall. Chem.* 939:121764.
(c) Chen, S.L.; Liu, X.Y.; Li, S.C.; Wu, C.; Li, Z.Y.; Li, T.T. 2021, Synthesis and crystal structure of a new Zn (II) complex with anti-leukemia activity. *Inorg. Nano-Met. Chem.* 51:224–229.

12 Gupta, K.C.; Sutar, A.K. 2008, Catalytic activity of Schiff base complexes. *Coord Chem Rev.* 252:1420–1450.

13 Wang, Z.G.; Ding, X.J.; Huang, Y.Y.; Yan, X.J.; Ding, B.; Li, Q.Z.; Xie, C.Z.; Xu, J.Y. 2020, The development of coumarin Schiff base system applied as highly selective fluorescent/colorimetric probes for Cu^{2+} and tumor biomarker glutathione detection. *Dyes Pigments* 175:108156.

14 (a) Gaur, S. 2003, Physico-chemical and biological properties of Mn(II), Co(II), Ni(II) and Cu(II) Chelates of Schiff bases. *Asian J. Chem.* 15 (1): 250–254.
(b) Genin, M.J.; Biles, C.; Keiser, B.J.; Poppe, S.M.; Swaney, S.M.; Tarapley, W.G.; Romeso, D.L.; Yage, Y. 2000, Novel 1,5-diphenylpyrazole nonnucleoside HIV-1 reverse transcriptase inhibitors with enhanced activity versus the delavirdine-resistant p236l mutant: Lead identification and SAR of 3- and 4-substituted derivatives. *J. Med. Chem.* 43:1034–1040.

15 (a) American Cancer Society. *Cancer Facts& Figures* 2016. Atlanta: American Cancer Society; 2016.
(b) Harold, V. 2006, The new era in cancer research. *Science* 312 (5777):1162–1165.

16 (a) Krohn, K. 2008, Anthracycline chemistry and biology II. Mode of action, clinical aspects and new drugs. In: Krohn, K, editor. *Top Curr Chem.* 283. Heidelberg: Springer; 136.
(b) Coley, H.M. 2008, Mechanisms and strategies to overcome chemotherapy resistance in metastatic breast cancer. *Cancer Treat Rev.* 34(4):378–390.

17 (a) Zhukova, O.; Dobrynin, I. 2000, Current results and perspectives of the use of human tumor cell lines for antitumor drug screening, *Vopr. Onkol.* 47:706–709.
(b) I. Kostova, 2006, Platinum complexes as anticancer agents, Recent Pat. *Anti-cancer drug Discov.* 1:1–22.

18 Nurmamat, M.; Yan, H.; Wang, R.; Zhao, H.; Li, Y.; Wang, X.; Nurmaimaiti, K.; Kurmanjiang, T.; Luo, D.; Baodi, J.; et al. 2021, Novel Copper (II) complex with a 4-acylpyrazolone derivative and coligand induce apoptosis in liver cancer cells. *ACS Med. Chem. Lett.* 12:467–476.

19 Yadamani, S.; Neamati, A.; Homayouni-Tabrizi, M.; Beyramabadi, S.A.; Yadamani, S.; Gharib, A.; Morsali, A.; Khashi, M. 2018, Treatment of the breast cancer by using low frequency electromagnetic fields and Mn (II) complex of a Schiff base derived from the pyridoxal. *Breast* 41:107–112.

20 Demir, B. S.; Gonul, İ.; Çelik, G.G.; İpekbayrak, S.; Saygideger, Y. 2020, Synthesis and anticancer activities of water soluble schiff base metal complexes, *Adyu J Sci.* 10(2): 441–454.

21 Talebi, A.; Salehi, M.; Khaleghian, A. 2021, Synthesis, characterization, electrochemical behavior, molecular simulation studies and in vitro toxicity assessment of new metal Schiff base complexes derived from 3-methoxy-2-hydroxy-benzaldehyde and allylamine. *Journal of Molecular Structure* 1246:131076.

22 Mukherjee, S.; Kumar Pal, C.; Kotakonda, M.; Joshi, M.; Shit, M.; Ghosh, P.; Choudhury, A.R.; Biswas, B. 2021, Solvent induced distortion in a square planar copper (II) complex containing an azo-functionalized Schiff base: Synthesis, crystal structure, *in-vitro* fungicidal and anti-proliferative, and catecholase activity. *Journal of Molecular Structure* 1245:131057.

23 Rajakkani, P.; Alagarraj, A.; Gurusamy Thangavelu, S.A. 2021, Tetraaza macrocyclic Schiff base metal complexes bearing pendant groups: Synthesis, characterization and bioactivity studies. *Inorganic Chemistry Communications* 134:108989.

24 Kavitha, R.; Reddy, C.V.R.; Sireesha, B. 2021, Synthesis, spectroscopic and biological activity evaluation of Ni (II), Cu (II) and Zn (II) complexes of Schiff base derived from pyridoxal and 4-fluorobenzohydrazide. *Nucleosides, Nucleotides and Nucleic Acids*, 40 (8): 845–866.

25 Kavitha, R.; Sireesha, B.; Venkata Ramana Reddy, C. 2016, Synthesis, Characterisation and Antibacterial Activity of Benzohydrazones Derived from 3-Hydroxy-5-Hydroxymethyl-2-Methylpyridine-4-Carboxaldehyde. *Heterocycl. Lett.* 6:741–747.

26 Justin Dhanaraj, C.; Jebapriya, M. 2020, Metal schiff base complexes of tridentate antipyrine based ligand: Synthesis, spectral characterisation, image analysis and biological studies. *Journal of Molecular Structure* 1220:128596.

27 Baoa, R-D.; Songb, X-Q.; Kongc, Y-J.; Lia, F-F.; Liaoa, W-H.; Zhoua, J.; Zhangd, J-H.; Zhaoa, Q-H.; Xub, J-Y.; Chene, C-S.; Xiea, M-J. 2020, A new Schiff base copper(II) complex induces cancer cell growth inhibition and apoptosis by multiple mechanisms. *Journal of Inorganic Biochemistry* 208: 111103.

28 Shi, S.; Yu, S.; Quan, L.; Mansoor, M.; Chen, Z.; Hu, H.; Liu, D.; Liang, Y.; Liang, F. 2020, Synthesis and antitumor activities of transition metal complexes of a bis-Schiff base of 2-hydroxy-1-naphthalenecarboxaldehyde. *Journal of Inorganic Biochemistry* 210:111173.

29 Ganguly, S.; Bauza, A.; Frontera, A.; Ghosh, 2019, Tri-nuclear copper-cadmium complexes of a N_2O_2-donor ligand with the variation of counter anions: Structural elucidation and theoretical study on inter-molecular interactions. *Inorg. Chim. Acta.* 492:142–149.

30 Aslana, H.G.; Akkocb, S.; Kokbudaka, Z. 2020, Anticancer activities of various new metal complexes prepared from a Schiff base on A549 cell line. *Inorganic Chemistry Communications* 111:107645.

31 Mbugua, S.N.; Nicole R. S. 2020, New palladium (II) and platinum (II) complexes based on pyrrole schiff bases: Synthesis, characterization, X-ray structure, and anticancer activity. *ACS Omega* 5:14942–14954.

32 Dasa, M.; Biswas, A.; Kundu, B.K.; Charmierc, M.A.J.; Mukherjeed, A.; Mobina, S. M.; Udayabhanub, G.; Mukhopadhyay, S. 2019, Enhanced pseudo-halide promoted corrosion inhibition by biologically active zinc (ii) schiff base complexes. *Chemical Engineering Journal*, 357:447–457.

33 Gwaram, N.S.; Ali, H.M.; Khaledi, H.; Abdulla, M.A.; Hadi, A.H.A.; Lin, T.K.; Ching, C.L.; Ooi, C.L. 2012, Antibacterial evaluation of some schiff bases derived from 2-Acetylpyridine and their metal complexes. *Molecules* 17: 5952–5971.

34 Manojkumar, Y.; Ambika, S.; Arulkumar, R.; Gowdhami, B.; Balaji, P.; Vignesh, G.; Arunachalam, S.; Venuvanalingam, P.; Thirumurugan, R.; Akbarsha, M.A. 2019, Synthesis, DNA and BSA binding, in vitro anti-proliferative and in vivo anti-angiogenic properties of some cobalt (III) Schiff base complexes. *New J. Chem.* 43:11391.

35 Xia, Y.; Liu, X.; Zhang, L.; Zhang, J.; Li, C.; Zhang, N.; Xu, H.; and Li, Y. 2019, A new Schiff base coordinated copper (II) compound induces apoptosis and inhibits tumor growth in gastric cancer. *Abrash Cell Int.* 19:81.

36 Andiappan, K.; Sanmugam, A.; Deivanayagam, E.; Karuppasamy, K.; Kim, H-S.; Vikraman, D. 2018, *In vitro* cytotoxicity activity of novel Schiff base ligand–lanthanide complexes. *Scientific Reports* 8:3054.

37 Arafath M.A.; Adam, F.; Razali, M.R.; Ahmed Hassan, L.E.; Ahamed, M.B.K.; Majid, A.M.S.A. 2017, Synthesis, characterization and anticancer studies of Ni (II), Pd (II) and Pt (II) complexes with Schiff

base derived from N-methylhydrazinecarbothioamide and 2-hydroxy-5-methoxy-3-nitrobenzaldehyde. *Journal of Molecular Structure* 1130:791–798.

38 Xiao1, Y.-J.; Diao, Q.-C.; Liang, Y.-H.; Zeng, K. 2017, Two novel Co (II) complexes with two different Schiff bases: Inhibiting growth of human skin cancer cells. *Brazilian Journal of Medical and Biological Research* 50(7):6390.

39 Gwaram, N. S. 2017, Synthesis and characterization of a Schiff base cobalt (III) complex and assessment of its anti-cancer activity. *Chem.Search Journal* 8(2):56–67.

40 Sathiyanarayanan, V.; Prasath, P.V.; Chandra Sekhar, P.; Ravichandran, K.; Easwaramoorthy, D.; Mohammad, F.; Al-Lohedan, H.A.; Oh, W.C.; Sagadevan, S. 2020, Docking and in vitro molecular biology studies of p-anisidine-appended 1-hydroxy-2-acetonapthanone Schiff base lanthanum (III) complexes. *RSC Adv.* 10:16457–16472.

41 Moubeena, S.A.M.; El-Shahata, M. F.; Abdel Aziza, A.A.; and Attiaa, A.S. 2021, Synthesis, characterization and biological evaluation of novel octahedral Ru (III) complexes containing pentadentate Schiff base ligands. *Current Chemistry Letters* 10:17–32.

42 Acharya, S.; Maji, M.; Chakraborty, M.P.; Bhattacharya, I.; Das, R.; Gupta, A.; and Mukherjee, A. 2021, Disruption of the microtubule network and inhibition of VEGFR2 phosphorylation by cytotoxic N, O-coordinated Pt (II) and Ru (II) complexes of trimethoxy aniline-based Schiff bases. *Inorg. Chem.* 60:3418–3430.

43 Ozdemir, O.; Gürkan, P.; Dilber. Y. S.; Mustafa Ark, Y.D. 2020, Novel palladium (II) complexes of N-(5-nitro-salicylidene)-Schiff bases: Synthesis, spectroscopic characterization and cytotoxicity investigation. *Journal of Molecular Structure* 1207:127852.

44 Güngor, O.; Gürkan, P. 2014, Synthesis and characterization of higher amino acid Schiff bases, as monosodium salts and neutral forms. Investigation of the intramolecular hydrogen bonding in all Schiff bases, antibacterial and antifungal activities of neutral forms. *J. Mol. Struct.* 1074:62–70.

45 Liao. W-H.; Song. X-Q.; Kong. Y-J.; Bao. R-D.; Li. F.F.; Qi-Hua Zhao. J.Z.; Xu. J.Y.; Ming-Jin Xie, N.X. 2021, A novel Schiff base cobalt (III) complex induces a synergistic effect on cervical cancer cells by arresting early apoptosis stage. *Biometals* 34:277–289.

46 Mehmet, E.; Alkı¸ R.; Kele¸ U.; Alan, Y.; Turan, N.; Buldurun, K. 2021, Cobalt and ruthenium complexes with pyrimidine based schiffbase: Synthesis, characterization, anticancer activities and electrochemotherapy efficiency. *Journal of Molecular Structure* 1226:129402.

47 Alyar, S.; S¸ Ozmen, U.O.; Adem, S.; Alyar, H.; Bilen, E.; Kaya, K. 2021, Synthesis, spectroscopic characterizations, carbonic anhydrase II inhibitory activity, anticancer activity and docking studies of new Schiff bases of sulfa drugs. *Journal of Molecular Structure* 1223:128911.

4 Biologically Active Schiff Base Metal Complexes as Antibacterial and Antifungal Agent

Prashant Tevatia, Sumit Kumar, and Priyansh Kumar Utsuk*
Department of Chemistry, Gurukula Kangri (Deemed to be University), Haridwar, India
*Corresponding author
Email: Prashant_tevatia@yahoo.co.in

CONTENTS

4.1 INTRODUCTION

Bacterial infection has been a major cause of illness and mortality in humans as well as in our useful domestic animals from the long past. Bacteria create illness by ejecting toxins into the body and the infection spreads as bacteria multiply. Bacteria can be found in practically any environment. Bacterial growth is inevitable, even in the most harsh and favorable situations. Bacteria are everywhere around us. Bacteria can be found inside the human body as well as on our skin, hair, and other surfaces. Microorganisms in our gut or stomach are beneficial bacteria that we need to live a healthy existence. Others that live on our skin and on our surfaces are mostly harmless, and their growth has no negative impact on our health. However, some bacteria strains are toxic to humans and cause a variety of infections and diseases.

When humans come into contact with a range of bacteria that are dangerous and cause infection or disease, they can cause a variety of infections and disorders. Often, they are classified as gram-positive and gram-negative bacteria. Some significant and common diseases caused by bacteria are Tetanus (by *Clostridium tetani*), Diphtheria (by *Corynebacterium diphtheria*), Syphilis (by *T. pallidum*), Leprosy (by *Mycobacterium leprae*), Pertussis (by *Bordetella pertussis*), Cholera (by *Vibrio cholera*), Gonorrhoea (by *N. gonorrhoeae*), Plague (by *Y. pestis*), Pulmonary Tuberculosis (by *M. tuberculosis*) and Salmonellosis (by *S. enteritis*). Even the most common bacterial illnesses might be lethal to humans in the past. Bacterial infections, which are today easily treatable, were

DOI: 10.1201/9781003291459-4

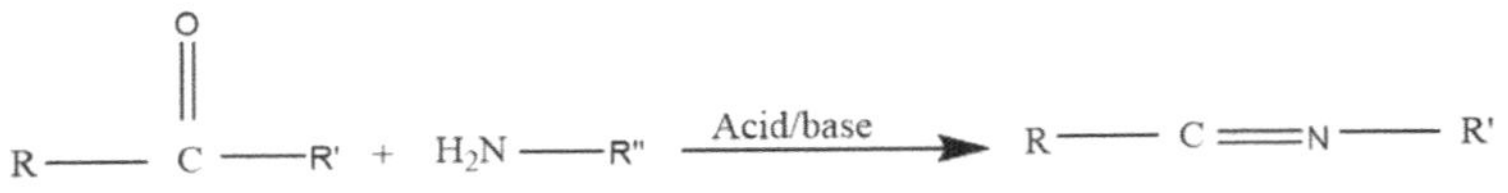

R=H, CH_3, Ph, C_2H_5

R"= H, Ph, CH_2

SCHEME 4.1 General route of Schiff base synthesis.

historically lethal and posed a threat to people all over the world. In the sphere of medicine, the development of vaccines and antibiotics was a great breakthrough. Antibiotics are the most important type of antibacterial agents. Antibiotics are produced naturally (by one microorganism fighting against another). The use of antibiotics became popular with the advent of penicillin in 1928. Antibacterial agents, other than antibiotics, are sulphonamides. Recently various Schiff bases have been found effective as antibacterial agents. Bacterial strains have evolved and become antibiotic-resistant; hence scientists are interested in developing newer, more effective antibacterial agents and Schiff base ligands and complexes have provided an effective and significant field in this quest.

Fungal infections or mycoses are also very common in humans caused by various species of fungi. These infections result in mild to severe diseases, more common in our external organs but many times in our internal organs also, due to ingestion or inhalation of spores of fungi. Mycoses are often classified as superficial, subcutaneous and systemic on the basis of body parts affected by fungi. Some common fungal diseases include superficial infections like Candidiasis (thrush), Dermatophytosis (ringworm), subcutaneous infections like eumycetoma and systemic diseases like Pneumocystis Pneumonia, Histoplasmosis and systemic Candidiasis, and so on. These fungal infections are immensely irritating and sometimes even fatal. Their treatment involves the use of topical (applied on affected part) as well as systemic antifungal medicines. Many well-known antifungal medicines such as Ketoconazole, Fluconazole, Albendazole, Clotrimazole, and so on, are based on Schiff base like azoles, imidazole, triazoles, and so on.

Under certain conditions, to form Schiff bases (which are named after Hugo Schiff), any primary amine can react with a ketone or an aldehyde [1]. Imines were originally prepared in the nineteenth century. A lot of imine manufacturing methods have been revealed since then. Schiff discusses a conventional synthesis that employs azeotropic distillation to combine a carbonyl molecule with an amine

A Schiff base, commonly known as azomethine or imine, is a nitrogen homologue of an aldehyde group or ketone group with an imine or azomethine group substituting for the carbonyl group (>CO). Metal complexes' chelating capabilities and biological applications have received a lot of attention, and they can be used as models for biologically significant species [2–4]. Schiff bases have been shown to have activities against fungi, bacteria, viruses and also have antimalarial, antiproliferative, anti-inflammatory and antipyretic properties [5–8]. The azomethine or imine groups found in a variety of natural-derived, non-natural and natural chemicals have been shown to be critical to their biological roles. The inclusion of such imine group in the compounds has been shown to be crucial for their biological actions. Ancistrocladidine (Figure 4.1) is a secondary metabolite produced by plants belonging to the Dioncophyllaceae and Ancistrocladacea families. Its molecular scaffold contains an imine group [9–10].

Multiple antibiotic resistance in bacteria has been linked the increase in death rates related to infectious diseases. The lack of effective remedies is primarily to blame for this problem [11]. It is undeniably a medical imperative to create novel antibacterial drugs with more effective mechanism of action [12]. Schiff bases have been proposed as antibacterial candidates. Among the several Schiff bases, azomethine ligands produced from benzimidazole are of particular interest for

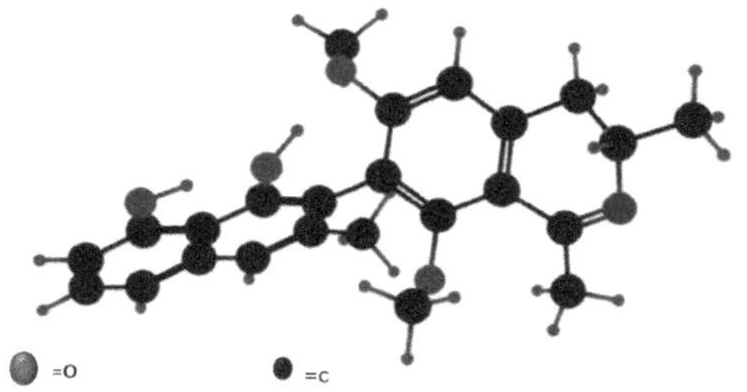

FIGURE 4.1 Structure of ancistrocladidine.

their anticancer, antifungal, diuretic, and antibacterial properties. The benzimidazole nucleus is a bioactive heterocyclic molecule whose derivatives have been proven to be effective in a variety of pharmacological and pharmacokinetic applications [13–15]. Albendazole, mebendazole, and thiabendazole are common anthelmintic medications [16]. Benzimidazole derivatives for nosocomial strains of *Stenotrophomonas malthophilia* were made using 2-thioalkyl and thioaryl substituted benzimidazole, as well as 5,6-dinitrobenzimidazole [17]. Infrared irradiation/no solvent, K-10/microwave, water suspension medium BF_4/molecular sieves, solid-state synthesis, Solvent-free/clay/microwave irradiation, $NaHSO_4.SiO_2$/microwave/solvent-free, silica/ultrasound irradiation and solvent-free/CaO/microwave are some of the examples of new methods and innovations studied in microwave irradiation and established among these developments because of their operational simplicity, increased reaction rate and high selectivity [18–21]. The Majetich and Rousell groups conducted the first independent studies of microwave irradiation because it removes the need of Dean-Stark apparatus and aromatic solvents for azeotropic water removal. Microwave irradiation is less harmful to the environment than earlier processes. Another advantage of this technology is that it allows for more efficient reactions in a shorter amount of time [22–24]. Infections of fungi are not normally limited to the surface tissues; in fact, deadly systemic fungal infections are on the rise. The fundamental cause of the increasing number of ill people, including the elderly on immunosuppressive medication, AIDS patients, cancer patient, and those undergoing treatment to restore blood cell production, is fungal infection. Stronger antifungal medications are needed, and various Schiff bases have been identified as prospective antifungal agents [25–27]. In this chapter we focus on antifungal and antibacterial activities of various acyclic and cyclic Schiff base complexes with metals.

4.2 MECHANISM OF ACTION OF ANTIMICROBIAL ACTIVITY

Antimicrobial agents' effects are investigated by looking into their mechanisms of resistance. When antimicrobial drugs act on critical microbial processes, they have a minimum or no influence on host function. Different antimicrobial drugs have different mechanisms of action. The modes of action of antimicrobial medications can be categorized depending on the structure of bacteria or the function that the drugs affect, and they include:

(1) Cell wall production is inhibited.
(2) Ribosome function is inhibited.
(3) Nucleic acid production is inhibited.
(4) Folate metabolism inhibition.
(5) Cell membrane function is inhibited.

To analyse or evaluate the in vitro antibacterial activity of an extract or a pure medicine, a range of laboratory methods can be performed. Disc diffusion and broth or agar dilution are the most

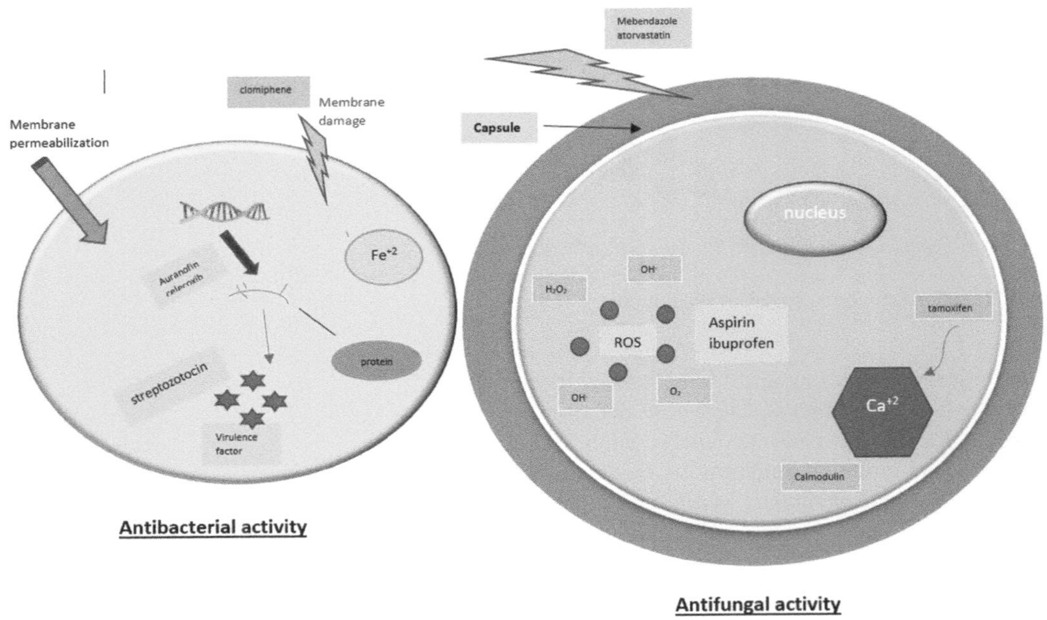

FIGURE 4.2 Mechanism of antimicrobial activity of schiff base metal complexes.

well-known and basic methods. Other procedures, such as the poisoned food technique, are used for antifungal testing. For further research into an agent's antimicrobial effect, flow cytofluorometric methods and the time-kill method are recommended because they provide information on the type of the inhibitory effect (bacteriostatic or bactericidal), the time or concentration-dependent nature of the inhibitory effect, and the cell damage inflicted on the test microorganism. The antimicrobial activities are discussed under the two main functions (1) antibacterial and (2) antifungal. The antibacterial and antifungal activity of an antibacterial drug is shown in Figure 4.2.

4.3 CYCLIC SCHIFF BASES METAL COMPLEXES

Template condensation of glutaric, succinic, malonic, and adipic acids with 2,6-diaminopyridine and diethylenetriamine resulted in macrocyclic Mn (II) complexes (Figure 4.3). The complexes were evaluated in vitro against a variety of hazardous bacterial strains including *P. cepacicola*, *S. aureus*, and *Xanthomonas campestris*, to see if they could limit growth. High solubility (due to which complexes are rapidly absorbed in cells of bacteria and fungi, resulting in activating the enzymes), particle fineness, metal ion size, and the addition of bulkier chemical moieties all contribute to the complexes' increased activity. From acetophenone, two Schiff bases were synthesized: (i) 4-ethyl-6-[(E)-1-(2 nitrophenyl)imino]ethylbenzene-1,3-diol, (ii) 4-ethyl-6-(E)-1-[(3-nitrophenyl)imino] ethylbenzene-1,3-diol. The antibacterial activity of the second was better in the polar solvent DMF against the tested bacteria. In 1,4-dioxane, Znis proved to have the best antibacterial activity than iron, whereas, in DMF, Ni showed the best antibacterial activity than Zn and Fe. The bacteria *P. vulgaris* proved to be the most resistant [28,29].

Four Cu^{+2} complexes has been reported by Shabbir et al., namely, bis((E)-2-((4-(4-biphenyloxy)phenylimino)methyl)phenolate) Cu^{+2}, (Copper(bis((E)-2-((4-(4-biphenyloxy) phenylimino)methyl)phenolate))$_2$), bis((E)-2-((4 phenoxyphenylimino)methyl)phenolate) Cu^{+} ^{2} and (Copper(bis((E)-2-((4 phenoxyphenylimino)methyl)phenolate))$_2$). Certain complexes are bioactive in these settings, according to biological studies.The Schiff bases were very active

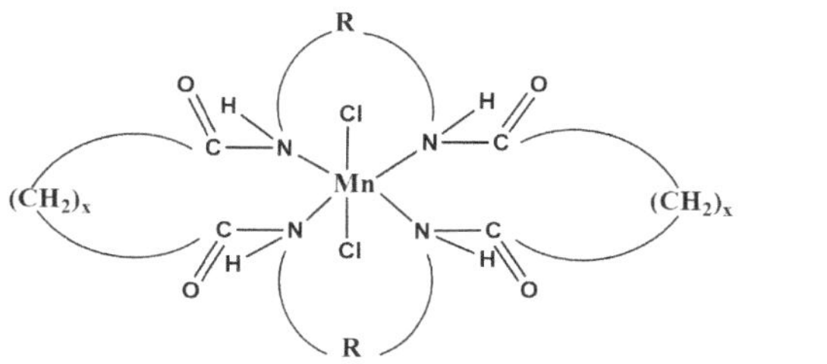

$L_1 = C_5H_3N$ and X=1 $L_2 = C_5H_3N$ and X=2 $L_3 = C_5H_3N$ and X=3
$L_4 = C_5H_3N$ and X=4 $L_5 = C_4H_9N$ and X=1 $L_6 = C_4H_9N$ and X=2
$L_7 = C_4H_9N$ and X=3 $L_8 = C_4H_9N$ and X=4

FIGURE 4.3 Macrocyclic Mn (II) complexes synthesized from dibasic acids.

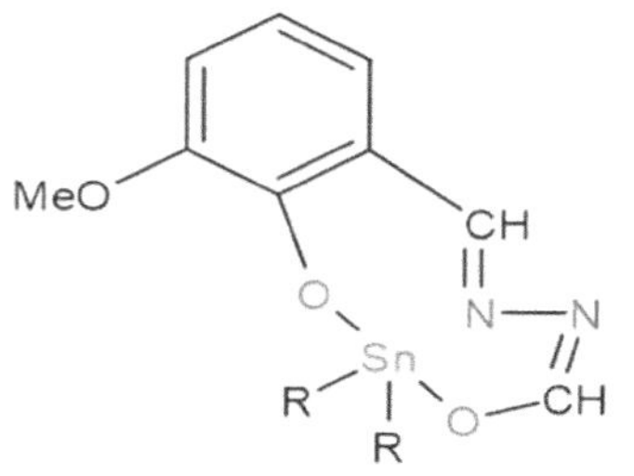

FIGURE 4.4 Structure of hydrazide ligand.

in shielding DNA from OH radicals. DNA protection enhanced as the concentration of the test substance increased. [30, 31]

Bioactivity of dimethyl di-n-octyl, diphenyl, n-butylchlorotin(IV), diethyl, di-tert-butyl, diethyl, derivatives of N-(2-hydroxy-3-methoxybenzylidene) form hydrazide ligand (Figure 4.4) was investigated. The stronger antibacterial property of organotin derivatives is most likely because of decreased polarity of the Sn^{+4} after complex formation with the ligand. This decrease in polarity is due to the delocalization of charge of tin metal with ligand groups as well as delocalization of electrons across the chelating ring. The tin Sn atom becomes more solubilized as a result, enabling complexes to infiltrate deeply into the lipid layer of the cell membrane [32].

Seven Schiff bases were produced by combining thiophene-2 carboxaldehyde, pyrrole-2-carboxaldehyde with 2-amino benzimidazole, 2-amino-4-methylphenol, 2-aminophenol and 2-amino-4-nitrophenol. Individual Schiff bases inhibited the development of the microorganisms studied to varying degrees [33]. With newly created physiologically active ligands, certain metal complexes of Zn^{+2}, Cu^{+2},Ni^{+2}, Co^{+2} were synthesized (Figure 4.5). Condensation of 4-amino-3-ethyl-5-mercapto-*s*-triazole and 2- acetylpyridine with 4-amino-5-mercapto-3-methyl-s-triazoleAntibacterial activity of heterocyclic bidentate Schiff bases was found to be significantly higher than that of several commercial antibiotics [34].

Co(II) complexes of N-salicylidene derivative (with amine and nitro group) with certain amino acids were synthesized. For the investigation of the biological effects of the Schiff base and its mixed ligand complexes, they were first scanned for a variety of microorganism strains. The observed results show that metal complexes have stronger antimicrobial properties than free ligands [35].

Condensation of isatin with ortho-phenylenediamine yielded a new class of complexes of the form $[M(TML)X]X_2$, where tetradentate ligand is TML type (M = $Cr^{(III)}$ and $Fe^{(III)}$, X= OAc^-, NO_3^-, and Cl^- (Figure 4.6). Antibacterial activity was also investigated in vitro on the complexes. Several of the complexes demonstrated impressive antibacterial activity against a variety of bacterial species [36].

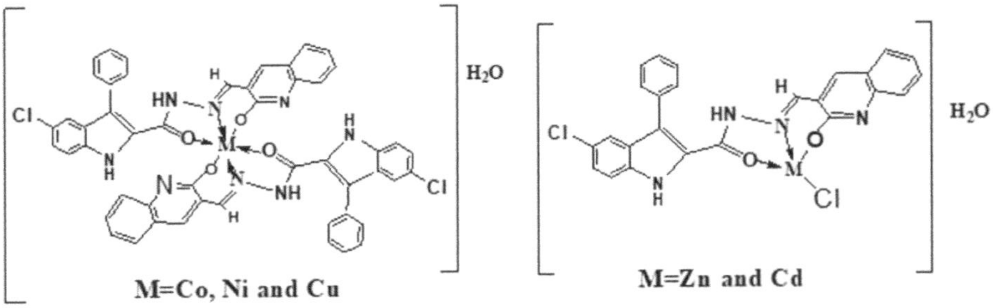

FIGURE 4.5 Structure if metal complexes of Zn^{+2}, Cu^{+2},Ni^{+2}, Co^{+2} and Cd.

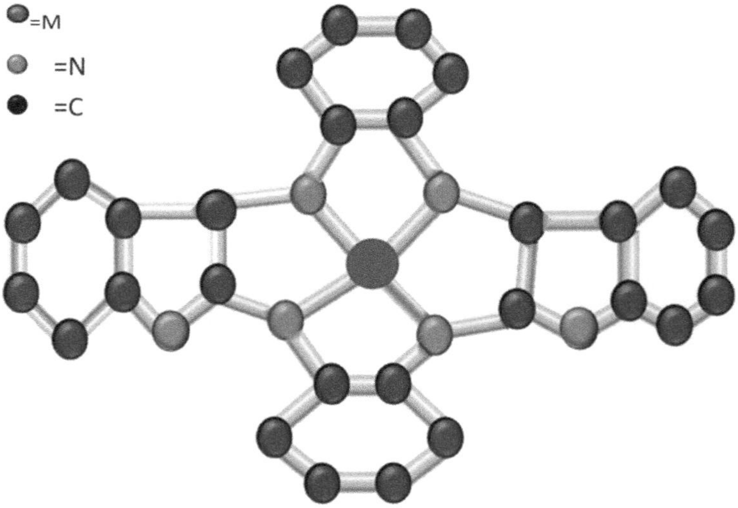

FIGURE 4.6 Indole-2,3-dione (isatin) macrocyclic derivative with M= Cr(III) and Fe (III).

4.4 ACYCLIC SCHIFF BASE METAL COMPLEXES

The ligand hard-soft Schiff base, which is made by reacting 4,6-diacetylresorcinol with thiocabohydrazide in a 1:2 molar ratio, has been utilized to make mono- and bimetallic acyclic and macrocyclic complexes. The prepared SBs ligand form acyclic binuclear complexes with Co^{+2}, Ni^{+2}, Cu^{+2}, Zn^{+2}, Mn^{+2} and Ru^{+3} chloride, $VO^{(IV)}$ sulphate, and $UO_2^{(VI)}$ nitrates, except for VO^{+4} and Ru^{+3}, which form acyclic mononuclear complexes. The antibacterial properties of the ligands and their complexes with metals were tested with gram-negative bacteria *P. fluorescens* and gram-positive bacteria *S. aureus*. The bulk of the complexes demonstrated just minor antibacterial activity against these infections [37]. Schiff base condensation of N,N-bis(3-aminopropyl)ethylene diamine, tris(2-aminoethyl) amine, 5-methylsalicylaldehyde, diethylenetriamine, triethylenetetramine, N,N-bis(aminopropyl) piperazine, and $GdNO_3.5H_2O$ (Figure 4.7) produced a new family of noncyclic mononuclear gadolinium(III) complex. Antibacterial activity was tested on all gadolinium (III) complexes against *P. aeruginosa*, *S. typhi*, *Klebsella*. Using concentrations 25, 50,75,100,125 µm in DMSO, it was found that $[Gd(III)L^1(NO_3)_2]$ along with $[Gd\ (III)L^5(NO_3)_2]$ showed maximum inhibitory activity against all microorganisms [38].

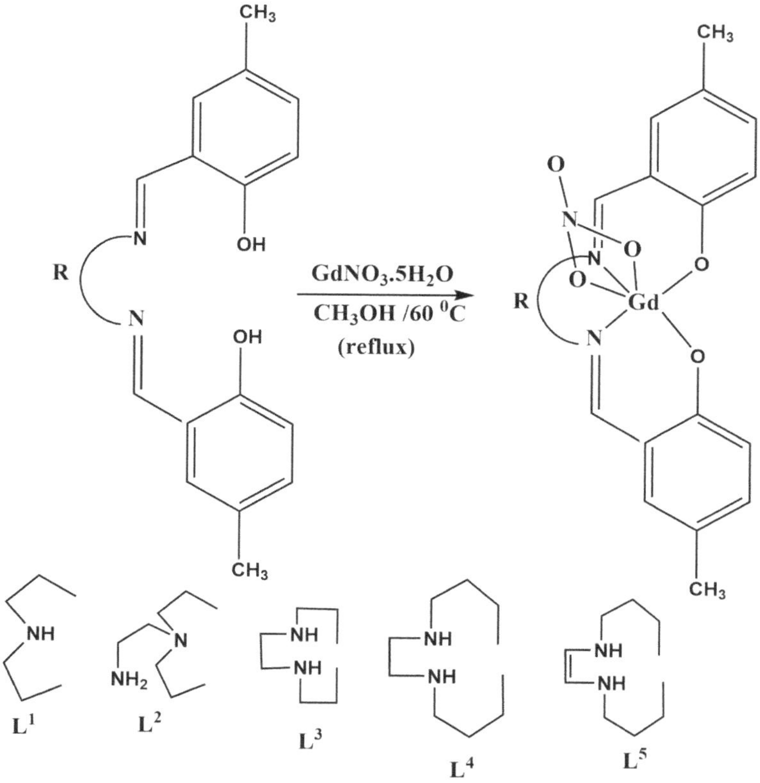

FIGURE 4.7 Synthesis of a family of noncyclic mononuclear gadolinium(III) complex.

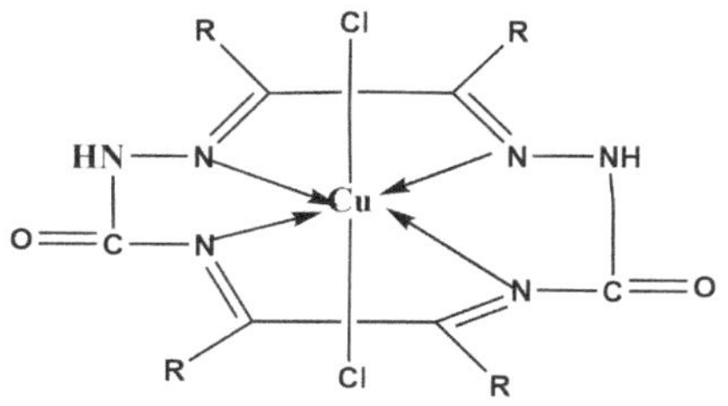

FIGURE 4.8 Proposed structure of Cu(II) complex.

The reaction of a precursor of diamine (thiosemicarbazide or semicarbazide) with diketone (glyoxal, diacetyl, benzil or 2,3-pentanedione) yielded macrocyclic ligands. These ligands' Copper (II) complexes have been synthesized. The antibacterial activity of the complexes against diverse bacteria was tested in vitro using the paper disc method, and the results show that binuclear complexes (Figure 4.8) are more powerful than free ligands [39].

4.5 SCHIFF BASE METAL COMPLEXES: ANTIFUNGAL PROPERTY

4.5.1 Cyclic Schiff Bases Metal Complexes

N-4-hydroxysalicylidene-glycylglycine and N-O-vanillal-glycylglycine have been synthesized, as well as their Mn(II), Co(II), Ni(II), and Cu(II) complexes (Figures 4.9 and 4.10). The results

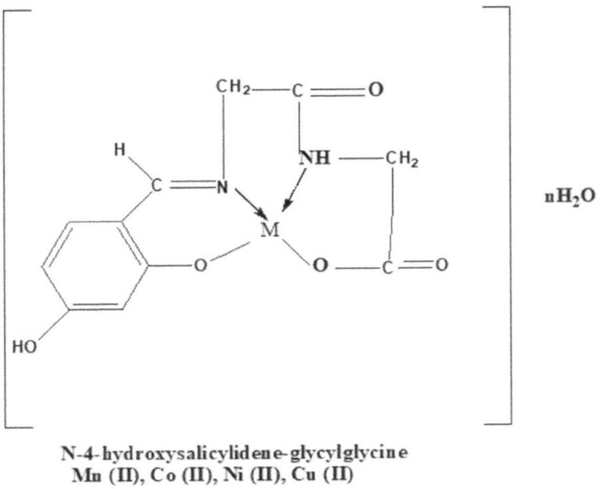

FIGURE 4.9 Mn(II), Co(II), Ni(II), and Cu(II) complexes derived from salicaldehyde.

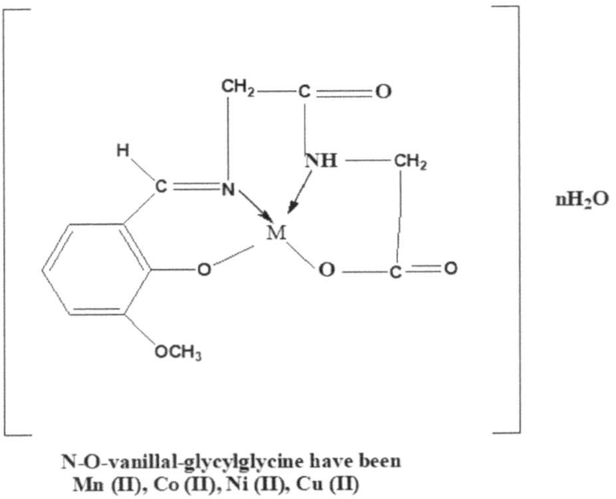

FIGURE 4.10 Metal complexes derived from vanillin.

reveal that under the test conditions, the ligand N-O-vanillal-glycylglycine.3H_2O and its complexes have a strong inhibitory effect on *Cryptococcus neoformans*. When the concentrations are high enough, all of the chemicals have a strong inhibitory effect on *Candida albicans*. Except for Cu' N-4-hydroxysalicylidene-glycylglycine.H_20 the ligand 4-hydroxysalicylidene-glycylglycine.H_20 and its complexes demonstrate high antifungal action to the two chosen fungi when the compound concentration is 200 ppm [40].

[RuHCl(CO)(PPh_3)$_2$(B)] reactions of $Ru^{(II)}$ complexes for bidentate Schiff base ligands produced by condensing salicylaldehyde with $C_6H_5NH_2$ and ortho, meta, para toluidine have been used with B = PPh_3, pyridine or piperidine (Figures 4.10 and 4.11). The SBs (Schiff base) and novel complexes were tested against the fungus *Aspergillus flavus* to see how effective they were. The ligand containing pyridine shows better antifungal activity. The pi-electron delocalization spread all over the chelate ring which increases the lipophilicity of Ruthenium and it supports the penetration through the lipid cell membrane [41].

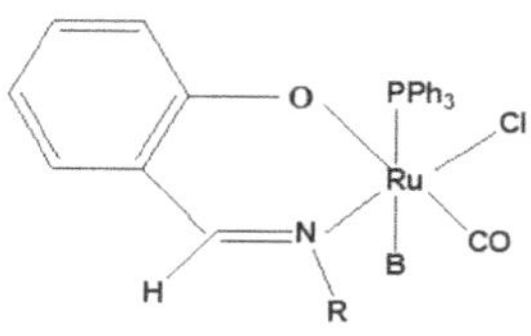

Stucture of Ru(II) complex B=PPh_3,Py
R=Ph,MeC_6H_4

FIGURE 4.11 Structure of Ru(II) complexes.

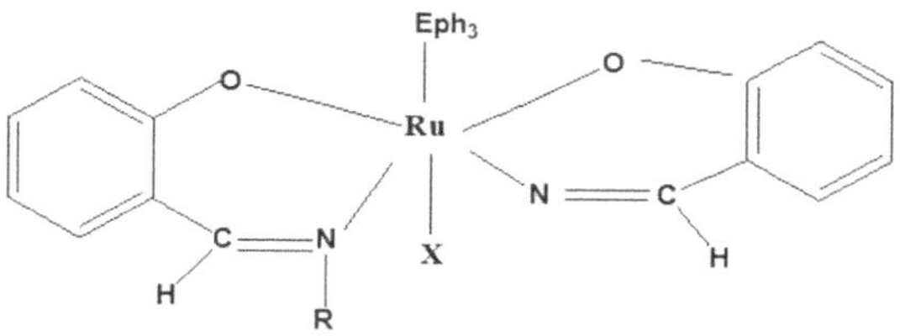

Structure of Ru(III) complexes E= P or As X=Cl or Br
R=-CH_3, -C_6H_{11} C_5H_9N

FIGURE 4.12 Structure of Ru(III) complexes.

The reaction of [RuX_3(Eph_3)] (E = phosphorous X = chlorine; E = arsenic X = chlorine or bromine) and [RuBr (PPh_3) (CH_3OH)] with SBs ligands obtained after condensation of salicyaldheyde with cyclohexylamine, methylamine have been carried out (Figure 4.11). At two different doses, the antifungal activity of SB ligands and their Ru chelates against *Aspergillus flavus and Fusarium species* was investigated in vitro [42].

By the reaction of 2-aminobenzoic acid with acetoacetanilido-4-aminoantipyrine in ethanol, a new diaza dioxa type Sb has been created and produced. Manganese(II), Nickle(II), Copper(II), Zinc(II), Cd(II), Hg(II), and VO(IV) were used to make metal complexes of the Schiff base in solid state. *C. albicans*, *A. flavus*, *Rhizoctonia bataticola*, *Rhizopus stolonifer*, *Aspergillus niger* and Trichoderma harizanum were used to investigate the compounds' antifungal activity in vitro. The metal complexes all outperformed the free ligand in terms of antifungal activity. The metal complexes' MIC was determined to be in the range of 10-31 g/ml [43]. Salicylaldehyde and *p*-amino acetanilide were combined in ethanol to create tridentate Schiff base, which was then treated with $CuCl_2$, $NiCl_2$, $CoCl_2$, and $ZnCl_2$ to form the complexes. *Candida albicans*, *Rhizoctonia bataticola*, *Trichoderma harizanum*, *Aspergillus flavus*, *Rhizopus stolonifer* and *Aspergillus niger* were evaluated against microorganisms cultivated on potato dextrose agar as media [44].

Co(II), Ni(II), Cu(II), and Mn(II) metal-ligand complexes, as well as SB formed from, thiophene-2 carboxy, furfur aldehyde, pyridine-2 aldehyde with vinyl aniline, were formed and evaluated for Candida *albicans and Candida* Kruse. [45]

Some novel SB ligands were synthesized by condensing 2-aminobenzoic and ortho-phthaldehyde and acid in a 1:2 ratio (Figure 4.13), and it has been demonstrated to be efficient against fungi (Candida *albicans and Aspergillus flavus*). [46]

A series of novel derivatives of derived benzothiazole's, benzoxazole benzimidazole, containing the pyrazole had been synthesized by reacting substituted phenylenediamine with 3-aryl-4-formyl pyrazole in the search for a new series of antifungal agents (Figures 4.14 and 4.15). The antifungal activity of the new compounds synthesized by this scheme, against *A. niger* and *C. albicans* strains, was tested in vitro [47].

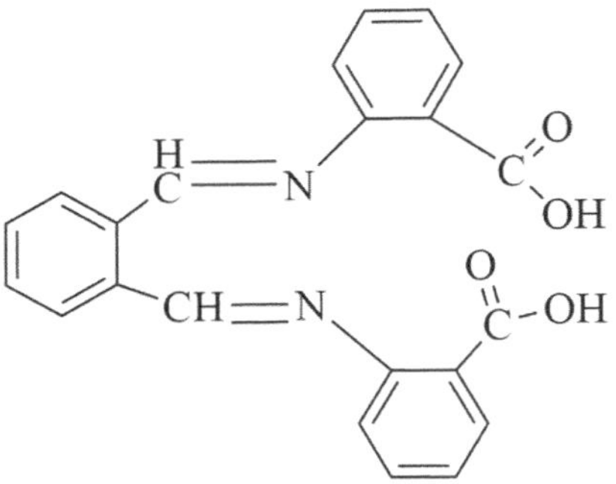

FIGURE 4.13 Structure of synthesised novel Schiff's base.

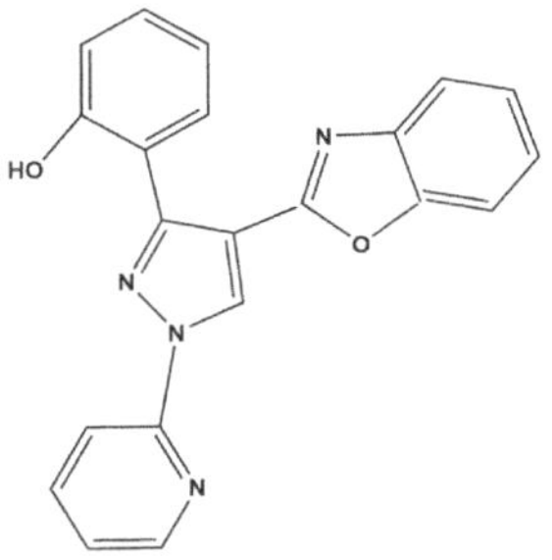

FIGURE 4.14 Synthesised novel derivatives of benzothiazole, benzoxazole and benzimidazole.

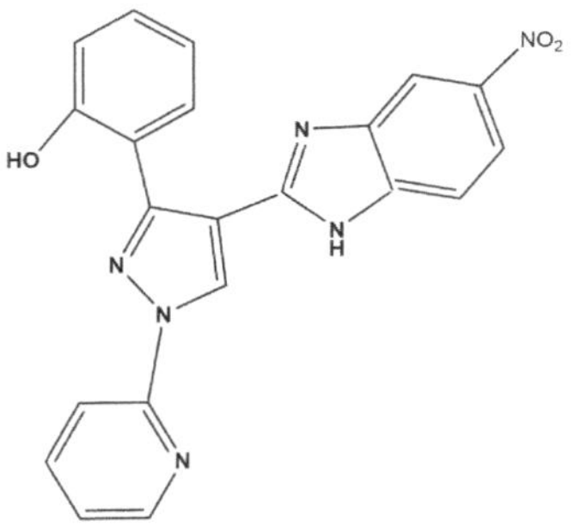

FIGURE 4.15 Derivatives of benzothiazole, benzoxazole and benzimidazole containg pyrazole.

By condensation of 2,3-butanedione(diacetyl) and 1,8-diaminonaphthalene with metal salts, a unique class of complexes of the type $[M(C_{28}H_{24}N_4)X]X_2$ where M = Chromium (III), iron (III), and manganese (III), X = chloride, nitrate, Oxoacetate have been synthesized. The antifungal property of all the complexes was checked against a variety of fungal species in vitro. The results were compared to the outcomes of a typical antifungal treatment [48].

The metal complexes were synthesized, with M = Nickle(II), Copper(II), X = chlorides, nitrates, sulphates, and the $(C_{19}H_{18}N_2O_2S)$ ligand of tetradentate aza oxo macrocyclic species. To investigate fungicidal examinations of synthesized metal complexes against *Aspergillus niger*, *Salmonellarolfsii*, *Fusarium oxysporum*, and *Alternaria brassicae*, the technique reported earlier was used. The activity against fungal strains of all metal complexes was excellent [49].

4.5.2 Acyclic Complexes Schiff Base Complexes

Transition metal complexes of Cobalt(II), Nickle(II) and Copper(II) with the ligand L (Figure 4.16) were synthesized. Antifungal and antibacterial properties of the ligand and its metal complexes were evaluated for three fungi, *Alternaria brassicae*, *Aspergillus niger*, and *Fusarium oxysporum* [50]. SBs were synthesized from 4-acetylbiphenyl and S-benzyldithiocarbalate [H_2-N-N-H-C(S)-S-CH_2-Ph] and their complexes of $Zn^{(II)}$, $Ni^{(II)}$, $Co^{(II)}$. The SBs and the synthesized complexes showed antifungal action against *phytoplithiru capsica* [51].

By combining a SB generated from 2,3-butanedione, 5-methyl-2,6-pyrimidine-dione with glycine and chlorides or acetates of metal as a template, new bimetallic complexes were formed where M = Copper(II), Nickle(II), Cobalt(II), or Zinc (II). The antifungal activity in vitro of the SBs and bimetallic complexes against *Candida parapsilosis* and *Candida albicans* was tested using the poisoned food technique [52]. Two novel uni-metallic and bimetallic complexes with SB ligand generated from the reaction of diethylenetriamine and 3-Acetylcoumarine in a 2:1 molar ratio were synthesized. The SB and complexes were investigated for antifungal efficacy against *Fusarium oxysporum*, a dangerous fungus [53]. The SB ligand synthesized from the reaction of thiocabohydrazide and 4,6-diacetylresorcinol and its Co(II), Ni(II), Cu(II), Zn(II) and Mn(II) complexes (Figure 4.17) have been investigated for antifungal activity against *Fusarium oxysporum in vitro* and were found mild active [54].

The SB [(S,Z)-2-((2-hydroxy-1-phenylethylidene) amino)-3-(4-hydroxyphenyl)propanoic acid] was synthesized and tested for antifungal properties against *Rhizopus stolonifer*, *Aspergillus niger*, *Aspergillus flavus and Alternaria alternata*. The activity was found to be more active against *Altarnaria alternata* than other fungi [55]. SB metal complexes (M= Cu^{II}, Ni^{II}, Zn^{II}, Co^{II}, Mn^{II}, Fe^{III} and Cr^{III}) of ligand N^4 -(7' -chloroquinoline-4'-ylamino)-N^1 -(2-hydroxy-benzylidene) thiosemicarbazone (Figure 4.18) were tested for antifungal action against two fungal strains (*Candida albicans and Fusarium solani*) and found to be physiologically active. The metal complex, on the other hand,

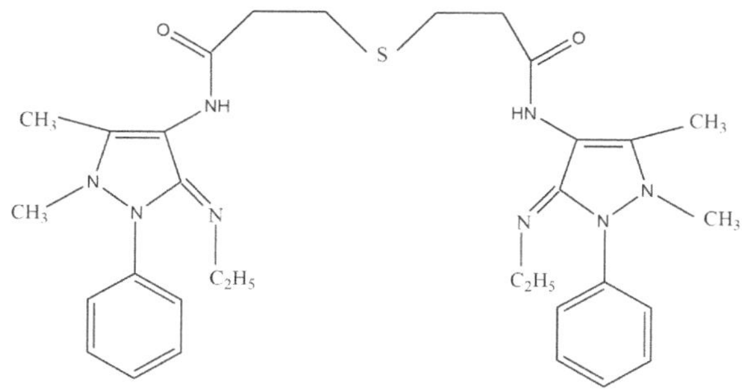

FIGURE 4.16 Structure of ligand 'L'.

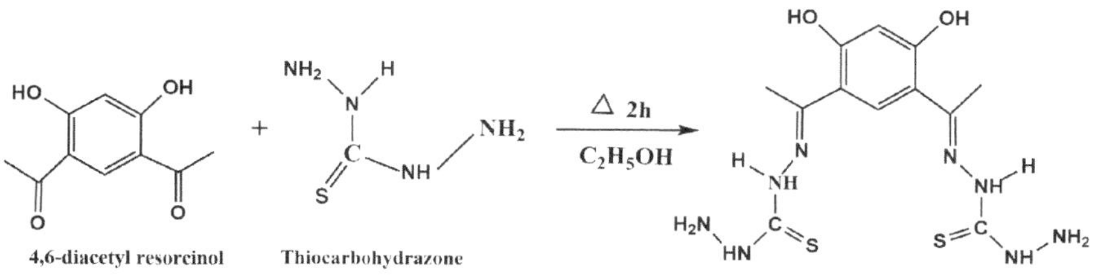

FIGURE 4.17 Synthesis of ligand by reacting thiocabohydrazide and 4,6-diacetylresorcinol.

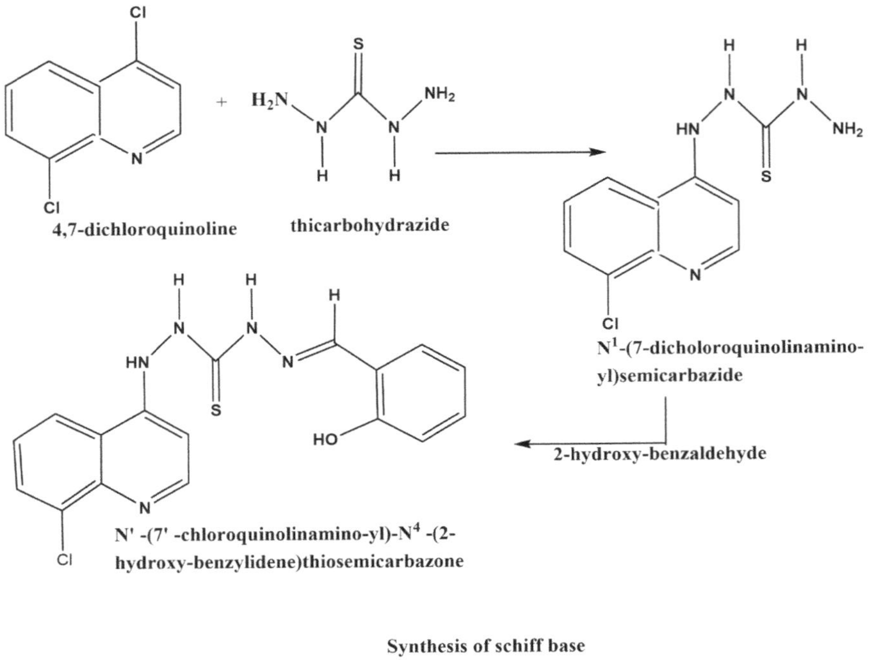

FIGURE 4.18 Synthesis of Schiff base.

showed remarkable antibacterial action, which is to be expected given that it contains sulphur and nitrogen atoms [56].

The synthesis of 11 Schiff bases has been completed. Piperonal (3,4methylenedioxybenzaldehyde) was combined with the appropriate aromatic primary amines to produce them. *Microsporum canis*, *Trichophyton rubrum*, *Epidermophyton floccosum* and *Microsporum gypsum* were the fungi employed [57]. A novel Schiff's base ligand (NO-donor), 1,4-diformylpiperazine bis(4-imino-2,3-dimethyl-1-phenyl-3- pyrazolin-5 one) generated by condensation of 1,4 diformylpiperzine with 4-aminoantipyrineif formed with metals (Nickle(II) Copper(II) and Cobalt(II)) was produced, which was generated by the acid catalyzed. The chemicals were tested against a variety of opportunistic diseases, including *A. niger*, *F. oxysporum* and *A. brassicae.* The complexes of copper meta were discovered to have the most antifungal activity [58].

4.6 CONCLUSION

Any primary amine can react with a ketone or an aldehyde under specific conditions to generate Schiff bases (named after Hugo Schiff) [1]. Imines were created for the first time in the nineteenth century. Since then, numerous imine production methods have been revealed. Schiff describes a common synthesis that combines a carbonyl molecule with an amine via azeotropic distillation. Schiff bases are compounds resembling aldehydes or ketones in which the carbonyl group has been replaced with an imine or azomethene group. In addition to their extensive industrial applications, they demonstrate a vast array of biological activity. This brief article provides an overview of the most promising antimalarial, antibacterial, antifungal, and antiviral Schiff bases. In addition, an overview of synthetic methods used to prepare Schiff bases is provided.

REFERENCES

1. Schiff, H. "Mitteilungen aus dem universitatslaboratorium in Pisa: Eineneue reihe organischer basen." (In German) Justus Liebigs Ann. Chem, 131, 118–119 [1864].
2. Abdel-Rahman, Laila H., Abu-DiefRafat, Ahmed M., El-Khatib, M., Mahdy, Shimaa, Amin Abdou Seleem, Abdel-Fatah, "New Cd(II), Mn(II) and Ag(I) Schiff Base Complexes : Synthesis, Characterization, DNA Binding and Antimicrobial Activity." International Journal of Nanomaterials and Chemistry, 2(3), 83–91 [2016].
3. Mokhles, M. Abd-Elzaher, "Spectroscopic characterization of some tetradentate Schiff bases and their complexes with nickel, copper and zinc." Journal of Chinese Chemical Society, 48, 153–158 [2001].
4. Ana O. de Souza, Fabio C. S. Galetti, Célio L. Silva, Beatriz Bicalho, Márcia M. Parma, Sebastião F. Fonseca, Anita J. Marsaioli, Angela C. L. B. Trindade, Rossimíriam P. Freitas Gil, Franciglauber S. Bezerra, Manoel Andrade-Neto, Maria C. F. de Oliveria, "Antimycobacterial and cytotoxicity activity of synthetic and natural compounds." Química Nova, 30, 7, 1563–1566 [2007].
5. Abdel-Rahman, Laila H., El-Khatib, Rafat M., Nassr, Lobna A.E., Abu-Dief, Ahmed M., El-Din Lashin, Fakhr, "Design, characterization, teratogenicity testing, antibacterial, antifungal and DNA interaction of few high spin Fe (II) Schiff base amino acid complexes." Molecular and Biomolecular Spectroscopy, 111, 266 [2013].
6. Harpstrite, Scott E., Collins Oksman, Silvia D., Goldberg, Daniel E., Sharma, Vijay, "Synthesis, characterization, and antimalarial activity of novel schiff-base-phenol and naphthalene-amine ligands." Medicinal Chemistry, 4, 392–395 [2008].
7. Alam, Mohammad Sayed, Choi, Jung-Hyun, Lee, Dong-Ung, "Synthesis of novel Schiff base analogues of 4-amino-1, 5-dimethyl-2-phenylpyrazol-3-one and their evaluation for antioxidant and anti-inflammatory activity." Bioorganic and Medicinal Chemistry, 20, 4103–4108 [2012].
8. Jain, Pallavi, Vijay K. Vishvakarma, Prashant Singh, Sulekh Chandra, Dinesh Kumar, and Namita Misra. "Co (II) and Ni (II) Complexes of a Heterocyclic Ligand: Synthesis, Characterization, Docking and Biological Activity." Iranian Journal of Science and Technology, Transactions A: Science, 1–13 [2022].
9. Fayez, Shaimaa, Cacciatore, Alessia, Sun, Sijia, Kim, Minjo, Ake Assi, Laurent, Feineis, Doris, Awale, Suresh, Bringmann, Gerhard, "Ancistrobrevidines A-C and related naphthylisoquinoline alkaloids with cytotoxic activities against HeLa and pancreatic cancer cells, from the liana Ancistrocladus abbreviates." Bioorganic and Medicinal Chemistry, 30, 115950 [2020].
10. Fayez, Shaimaa, Bruhn, Torsten, Feineis, Doris, AkeAssi, Laurent, Awale, Suresh, Bringman, Gerhard, "Ancistrosecolines A–F, Unprecedented seco-Naphthylisoquinoline Alkaloids from the Roots of Ancistrocladus abbreviatus, with Apoptosis-Inducing Potential against HeLa Cancer Cells." Pubs.acs.org/jnp 25 [2019].
11. Alekshun, M.N. and Levy, S.B. "Molecular mechanisms of antibacterial multidrug resistance." Cell, 128(6), 1037–1050 [2007].
12. Fernando Baquero, Fernando, "Gram-positive resistance: challenge for the development of new antibiotics." Journal of Antimicrobial Chemotherapy, 39, Suppl. A, 1–6 [1997].
13. Garuti, Laura, Roberti, Marinella, Cermelli, Claudio, "Synthesis and antiviral activity of some N-benzenesulphonylbenzimidazoles." Bioorganic and Medicinal Chemistry, 9, 2525–253 [1999].
14. Krishnanjaneyulu, Immadisetty Sri, Saravanan, Govindaraj, Vamsi, Janga, Supriya, Pamidipamula, Bhavana, Jarugula Udaya, Kumar Mittineni Venkata Sunil, "Synthesis, characterization and antimicrobial activity of some novel benzimidazole derivatives." Journal of Advanced Pharmaceutical Technology & Research 5 (1), 21–27 [2014].
15. Yellajyosula Lakshmi, Narasimha Murty, Guduru Durga, Anjli Jha, "Synthesis, characterization, and antimicrobial activity of some new 2-diazo-benzimidazole derivatives and their Ni(II), Cu(II), and Ag(I) complexes." Medicinal Chemistry Research, 22, 2266–2272 [2013].
16. Koehler, Peter, "The biochemical basis of anthelmintic action and resistance." International Journal for Parasitology, 31s, 336–345 [2001].
17. Kazimierczuk, Zygmunt, Upcroft, Jacqueline A., Upcroft, Peter, Gorska, Agata, Starosciak, Bohdan, Laudy, Agnieszka, "Synthesis, antiprotozoal and antibacterial activity of nitro-and halogeno-substituted benzimidazole derivatives." Acta Biochemical Polonica, 49, 1, 185–195 [2001].

18. Schmeyers, J., Toda, F., Boy, J., & Kaupp, G. "Quantitative solid–solid synthesis of azomethines." Journal of the Chemical Society, Perkin Transactions, 2, (4), 989–994 [1998].
19. Varma, Rajender S., Rajender Dahiya, and Sudhir Kumar. "Microwave-assisted Henry reaction: Solventless synthesis of conjugated nitroalkenes." Tetrahedron letters, 38, 29, 5131–5134 [1997].
20. Tanaka, Koichi, and Ryusuke Shiraishi. "Clean and efficient condensation reactions of aldehydes and amines in a water suspension medium." Green Chemistry, 2(6), 272–273 [2000].
21. Vazquez, Miguel A., Landa, Miguel, Reyes, Leonor, Miranda, Rena, Tamariz, Joaquın, Delgado, Fransisco. "Infrared irradiation: effective promoter in the formation of n-benzylideneanilines in the absence of solvent." Synthetic Communications, 34, 2705–2718 [2013].
22. Gopalakrishnan, M., Sureshkumar, P.. Kanagaraan, V. "New environmentally- friendly solvent free synthesis of imines using calcium oxide under microwave irradiation." Research on Chemical Intermidates, 33, 541–548 [2007].
23. Langa, Fernando, La Cruz, Pilar, De, De La Hoz, Antonio, Angeldiaz-Ortiz and Barra, Enriquediez, "Microwave irradiation more then just a method for a accelerating reaction." Contemporary Organic Synthesis [1997].
24. Hassan, Ali M., Heakal, Bassem H., Younis, Ahmed, Bedair, M.A., Elbialy, Zaghloul I., Abdelsalam Mohamed, Moataz Mohamed. "Synthesis of some triazole Schiff base derivatives and their metal complexes under microwave irradiation and evolution of their coorsion inhibition and biological activity." Egyptian Journal of Chemistry, 62, 1603–1624 [2019].
25. Nucci, Marcio, Marr, Kieren A. "Emerging fungal diseases." Clinical Infectious Diseases, 41(4), 521–526 [2005].
26. Martins, C. V. B., da Silva, D. L., Neres, A. T. M., Magalhaes, T. F. F., Watanabe, G. A., Modolo, L. V., Sabino, A. A., de Fátima, A., de Resende, M. A., "Curcumin as a promising antifungal of clinical interest." Journal of Antimicrobial Chemotherapy, 63(2), 337–339 [2009].
27. Martins, C.V.B., de Resende, M.A., da Silva, D.L., Magalhaes, T.F.F., Modolo, L.V., Pilli, R.A., de Fátima, A., "In vitro studies anticanadidal activity of goniothalamin enantiomers." Journal of Applied Microbiology, 107(4), 1279–1286 [2009].
28. Chaudhary, Ashu, Bansal, Nidhi, Gajraj, A., Singh, R.V., "Antifertility, antibacterial, antifungal and percent diseaseincidence aspects of macrocyclic complexes of manganese(II)." Journal of Inorganic Biochemistry, 96, 393–400 [2003].
29. Nair, R., Shah, A., Baluja, S., Chand, S., "Synthesis and antibacterial activity of some Schiff base complexes." J. Serb. Chem. Soc., 71(7) 733–744 [2006].
30. Shabbira, Muhammad, Akhtera, Zareen, Ismailb, Hammad, "Synthetic bioactive novel ether based Schiff base and their copper(II) complexes." Journal of Molecular Structure, 1146, 57–61 [2017].
31. Shabbir, Muhammad, Akhter, Zareen, Raithby, Paul R., Thomas, Lynne H., Ismail, Hammad, Arshad, Faiza, Mirza, Bushra, Teat, Simon J., Mahmood, Khalid, "Synthesisc, characterisation and biological properties of novel ON donor bidentate Schiff bases and their copper(II) complexes." Journal of Coordination Chemistry, 70(14), 2463–2478 [2017].
32. Shujah, Shaukat, Niaz Muhammad, Afzal Shah, Saqib Ali, Nasir Khalid, and Auke Meetsma. "Bioactive hepta-and penta-coordinated supramolecular diorganotin (IV) Schiff bases." Journal of Organometallic Chemistry, 741–742, 59–66 [2013].
33. Shanty, Angamaly Antony, Jessica Elizabeth Philip, Eeettinilkunnathil Jose Sneha, Maliyeckal R. Prathapachandra Kurup, Sreedharannair Balachandran, and Puzhavoorparambil Velayudhan Mohanan. "Synthesis, characterization and biological studies of Schiff bases derived from heterocyclic moiety." Bioorganic Chemistry, 70 (2017): 67–73 [2016].
34. Singh K, Barwa MS, and Tyagi P, "Synthesis, chacterization and biological studies of Co(II), Ni(II), Cu(II) and Zn(II) complexes with bidentate Schiff bases derived by heterocyclic ketone." European Journal of Medicinal Chemistry, 41, 147–153 [2006].
35. A. R. Patil, K. Donde, S. Raut, V. Patil, R. Lokhande. "Synthesis, characterization and biological activity of mixed ligand Co(II) complexes of schiff base 2-amino-4-nitrophenol-n-salicylidene with some amino acids." Journal of Chemical and Pharmaceutical Research, 4, 1413–25 [2012].
36. Singh, D.P., Grover, Vidhi., Kumar, Krishan, Jain. Kiran. "Metal Ion Prompted Macrocyclic Complexes Derived from Indole-2,3-dione (isatin) and O-phenylenediamine with their Spectroscopic and Antibacterial Studies." Acta Chem. Solv., 57, 775–780 [2011].

37. Abou-Hussein, Azza A.A., Linert Wolfgang "Synthesis, spectroscopic and biological activities studies of acyclic and macrocyclic mono and binuclear metal complexes containing a hard-soft Schiff base." Spectrochimica acta Part A: Molecular and Biomolecular Spectroscopy, 95, 596–509 [2012].
38. Vijayaraj, A., Prabu, R., Suresh, R., Kumari, R.S., Kaviyarasan, V. and Narayanan, V., "Spectral, Electrochemical, Fluorescence, Kinetic and Anti-microbial Studies of Acyclic Schiff-base Gadolinium (III) Complexes." Bull. Korean Chem. Soc., 33(11), 3581 [2012].
39. Sulekh Chandra, Sangeetika and Shalini Thakur. "Electronic, epr, cyclic voltammetric and biological activities of copper (II) complexes with macrocyclic ligands." Transition Metal Chemistry, 29, 925–935 [2004].
40. Zishen, Wu, Zhiping, Lu, Zhenhunan, Yen, "Synthesis, characterization and antifungal activity of glycylglycine Schiff base complexes of 3d transition metals." Transition Metal Chemistry, 18, 291–294 [1993].
41. Dharmaraj, Nallasamy, Periasamy Viswanathamurthi, and Karuppannan Natarajan. "Ruthenium (II) complexes containing bidentate Schiff bases and their antifungal activity." Transition Metal Chemistry 26, 105–109 [2001].
42. Ramesh, R., Maheswaran, S., "Synthesis, spectra, dioxygen affinity and antifungal activity of Ru(III) Schiff base complexes." Journal of Inorganic Biochemistry, 96, 457–462 [2003].
43. Raman, N., Sakthivel, A., Rajasekaran, K., "Synthesis and spectral characterization of antifungal sensitive Schiff base transition metal complexes." Mycobiology, 35(3), 150–153 [2007].
44. Sakthivel, A., Thalavaipandian, A., Raman, N., Thangagiri, B., "Synthesis, characterization and antifungal activity of transition metal (II) complexes of Schiff base derived from *p*-aminoacetanilide and salicylaldehyde." International Journal of Current Science and Research, 3, 253–1260 [2017].
45. Pratibha, M.S.J., Vatsala, P., Uma, V., "Biologically active Co (II), Ni (II), Cu (II) and Mn(II) Complexes of Schiff Bases derived from Vinyl aniline and Heterocyclic Aldehydes." Int. J. Chem. Techno. Res., 1, 225–32 [2009].
46. Abdallah, Sayed M., Zayed, M.A., Mohamed, Gehad G., "Synthesis and spectroscopic characterization of new tetradentate Schiff base and its coordination compounds of NOON donor atoms and their antibacterial and antifungal activity." Arabian Journal of Chemistry, 3, 103–113 [2013].
47. Padalkar, Vikas S., Borse, Bhushan N., Gupta, Vinod D., Phatangare, Kiran R., Patil, Vikas S., Nagaianj, S. "Synthesis and Antimicrobial Activities of Novel 2-[substituted1H-pyrazol-4-yl] Benzothiazoles, Benzoxazoles." Heterocyclic Chemistry [2016].
48. Singh, D.P., Kumar, K. Macrocyclic complexes: synthesis and characterization. Journal of the Serbian Chemical Society, 75(4): 475–482 [2010].
49. Singh, Jugmendra, Kanojia, Rajni, TyagiInternational Mukesh, "Template synthesis and spectral studies of Ni(II) and Cu(II) transition metal complexes of tetradentate aza-oxo(N_2O_2) macrocyclic ligand." Journal of Pharmaceutical Sciences and Research, 5(9): 3903–3911 [2014].
50. Chandra, Sulekh, Sharma, Deepali Jain Amit Kumar, Sharma, Pratibha, "Coordination Modes of a Schiff Base Pentadentate Derivative of 4-Aminoantipyrine with Cobalt(II), Nickel(II) and Copper(II) Metal Ions: Synthesis, Spectroscopic and Antimicrobial Studies." Molecules, 14, 174–190 [2009].
51. Bi, Siwei, Li, Guizhi, "Synthesis, characterization and antifungal activity of some transition metal complexes of Schiff base derived from 4-acetylbiphenyl and S-benzyldithiocarbazate." Synthesis and Reactivity in Inorganic Metal-Organic Chemistry, 29(10), 1829–1841 [1999].
52. Srivastva, Abhay Nanda, Singh, Netra Pal, Shrivastaw, Chandra Kiran, "In vitro antibacyerial and antifungal activities of binuclear transition metal complexes of ONNO Schiff base and 5,methyl, 2,6-pyrimidine-dioneand their spectroscopic validation." Arabian Journal of Chemistry [2014].
53. Abou-Husseina, A.A., Linert, Wolfgang, "Synthesis, spectroscopic studies and inhibitory activity against bactria and fungi of acyclic and macrocyclic transition metal complexes containing a triamine coumarine Schiff base ligand." Spectrochemical Acta Part A: Molecular and Biomolecular Spectroscopy, 141, 223–232 [2015].
54. Azza, A.A., Abou-Hussein, A., "Synthesis, spectroscopic and biological activities studies of acyclic and macrocyclic mono and binuclearmetal complexes containinga hard soft shiff base." Molecular and Biomolecular Spectroscopy, 95, 596–609 [2012].
55. Miloud, M.M., El-ajaily, M.M., Al-noor, T.H., Al-barki N.S., "Antifungal activity of some mixed ligand complexes incorporating Schiff bases." Journal of Bacteriology and Mycology, 7(1), 1122 [2020].

56. Khlood, S., Melha, Abou, "In-vitro antibacterial, antifungal activity of some transition metal complexes of thiosemicarbazone Schiff base (HL) derived from N4 -(70 -chloroquinolin-40 -ylamino) thiosemicarbazide." Journal of Enzyme Inhibition and Medicinal Chemistry, 23(4), 493–503 [2008].
57. Echevarria, Aurea, Maria da Graça Nascimento, Vanilde Gerônimo, Joseph Miller, and Astréa Giesbrecht. "NMR spectroscopy, hammett correlations and biological activity of some Schiff bases derived from piperonal." Journal of the Brazilian Chemical Society, 10(1), 60–64 [1999].
58. Sharmaa, Amit Kumar, Chandra, Sulekh, Sharma, A.K., Chandra, S., "Spectroscopic and mycological studies of Co(II), Ni(II) and Cu(II) complexes with 4-aminoantipyrine derivative." Spectrochemical Acta Part A, 81, 424–430 [2011].

5 Utilization of Schiff Base Metal Complexes as Antiviral and Antiparasitic Agent

Shaina Joarder[1], Divyam Bansal[1], Madhur Babu Singh[1,2], and Kamlesh Kumari[3]*

[1] Department of Chemistry, Atma Ram Sanatan Dharma College, University of Delhi, Delhi, India
[2] Department of Chemistry, Faculty of Engineering and Technology, SRM Institute of Science and Technology, Delhi-NCR Campus, Uttar Pradesh, India
[3] Department of Zoology, University of Delhi, Delhi, India
*Corresponding author
Email: biotechnano@gmail.com

CONTENTS

5.1 INTRODUCTION

Schiff bases have emerged as an essential class of compounds in bio-inorganic chemistry since their discovery in 1864 by Hugo Joseph Schiff [1]. Any primary amine acting as an effective chelating framework can react with an aldehyde or a ketone to produce a Schiff base under specific conditions (constant heat and presence of a chemical catalyst). A Schiff base can be described as a nitrogen homolog of a ketone or an aldehyde where the carbonyl group (CO) has been exchanged for an azomethine or imine group [2]. Various natural and artificial sources, as well as laboratories, can be used to create compounds with azomethine functional groups. The mechanism of formation of Schiff base involves a tetrahedral intermediate as a result of the amine's nucleophilic attack on the electrophile site during the synthesis of a Schiff base as shown in Figure 5.1. Therefore, regardless of the pathways or techniques used, this process results in the escape of water and the production of the imine group [3]. Typically, these reactions take place in acidic media. Most of the Schiff bases happen to be crystalline solids, mildly basic, and a few of them combine along strong acids to generate insoluble salts. Amino acid Schiff bases can be utilized as ligands or intermediates in the synthesis of various metal complexes [4].

It is known that metal ions play a significant role in many essential biological processes and are essential to life. They are crucial cofactors of numerous enzymes and help control electron transfer, electrolyte balance, and oxygen transport. Additionally, they have been linked to controlling the immune system and the host's defence against harmful invaders. Metals have a wide range

DOI: 10.1201/9781003291459-5

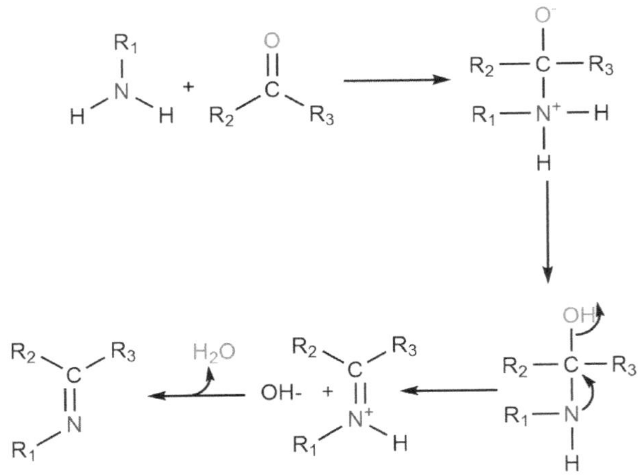

FIGURE 5.1 Mechanism for formation of Schiff base.

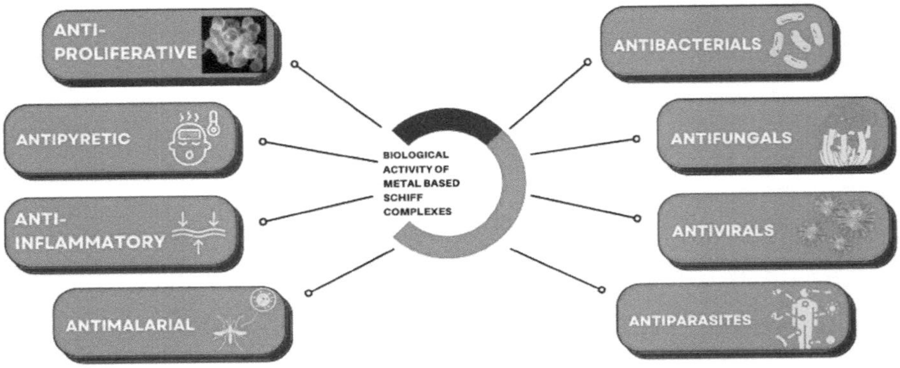

FIGURE 5.2 Various biological applications of metal-based Schiff base complexes.

of features, including various coordination numbers, redox states, geometries, as well as thermodynamic behavior and reactivity, which has led to their significant usage in medicine for the diagnosis and treatment of many disorders [5] [6].

Schiff bases have found numerous biological applications due to the presence of the double bond between nitrogen and carbon in their structure. The advantageous characteristics of Schiff base complexes extend beyond the azomethine link and include many scaffolds that have strengthened the structure, such as substituted aromatics and aromatic compounds. The presence of hydrogen bonding between the existing atoms and present functional groups like the hydrogen bonding amongst OH group and N atom of azomethine part and the presence of nucleophile group can be contributing factors for the resultant biological effects of Schiff bases [3].

Schiff base ligands are simply manufactured and can combine with practically any metal ion to produce complexes. The curiosity towards complexes of Schiff bases has expanded since it is now understood that several of these may act as mock-ups for physiologically significant species [2]. The Schiff bases have been found applicable in coordinating the transition metal ions and creating diverse enriched metal complexes with numerous antibacterial [7], antiproliferative [8], antiviral, antifungal [9], anticancer [10], and even antidiabetic [11] activities (Figure 5.2). They possess effective chelating frameworks as well as flexi-dentate capabilities; in particular, the soft ligands make it easier to synthesize, modify, and replace their structural components, leading to flexible chains [5]. The biological activities of metal complexes are greater than those of their corresponding ligands. Due to their optical nonlinearity features, electron-donating capacity, stability, catalytic, photochromic, and biological activities, Schiff base complexes are of great interest. The coordination of Schiff bases to metal ions serves as the foundation for their activity [12]. In this chapter we will gain an insight into the antiviral and antiparasitic effectiveness of the metal-based Schiff base complexes.

5.2 ANTIVIRAL

Viruses are sub-microscopic agents that are responsible for an array of diseases and illnesses caused in plants as well as humans. An antiviral drug is an agent utilized for treating the diseases caused by such viruses. Nucleic acids (either RNA or DNA) with a protein covering make up viruses. Viruses are obligatory parasites, meaning they must invade a cell for replication to take place, because they lack the enzymes necessary to produce biological components. The host cell is given instructions to generate viral components by the virus' nucleic acid, which results in a virus capable of causing infection. In certain circumstances, such as herpes infections, the nucleic acid of the viral agent may stay in the host cell not leading to virus replication and harm to the host (viral latency) [13]. In other instances, the host cell's creation of a virus may result in the cell's demise.

Treatment of a viral disease is challenging since the drug must be competent to stop the virus without significantly harming the host cell because the virus needs a host cell to support its reproduction. To prevent a virus from reproducing, an antiviral mediator must act towards one of five key points in the process:

1. Attachment and permeation into the host cell;
2. Decoating of the virus (e.g., release of the viral DNA or RNA by removal of the protein surface);
3. Synthesis of fresh viral constituents by host cell under the control of the viral DNA;
4. Gathering of the components into a new virus;
5. Discharge of the virus from infected host cell.

Even if there are numerous therapeutic alternatives for the treatment of viral infections nowadays, the increase of mutations has made the available antiviral medications less effective [14]. The Cucumber mosaic virus responds effectively to the Silver coordinated Schiff base complex [4]. Due to the unusual properties of the metal centers now utilized in medicinal chemistry, metallodrugs may present a favorable opportunity to accomplish these objectives [15].

There have been a few previous studies on the antiviral potential of Schiff bases. In a work by Sriram et al., EC_{50} and CC_{50} (the cytotoxic concentration of chemical (M) necessary to lower the survival rate by 50% of pretend infected CEM cells) (ability of the compounds to constrain HIV-1 replication in the CEM cell lines) were calculated. Excellent anti-HIV activity was shown by abacavir and its synthetic ligands; in particular, compounds 2.43 to 2.51 were found to be quite effective. Compared to ketone analogs, substituted benzaldehyde derivatives were more active. With an EC_{50} of 0.05 M, CC_{50} > 100 M, a selectivity index (SI) > 2000, and an observation that it was 32 times extra effective than the parent drug (with EC_{50}, 1.6 M), compound 2.48 demonstrated the strongest inhibitory property. Additionally, it was discovered that the ligands' pharmacological effects were greatly enhanced by their lipophilicity [16].

According to Al-Masoudi et al., utilizing the MT-4/MTT assay, they were able to synthesize certain Schiff bases with antiviral activity counter to HIV-2 (strain ROD) and HIV-1 (strain IIIB) in human T-lymphocyte cells (MT-4). EC_{50} values were determined providing 50% defense in MT-4 cell lines in contradiction of cytopathic influence of HIV. It was discovered that the investigated strains were responsive to the selected ligands and their Schiff base complexes, with metal-based ligands showing the greatest activity. The most effective anti-HIV drugs had an index selectivity of 9 and an EC_{50} of 10.2 g/mL against HIV-2 in cell culture. Since gold possesses some antibacterial properties and Tweedy's theory could have also affected the permeability of membrane, the gold complex, compound 2.42, demonstrated stronger antiviral action compared to other compounds [17].

By forming 3-amino-2-phenylquinazoline-4(3H)-ones, Kumar et al. created 3-(benzylideneamino)-2-phenylquinazoline-4(3H)-ones Schiff bases including different substituted carbonyl compounds. (KOS) Herpes simplex virus-1, herpes simplex virus-2, vesicular stomatitis, vaccinia, reovirus1, herpes simplex virus-1 TK- KOS ACVr, Sindbis virus, para influenza-3 virus, and seven other viruses were all subjected to antiviral activity. The effectiveness of the ligands to modify each viral representative's respiration process was evaluated in comparison to para influenza-3 virus, brivudine, cidofovir, ribavirin, and ganciclovir, which were employed as reference materials. The examined microorganisms responded to the test compounds in various ways, with some showing greater sensitivity to the produced Schiff bases compared to that of the controls. In human embryonic lung cell culture, the herpes simplex viruses 1, 2, and 1 as well as the vaccinia virus were extremely profound to compound 2.43 (Figure 5.6), displaying 50% cell decease at EC_{50} ranging 0.8–30 g/mL larger compared to the positive controls (with EC_{50}, 0.05–250 g/mL). In comparison to the standards, the methoxy-substituted (2.44) and unsubstituted (2.45) analogs of compound 2.43 also displayed stronger antiviral activities [18].

The coronavirus is a broad family of viruses that includes the SARS-CoV-2. Both humans and some animals can contract these viruses. In 2019, SARS-CoV-2 was first identified as a human pathogen. The virus is believed to transmit from person to person via droplets emitted during coughing, sneezing, or talking. It can also be contracted by contacting a surface that has the virus on it and then touching the mouth, nose, or eyes. Treatments for COVID-19 and SARS-CoV-2 infection prevention are the subjects of research.

El-Gammal et al. recently studied Cobalt, Nickel, and Copper metal (II) complexes with a novel Schiff base [HL or (E)-4-((3-cyano-4,6-dimethyl pyridin-2-yl) amino)-N'-(1-(pyridin-2-yl) ethylidene) benzohydrazide)]. Molecular docking was performed to demonstrate the action of HL against COVID-19 by interacting with SARS-CoV-2 major protease, which was retrieved with PDB ID: 6 W41 from the RCSB protein data bank. Future investigations and clinical tests will therefore focus on HL as a viable COVID-19 sickness therapeutic medication. According to the research, ligands and receptor protein 6 W41 might be organized. Docking analysis showed that ligands and the receptor 6 W41 interacted favorably [19].

5.3 ANTIPARASITIC

Parasites is the term used for an organism that coexists with another living creature in order to gain nutrition, grow, or reproduce—often in a way that is harmful to the host directly or indirectly. A wide range of different organisms belonging to bacteria, viruses, fungi, protozoans, helminths, and so on are known to cause diseases in humans. Worldwide, billions of infections are brought on by parasites. They can enter our bodies through various means, and multiply and interfere with vital activities in our bodies which lead to morphological and functional damage. Infections with parasites can take many different forms. Malaria is among the most widespread parasite illnesses. Antiparasitic medications can function via three pathways:

1. Eliminating the parasite or its eggs;
2. Halting the parasite's growth;
3. Causing the parasites to become paralyzed so that they are unable to affix to the host.

An array of various diseases is caused by the organisms having a parasitic nature. *Staphylococcus aureus* is a round-shaped gram-positive bacterium which, being an opportunistic pathogen, causes

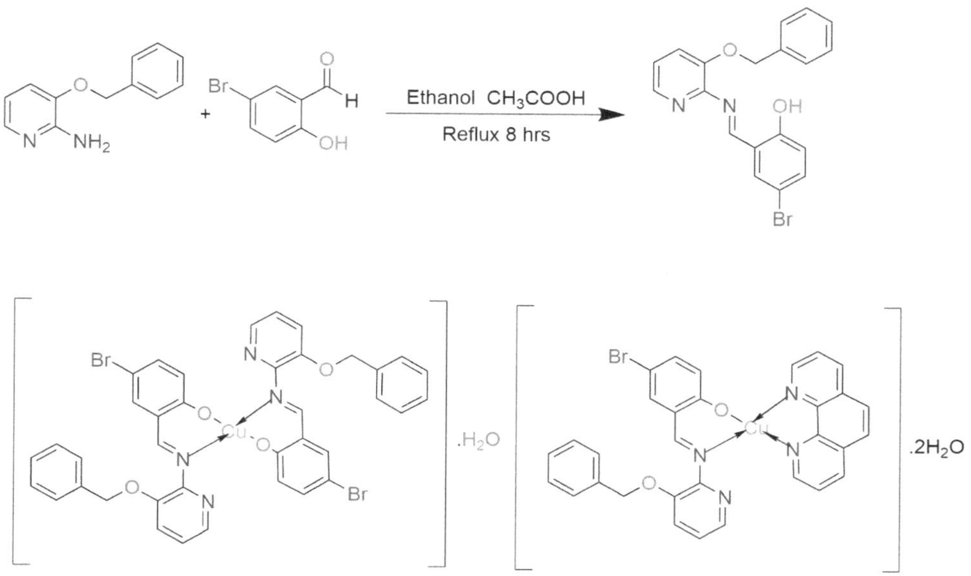

FIGURE 5.3 Representation of formation of ligand HL and Proposed structures of Schiff base complexes C1 and C2 respectively [20].

a wide number of clinical ailments in humans including community-acquired and nosocomial infections. *Salmonella Typhi* takes only humans as their hosts. Typhoid fever patients have the germs in their blood and intestines. Long-lasting high fever, exhaustion, headaches, nausea, constipation or diarrhea, abdominal pain, and rashes are other symptoms. Plasmodium parasites, which individuals contract through the bites of infected female *Anopheles* mosquitoes, are the source of the acute fever sickness known as malaria. Malaria can result in serious health issues such convulsions, brain damage, difficulty breathing, organ failure, and death if it is not treated. A yeast infection called candidiasis is a fungal infection. It poses a significant risk to world health. People who have a long-term illness with noticeable esophageal dilatation or who have a skin fungal infection are most at risk. *Aspergillus niger* frequently leads to aspergillosis. Otomycosis, a frequent fungal-related ear infection, has been reported to induce skin and lung infections. It frequently leads to aspergillosis. Otomycosis, a frequent fungal-related ear infection, has been reported to induce skin and lung infections. A number of different metal-based Schiff bases have therefore been studied to assess their impact on the growth of such organisms known to be harmful.

Sharma et al. made Copper(II) complexes with Schiff bases as shown in Figure 5.3. The bioactivity of Schiff base and its resultant Cu(II) complexes as antibacterial (on *Salmonella typhi* and *Staphylococci aureus*), antifungal (*Aspergillus niger* and *Candida albicans*), and anthelmintic agents have been examined in vitro. It was discovered that the biological action of the Cu(II) complexes is superior to that of its Schiff base ligand HL [20].

Manjunath et al. investigated the metal (II) complexes' (structure as depicted in Figure 5.4) antibacterial (*Escherichia coli*, *Salmonella typhi*, *Pseudomonas aeruginosa*, and *Staphylococcus aureus*), antifungal (*Cladosporium*, *Aspergillus flavus* and, *Aspergillus niger*), antithelmintic and DNA cleaving characteristics. The findings indicated that several of the produced chemicals may have antibacterial potential. Co(II) and Ni(II) complexes were discovered to have effective results when the produced compounds were evaluated for their anthelmintic activity [21].

Onuorah et al. created and characterized complexes of the Schiff base of Fe(II), Zn(II), Mn(II), Co(II), and Cu(II) as depicted in Figure 5.5. These metal complexes' capacity to prevent the

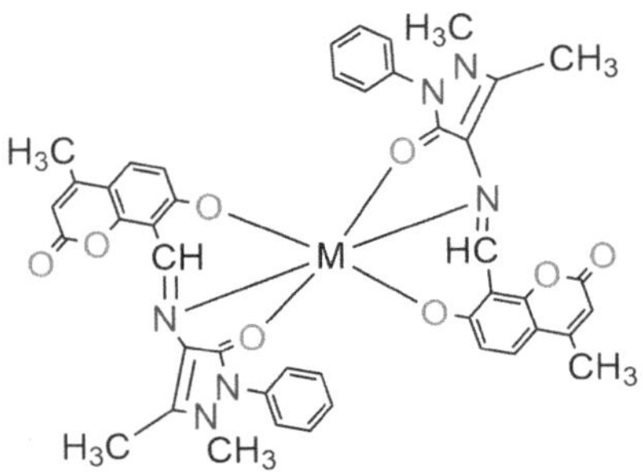

FIGURE 5.4 Structure of metal (II) Complexes [21].

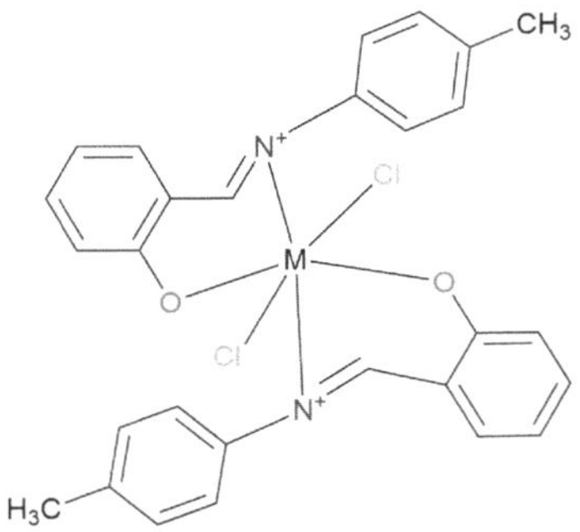

FIGURE 5.5 Proposed structure of Metal(II) based Schiff base complexes [22].

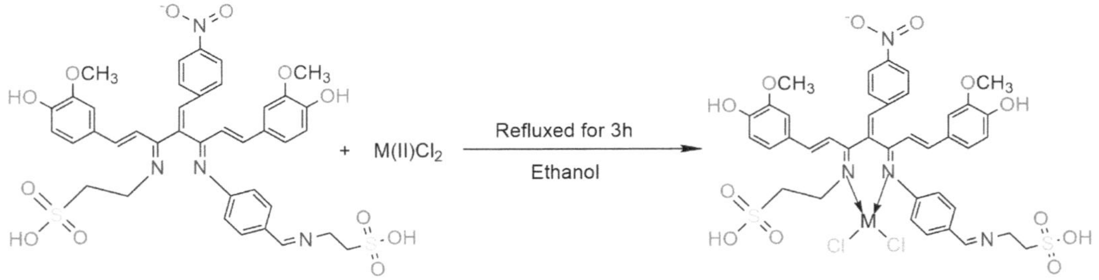

FIGURE 5.6 Proposed reaction for Metal (II) Schiff base complex formation of Cu, Ni, Zn, and Co [23].

progression of pathogenic microorganisms like bacteria (*Salmonella typhi*, *Staphylococcus aureus*) and fungi (*Aspergillus niger and Candida albicans*) was compared with the standard medications. Results of antibacterial activity have shown that maximum complexes are more effective than ordinary medicines against isolated microorganisms [22].

Thiravidamani et al. described the liquid-state synthesis of four novel 1,7-bis(4-hydroxy-3-methoxy phenyl)-1,6-heptadiene-3,5-dione ternary metal (II) complexes-based Schiff base from taurine as shown in Figure 5.6. The copper, nickel, zinc, and cobalt metal complexes as synthesized were characterized. The well diffusion method was used to test the ligands and produced metal complexes for (in vitro) antibacterial activity towards fungal and bacterial species like *Staphylococcus aureus*, *Pseudomonas aeruginosa*, *Staphylococcus epidermidis*, and *Escherichia coli* [23].

For a variety of biological activities, the zinc (II) and iron (III) metal complexes (Figure 5.8) of Schiff base ligands (Figure 5.7) were compared by Naureen et al. Antibacterial and antifungal assays

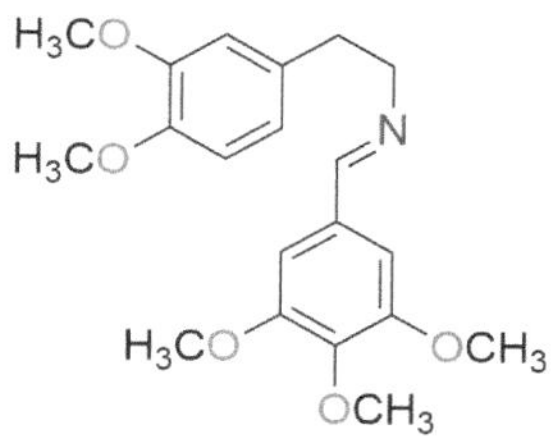

FIGURE 5.7 Structure of ligand used (L1) [24].

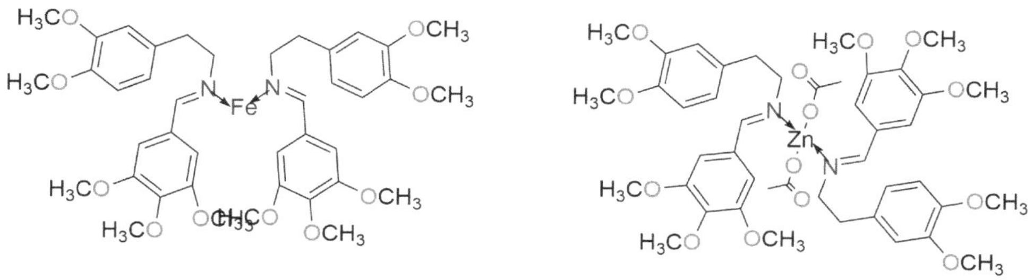

FIGURE 5.8 Proposed structures of Fe(III) and Zn(II) based Schiff based metal complexes [24].

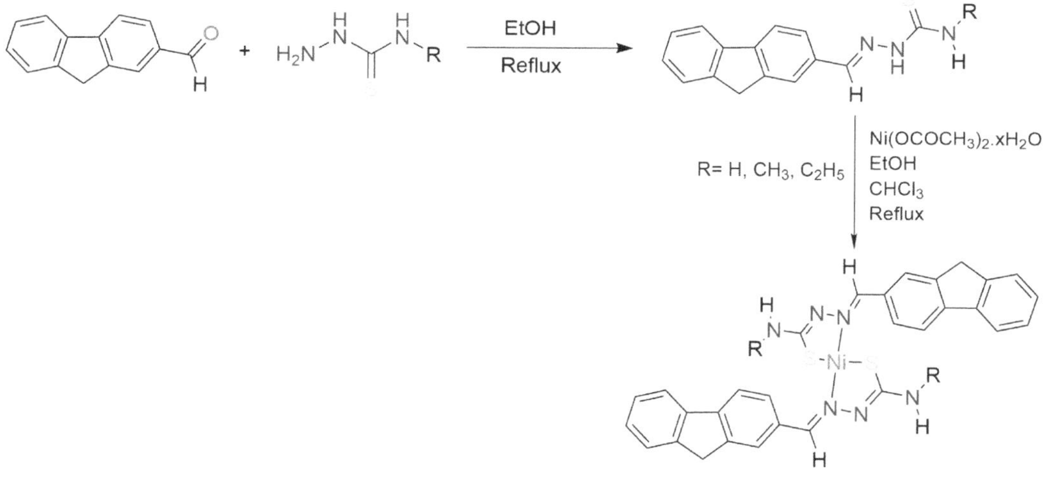

FIGURE 5.9 Proposed reaction for Ni(II) metal complex formation [25].

were performed as part of pharmacological activity. The substances have demonstrated antibacterial and antifungal properties, however the metal complexes—particularly all zinc (II) complexes—exhibited superior properties than the original ligands [24].

Savir et al. studied various Schiff base metal complexes (Figure 5.9) in different forms. As a result of their coordination to the Ni(II) ion by means of their N and S atoms, the thiosemicarbazone ligands appeared to function as bidentate ligands, according to the data. The nickel complexes showed moderate in vitro antimalarial efficacy among the six investigated drugs. It was noticed that the compound's antimalarial activity rises with the size of the substituent group [25].

The anti-*Ttrypanosoma cruzi* activities of the new VIVO-complexes and the previously created salicylaldimine compounds were assessed by Scalese et al. in vitro towards the *T. cruzi* epimastigote

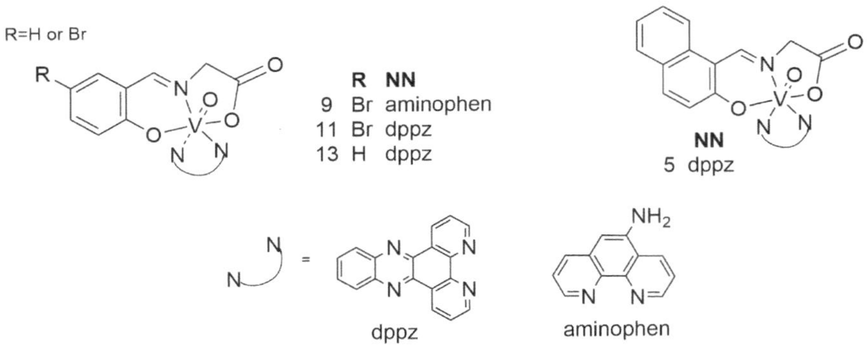

FIGURE 5.10 Proposed structures of complexes 5,9,11 and 13 [26].

strain, CL Brener strain. VERO cells were used as a mammalian model to examine their selectivities toward this parasite. The IC_{50} values for almost all complexes were in the lower micromolar range (within 1.4 and 4.7 M). Along with researching how vanadium compounds affect parasite endurance and reproduction, the mechanism(s) causing parasite decease when exposed to these chemicals were also understood. Table 5.1 lists the four compounds chosen for this purpose (Figure 5.10), including 5, [VO(naphthgly-2H)(dppz)], 9, [VO(5Brsalgly-2H)(aminophen)], 11, [VO(5Brsalgly-2H)(dppz)] and, 13, [VO(salgly-2H)(dppz)] which all showed great activity on T. cruzi and good selectivities against the parasite. In Trypanosoma cruzi, vanadium compounds cause apoptosis and alter the potential of the mitochondrial membrane [26].

A significant fraction of the global population is susceptible to helminth infections, which are among the most prevalent infections in humans [27]. They are a serious hazard to public health in underdeveloped nations and raise the risk of pneumonia, eosinophilia, anemia, and malnutrition. Many helminths dwell in the gastrointestinal system, while others also reside in tissues or their larvae move in to tissues. Anthelmintics are medications that either kill or expel infesting helminths. By denying the host of food, causing blood loss, organ damage, lymphatic or intestinal obstruction, or by secreting poisons, they cause harm to the host. Although rarely lethal, helminthiasis is a leading source of morbidity. Maia et al. examined that the Ni complex exhibited a potent in vitro action against *L. amazonensis* promastigotes (Leishmania specie). This activity may be connected to the complex's capacity to interface with parasite DNA, which would facilitate the suppression of DNA replication by simultaneous interface of intercalation and synchronization with DNA, a mechanism shared by other metallic complexes [28].

One of the most frequent types of intestinal parasites within humans are intestinal worms, commonly referred to as parasitic worms. They are most frequently seen in subtropical and tropical areas. The majority of intestinal worm infestations result in minor illness and are manageable with medicine. Roundworms, which cause ascariasis, pinworm, and hookworm infections; flatworms, which include tapeworms and flukes; and hookworms are the most prevalent types of intestinal worms that may harm people. Eating raw or undercooked from an infected animal, such as a pig, cow, or fish, is one way to contract intestinal worms. Consuming contaminated soil, drinking polluted water, touching contaminated faeces, and practicing poor sanitation and hygiene are additional factors that might result in intestinal worm infection [29]. The parasite enters your intestines once you've swallowed the contaminated item. *Pheretima posthuma* was chosen because of its morphological and physiological similarity to the human intestinal roundworm parasite.

Using adult Indian earthworms, Cu(II) complexes C1 and C2 were tested for their anthelmintic activity in vitro (*Pheretima posthuma*) by Sharma et al. By measuring how long it takes for worms to completely paralyze and die, the compounds were assessed for their action. Mortality and loss of motility against *P. Posthuma* are more pronounced at higher concentrations. The findings demonstrate

TABLE 5.1
Antibacterial, Anti-Protozoal and Antifungal Activities of Some Metal-Based Schiff Base Complexes

S.no.	Compound with Empirical Formula	Organism Tested	Disease Caused by Organism	Values	Reference
1	C1: $Cu(C_{38}H_{30}Br_2CuN_4O_5)_2$ C2: $Cu(C_{31}H_{26}BrCuN_4O_4)_2$	*Staphylococcus aureus*	Pneumonia, impetigo, meningitis, cellulitis, osteomyelitis, carbuncles, endocarditis, abscesses, toxic shock syndrome, bacteremia, and sepsis	**MIC (µg/mL)** **Zone of inhibition (mm)** 69 21 59 30	[20]
	$Co(C_{22}H_{18}N_3O_4)_2$ $Ni(C_{22}H_{18}N_3O_4)_2$ $Cu(C_{22}H_{18}N_3O_4)_2$			**Zone of inhibition in % at 100 mg mL^{-1}** 59 82 55	[21]
	$Mn(HL)_2Cl_2$ $Fe(HL)_2Cl_2$ $Cu(HL)_2Cl_2$ $Zn(HL)_2SO_4$ $Co(HL)_2Cl_2$			**Zone of inhibition (mm)** 10.58 14.40 17.00 9.20 7.80	[22]
	$C_{39}H_{38}Cl_2N_4O_{12}S_2Co$ $C_{39}H_{38}Cl_2N_4O_{12}S_2Ni$ $C_{39}H_{38}Cl_2N_4O_{12}S_2Cu$ $C_{39}H_{38}Cl_2N_4O_{12}S_2Zn$			**MIC (×10^4 µM)** 8.7 10.5 13.2 14.3	[23]
	$(L1)_2FeCl_3$ $(L1)_2Zn(Ac)_2$			**Zone of inhibition (mm)** 16 11	[24]
2	$Cu(C_{38}H_{30}Br_2CuN_4O_5)_2$ $Cu(C_{31}H_{26}BrCuN_4O_4)_2$	*Salmonella typhi*	Typhoid and paratyphoid fever	**MIC (µg/mL)** **Zone of inhibition (mm)** 62 19 65 21	[20]
	$Mn(HL)_2Cl_2$ $Fe(HL)_2Cl_2$ $Cu(HL)_2Cl_2$ $Zn(HL)_2SO_4$ $Co(HL)_2Cl_2$			**Zone of inhibition (mm)** Resistant 7.93 11.29 10.40 7.61	[22]
	$C_{39}H_{38}Cl_2N_4O_{12}S_2Co$ $C_{39}H_{38}Cl_2N_4O_{12}S_2Ni$ $C_{39}H_{38}Cl_2N_4O_{12}S_2Cu$ $C_{39}H_{38}Cl_2N_4O_{12}S_2Zn$			**MIC (×10^4 µM)** 12.4 11.5 13.5 14.6	[23]
3	$C_{30}H_{24}N_6NiS_2$ $C_{32}H_{26}N_6NiS_2$ $C_{34}H_{34}N_6NiS_2$	*Plasmodium falciparum*	Falciparum malaria	**$IC_{50\pm SD}$ (mM)** >25 23.79 ± 1.09 2.29 ± 2.19	[25]

(*continued*)

TABLE 5.1 (Continued)
Antibacterial, Anti-Protozoal and Antifungal Activities of Some Metal-Based Schiff Base Complexes

S.no.	Compound with Empirical Formula	Organism Tested	Disease Caused by Organism	Values	Reference
4	$Cu(C_{38}H_{30}Br_2CuN_4O_5)_2$ $Cu(C_{31}H_{26}BrCuN_4O_4)_2$	*Candida albicans*	Candidiasis	**MIC (μg/mL)** 82 87 Zone of inhibition (mm) 17 23	[20]
	$C_{39}H_{38}Cl_2N_4O_{12}S_2Co$ $C_{39}H_{38}Cl_2N_4O_{12}S_2Ni$ $C_{39}H_{38}Cl_2N_4O_{12}S_2Cu$ $C_{39}H_{38}Cl_2N_4O_{12}S_2Zn$			**MIC (×10⁴ μM)** 9.3 11.4 12.3 13.6	[23]
5	$Cu(C_{38}H_{30}Br_2CuN_4O_5)_2$ $Cu(C_{31}H_{26}BrCuN_4O_4)_2$	*Aspergillus niger*	Aspergillosis	**MIC (μg/mL)** 71 69 **Zone of inhibition (mm)** 19 20	[20]
	$Co(C_{22}H_{18}N_3O_4)_2$ $Ni(C_{22}H_{18}N_3O_4)_2$ $Cu(C_{22}H_{18}N_3O_4)_2$			**Zone of inhibition in % at 100 mg mL^{-1}** 65 46 81	[21]
	$C_{39}H_{38}Cl_2N_4O_{12}S_2Co$ $C_{39}H_{38}C_{12}N_4O_{12}S_2Ni$ $C_{39}H_{38}Cl_2N_4O_{12}S_2Cu$ $C_{39}H_{38}Cl_2N_4O_{12}S_2Zn$			**MIC (×10⁴ μM)** 9.4 10.5 13.9 14.2	[23]
6	$Co(C_{22}H_{18}N_3O_4)_2$ $Ni(C_{22}H_{18}N_3O4)_2$ $Cu(C_{22}H_{18}N_3O_4)_2$	*Cladosporium*	Subcutaneous abscesses and CNS infections	**Zone of inhibition in % at 100 mg mL^{-1}** 71 82 52	[21]
7		*Aspergillus flavus*	Chronic granulomatous sinusitis, wound infections, cutaneous aspergillosis, keratitis, and osteomyelitis ensuing trauma and inoculation	**Zone of inhibition in % at 100 mg mL^{-1}** 70 83 54	
8	$C_{39}H_{38}Cl_2N_4O_{12}S_2Co$ $C_{39}H_{38}Cl_2N_4O_{12}S_2Ni$ $C_{39}H_{38}Cl_2N_4O_{12}S_2Cu$ $C_{39}H_{38}Cl_2N_4O_{12}S_2Zn$	*Fusariumsolani*	Several crop diseases	**MIC (×10⁴ μM)** 9.9 10.4 11.8 13.6	[23]
9		*Curvularialunata*	Onychomycosis, dialysis-associated peritonitis, eumycotic mycetoma, mycotic keratitis and sinusitis	**MIC (×10⁴ μM)** 10.3 11.2 126 14.7	

TABLE 5.1 (Continued)
Antibacterial, Anti-Protozoal and Antifungal Activities of Some Metal-Based Schiff Base Complexes

S.no.	Compound with Empirical Formula	Organism Tested	Disease Caused by Organism	Values	Reference
10	[VO(naphthgly-2H)(dppz)]·2.5 H_2O [VO(5Brsalgly-2H)(dppz)]·1/2H_2O [VO(salgly-2H)(dppz)]·H_2O [VO(5Brsalgly-2H)(aminophen)]·2H_2O	*Trypanosoma cruzi*	Chagas disease (in Central and South America and Mexico) transmitted from triatomine bug	**IC_{50} ± SD (μM)** 1.41 ± 0.14 1.69 ± 0.49 1.48 ± 0.28 2.09 ± 0.08	[26]

TABLE 5.2
Anthelmintic Action of the Schiff Bases Metal Complexes towards *Pheretima prosthuma*

S.no.	Compound with Empirical Formula	Paralysis Time taken (min)	Death Time taken (min)	Reference
1	$Cu(C_{38}H_{30}Br_2CuN_4O_5)_2$	19	23	[20]
	$Cu(C_{31}H_{26}BrCuN_4O_4)_2$	18	25	
2	$Co(C_{22}H_{18}N_3O_4)_2$	7	13	[21]
	$Ni(C_{22}H_{18}N_3O4)_2$	10	15	
	$Cu(C_{22}H_{18}N_3O_4)_2$	7	10	
3	$Co(C_{18}H_{15}NO_4)_2$	4.30 ± 0.04	7.72 ± 0.03	[30]
	$Co(C_{18}H_{12}$- $NO_4F_3)_2$	3.43 ± 0.01	6.51 ± 0.20	
	$Ni(C_{18}H_{15}NO_4)_2$	3.29 ± 0.02	6.91 ± 0.01	
	$Ni(C_{18}H_{12}$- $NO_4F_3)_2$	54 ± 0.01	7.25 ± 0.01	
	$Cu(C_{18}H_{15}NO_4)_2$	3.25 ± 0.00	6.93 ± 0.10	
	$Cu(C_{18}H_{12}$- $NO_4F_3)_2$	3.20 ± 0.04	3.20 ± 0.04	

that these metal complexes produce worm paralysis and decease, proving that the Cu(II) complexes in Table 5.2 are effective anthelmintic agents. In comparison to ligand and the common medication albendazole, Cu(II) complexes are shown to be more activated. The complexes' biochemical method of anthelmintic activity may involve interfering with metabolic functions or the neuromuscular physiology of the parasites. Generally, the change of the parasite's motor activity and/or inhibition of energy metabolism may be the conceivable mechanisms by which complexes exert their anthelmintic effects [20].

Manjunath et al. proposed the findings of the produced compounds' anthelmintic action are shown in Figure 5.11. Some of the synthetic compounds showed effective anthelmintic properties. The Schiff bases were extremely active. There have been encouraging developments with the Co(II) and Ni(II) metal complexes. The chemicals that were presented exhibited intriguing anthelmintic action, making them potentially useful for in-vivo investigations when compared to commonly used medications [21].

Co(II), Cu(II), and Ni(II) metal complexes with Schiff base H2LI [$C_{18}H_{15}NO_4$] and Schiff base H2LII [$C_{18}H_{12}$- NO_4F_3] were synthesized by Patil et al. by combining o-toluidine/3-aminobenzotrifluoride with 6-formyl-7,8-dihydroxy-4-methylcoumarin (Figure 5.12 and 5.13). It was discovered that the

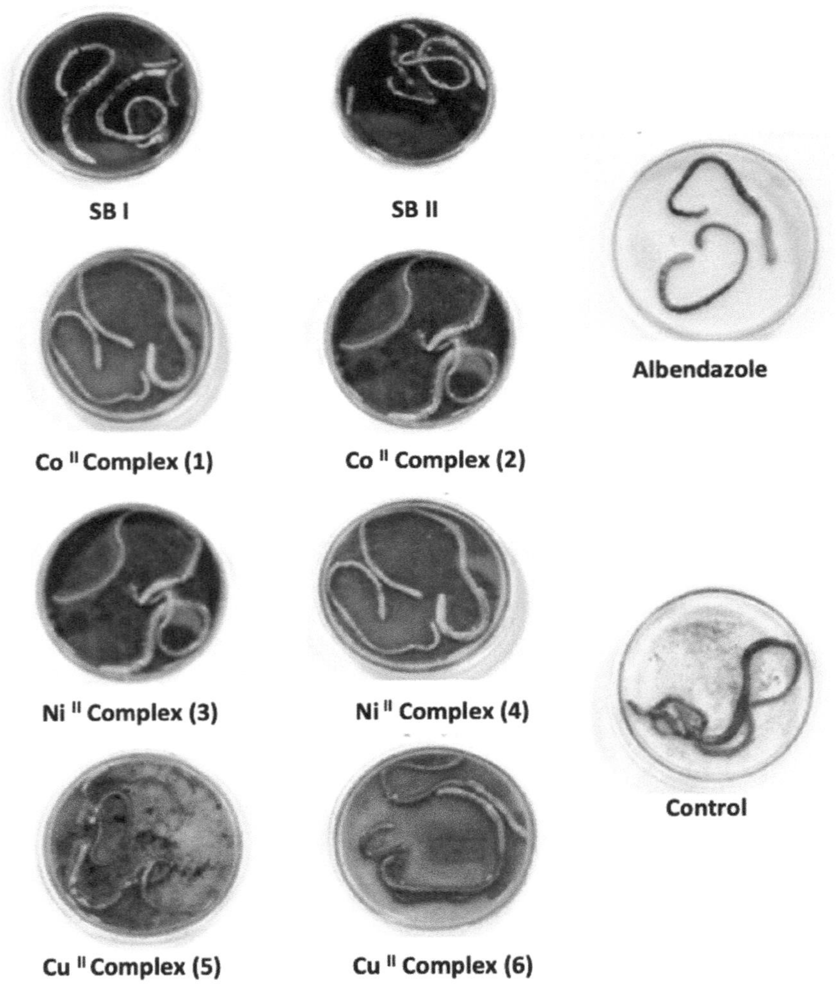

FIGURE 5.11 Anthelmintic Activity of Schiff bases along with their metal complexes [21].

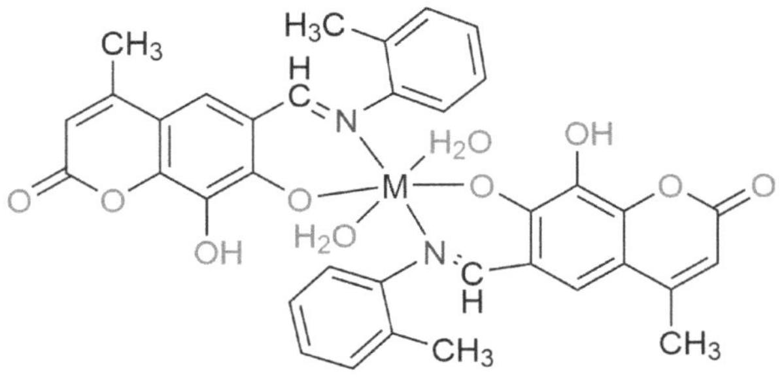

FIGURE 5.12 Proposed structures for complexes Co(II) (1), Ni(II) (3) and Cu(II) (5) [30].

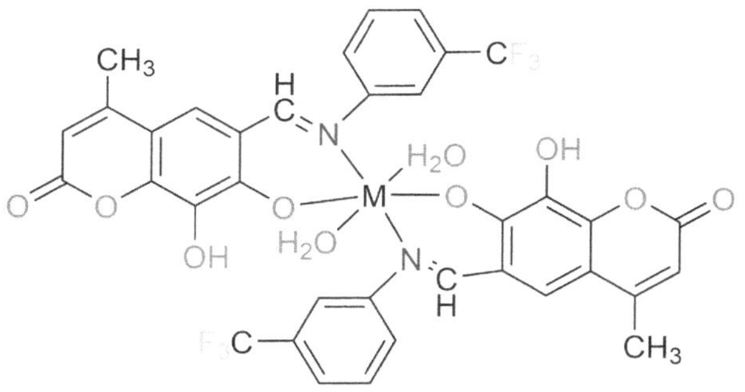

FIGURE 5.13 Proposed structures for complexes Co(II) (2), Ni(II) (4) and Cu(II) (6) [30].

Schiff bases and their corresponding metal complexes both possessed strong anthelmintic properties. The activity of the two ligands was comparable, with Schiff base H2LII being more active than H2LI. Compared to ligands, all metal complexes exhibited strong anthelmintic action. The complex with Schiff base of H2LII was found have the most potent antithelmintic activity [30].

5.4 CONCLUSION

Schiff bases have emerged as an essential class of compounds in bio-inorganic chemistry. The Schiff bases have been found to be applicable in coordinating the transition metal ions and creating diverse enriched metal complexes with several antiviral, antibacterial, and anticancer activities. A vast range of different organisms belonging to bacteria, viruses, fungi, protozoans, helminths, and so on, are known to cause disease in humans. Worldwide, billions of infections are caused on by parasites. The metal-based Schiff bases, according to the studies performed by various scientists, have proved to be more effective than their original Schiff bases and ligands. There have been few studies on the antiviral prospect of these compounds due the biological characteristics of viruses. These complexes have proved to be potent antiparasitic agents against different types of plants and human pathogens, generating possible pharmacological applications in the future.

REFERENCES

[1] T. T. Tidwell, "Hugo (Ugo) Schiff, Schiff bases, and a century of beta-lactam synthesis," *Angew. Chem. Int. Ed. Engl.*, vol. 47, no. 6, pp. 1016–1020, 2008, doi: 10.1002/ANIE.200702965.

[2] M. Munjal, "Biological activity of transition metal complexes incorporating Schiff bases: A review," vol. 6, no. 2, pp. 354–360, 2017.

[3] T. Y. Fonkui, M. I. Ikhile, D. T. Ndinteh, and P. B. Njobeh, "Microbial activity of some heterocyclic schiff bases and metal complexes: A review," *Trop. J. Pharm. Res.*, vol. 17, no. 12, pp. 2507–2518, 2018, doi: 10.4314/tjpr.v17i12.29.

[4] J. M. Mir, S. A. Majid, and A. H. Shalla, "Enhancement of Schiff base biological efficacy by metal coordination and introduction of metallic compounds as anticovid candidates: A simple overview," *Rev. Inorg. Chem.*, vol. 41, no. 4, pp. 199–211, 2021, doi: 10.1515/revic-2020-0020.

[5] C. Abate, F. Carnamucio, O. Giuffr, and C. Foti, "Metal-Based Compounds in Antiviral Therapy," Biomolecules, vol. 12, no. 7, p. 933, 2022, doi:10.3390/biom12070933.

[6] A. O. Rataan, S. M. Geary, Y. Zakharia, Y. M. Rustum, and A. K. Salem, "Potential role of selenium in the treatment of cancer and viral infections," *Int. J. Mol. Sci.* vol. 23, no. 4, p. 2215, Feb. 2022, doi: 10.3390/IJMS23042215.

[7] Z. T. Omar, S. Jadhav, S. Shejul, P. Chavan, R. Pathrikar, and M. Rai, "Synthesis, magnetic moment, antibacterial, and antifungal studies of INH incorporating Schiff base metal complexes," *Polycycl. Aromat. Compd.*, vol. 0, no. 0, pp. 1–14, 2022, doi: 10.1080/10406638.2022.2077776.

[8] T. S. Ababneh, M. E. Khateeb, A. K. Tanash, M. J. A. Shammout, T.M.A. Jazzai, M. Alomari, S. Daoud and W. H. Talib, "Synthesis, computational, anticancerous and antiproliferative effects of some copper, manganese and zinc complexes with ligands derived from symmetrical 2,2′-diamino-4,4′-dimethyl-1,1′-biphenyl-salicylaldehyde," *Polish J. Chem. Technol.*, vol. 23, no. 1, pp. 7–15, Mar. 2021, doi: 10.2478/PJCT-2021-0002.

[9] Z. T. Omar, S. Jadhav, R. Pathrikar, S. Shejul, and M. Rai, "Synthesis, magnetic susceptibility, thermodynamic study and bio-evaluation of transition metal complexes of new Schiff base incorporating INH pharmacophore," *Polycycl. Aromat. Compd.*, vol. 0, no. 0, pp. 1–15, 2021, doi: 10.1080/10406638.2021.2015397.

[10] K. Kar, D. Ghosh, B. Kabi, and A. Chandra, "A concise review on cobalt Schiff base complexes as anticancer agents," *Polyhedron*, vol. 222, p. 115890, Aug. 2022, doi: 10.1016/J.POLY.2022.115890.

[11] A. Sudha and S. J. Askar ali, "Investigation of new Schiff base transition metal (II) complexes theoretical, antidiabetic and molecular docking studies," *J. Mol. Struct.*, vol. 1259, p. 132700, Jul. 2022, doi: 10.1016/J.MOLSTRUC.2022.132700.

[12] A. Arunadevi and N. Raman, "Biological response of Schiff base metal complexes incorporating amino acids–a short review," *J. Coord. Chem.*, vol. 73, no. 15, pp. 2095–2116, Aug. 2020, doi: 10.1080/00958972.2020.1824293/SUPPL_FILE/GCOO_A_1824293_SM2131.DOCX.

[13] L. L. Dickey, L. J. Martins, V. Planelles, and T. M. Hanley, "HIV-1-induced type I IFNs promote viral latency in macrophages," *J. Leukoc. Biol.*, 2022, doi: 10.1002/JLB.4MA0422-616R.

[14] X. Liu, X. Liu, J. Zhou, Y. Dong, W. Jiang, and W. Jiang, "Rampant C-to-U deamination accounts for the intrinsically high mutation rate in SARS-CoV-2 spike gene," *RNA*, vol. 28, no. 7, pp. 917–926, Jul. 2022, doi: 10.1261/RNA.079160.122.

[15] D. Cirri, A. Pratesi, T. Marzo, and L. Messori, "Metallo therapeutics for COVID-19. Exploiting metal-based compounds for the discovery of new antiviral drugs," https://doi.org/10.1080/17460441.2020.1819236, vol. 16, no. 1, pp. 39–46, 2020, doi: 10.1080/17460441.2020.1819236.

[16] D. Sriram, P. Yogeeswari, N. S. Myneedu, and V. Saraswat, "Abacavir prodrugs: Microwave-assisted synthesis and their evaluation of anti-HIV activities," *Bioorg. Med. Chem. Lett.*, vol. 16, no. 8, pp. 2127–2129, Apr. 2006, doi: 10.1016/J.BMCL.2006.01.050.

[17] N. A. Al-Masoudi, N. M. Aziz, and A. T. Mohammed, "Synthesis and in vitro anti-hiv activity of some new Schiff base ligands derived from 5-amino-4-phenyl-4H-1,2,4-triazole-3-thiol and their metal complexes," http://dx.doi.org/10.1080/10426500802591630, vol. 184, no. 11, pp. 2891–2901, Nov. 2009, doi: 10.1080/10426500802591630.

[18] K. S. Kumar, S. Ganguly, R. Veerasamy, and E. De Clercq, "Synthesis, antiviral activity and cytotoxicity evaluation of Schiff bases of some 2-phenyl quinazoline-4(3)H-ones," *Eur. J. Med. Chem.*, vol. 45, no. 11, pp. 5474–5479, Nov. 2010, doi: 10.1016/J.EJMECH.2010.07.058.

[19] O. A. El-Gammal, A. A. El-Bindary, F. Sh. Mohamed, G. N. Rezk, and M. A. El-Bindary, "Synthesis, characterization, design, molecular docking, anti COVID-19 activity, DFT calculations of novel Schiff base with some transition metal complexes," *J. Mol. Liq.*, vol. 346, p. 117850, 2022, doi: 10.1016/j.molliq.2021.117850.

[20] D. Sharma and H. D. Revanasiddappa, "Synthesis, spectral characterization, in vitro antimicrobial and anthelmintic evaluations of cu(ii) complexes with a new Schiff base," *Orbital Electron. J. Chem.*, vol. 10, no. 7, 2018, doi: 10.17807/orbital.v10i7.1168.

[21] M. Manjunath, A. D. Kulkarni, G. B. Bagihalli, S. Malladi, and S. A. Patil, "Bio-important antipyrine derived Schiff bases and their transition metal complexes: Synthesis, spectroscopic characterization, antimicrobial, anthelmintic and DNA cleavage investigation," *J. Mol. Struct.*, vol. 1127, pp. 314–321, 2017, doi: 10.1016/j.molstruc.2016.07.123.

[22] N. J. Onuorah, N. P. Yahaya, W. L. Danbature, and M. Umar, "Synthesis, characterization and evaluation of antimicrobial potency of Fe(II), Mn(II), Cu(II), Co(II) and Zn(II) ion complexes with Schiff base, derived from salicylaldehyde and p.toluidine," *World J. Innov. Res.*, vol. 8, no. 2, pp. 16–21, 2020, doi: 10.31871/wjir.8.2.34.

[23] C. Thiravidamani and N. Tarannum, "Evaluation of DNA intercalation study and biological profile of a series of Schiff base metal(II) complexes derived from amino acid," *Inorg. Nano-Metal Chem.*, vol. 51, no. 8, pp. 1005–1016, 2021, doi: 10.1080/24701556.2020.1813770.

[24] B. Naureen, G. A. Miana, K. Shahid, M. Asghar, S. Tanveer, and A. Sarwar, "Iron (III) and zinc (II) monodentate Schiff base metal complexes: Synthesis, characterisation and biological activities," *J. Mol. Struct.*, vol. 1231, p. 129946, 2021, doi: 10.1016/j.molstruc.2021.129946.

[25] S. Savir, Z. J. Wei, J. W. K. Liew, I. Vythilingam, Y. A. L. Lim, H. M. Saad, K. S. Sim and K. W. Tan, "Synthesis, cytotoxicity and antimalarial activities of thiosemicarbazones and their nickel (II) complexes," *J. Mol. Struct.*, vol. 1211, p. 128090, 2020, doi: 10.1016/j.molstruc.2020.128090.

[26] G. Scalese, et al., "Heteroleptic oxidovanadium (IV) complexes of 2-hydroxynaphtylaldimine and polypyridyl ligands against Trypanosoma cruzi and prostate cancer cells," *J. Inorg. Biochem.*, vol. 175, no. June, pp. 154–166, 2017, doi: 10.1016/j.jinorgbio.2017.07.014.

[27] N. Salam and S. Azam, "Prevalence and distribution of soil-transmitted helminth infections in India," *BMC Public Health*, vol. 17, no. 1, pp. 1–12, Feb. 2017, doi: 10.1186/S12889-017-4113-2/TABLES/2.

[28] D. O. Maia, V. F. Santos, C. R. S. Barbosa, Y. N. Froes, D. F. Muniz, A. L. Santos, M. H. C. Santos, R. R. S. Silva, C. G. L. Silva, R. O. S. Souza, J. C. S. Sousa, H. D. M. Coutinho and C. S. Teixeira, "Nickel (II) chloride schiff base complex: Synthesis, characterization, toxicity, antibacterial and leishmanicidal activity," *Chem. Biol. Interact.*, vol. 351, no. October 2021, 2022, doi: 10.1016/j.cbi.2021.109714.

[29] M. Seid Id, T. Yohanes, Y. Goshu, K. Jemal, and M. Siraj, "The effect of compliance to Hand hygiene during COVID-19 on intestinal parasitic infection and intensity of soil transmitted helminthes, among patients attending general hospital, southern Ethiopia: Observational study," *PLoS One*, vol. 17, no. 6, p. e0270378, 2022, doi: 10.1371/JOURNAL.PONE.0270378.

[30] S. A. Patil, C. T. Prabhakara, B. M. Halasangi, S. S. Toragalmath, and P. S. Badami, "Spectrochimica Acta Part A: Molecular and Biomolecular Spectroscopy DNA cleavage, antibacterial, antifungal and anthelmintic studies of Co (II), Ni (II) and Cu (II) complexes of coumarin Schiff bases: Synthesis and spectral approach," *Spectrochim. ACTA PART A Mol. Biomol. Spectrosc.*, vol. 137, pp. 641–651, 2015, doi: 10.1016/j.saa.2014.08.028.

6 Medicinal Activities of Anti-Inflammatory Schiff Base Metal Complexes

Alka[1,2], *Pallavi Jain*[1*], *and Seema Gautam*[2]
[1]Department of Chemistry, SRM-IST, Delhi-NCR Campus, Modinagar, Ghaziabad, India
[2]Department of Chemistry, Deshbandhu College, University of Delhi, Kalkaji, India
[*]E-mail id: pallavij@srmist.edu.in

CONTENTS

6.1 INTRODUCTION

The substances with an azomethine or imine (-C=N-) functional group are known as Schiff bases. In 1960, Hugo Schiff was the first to describe them as the condensation products of primary amines with aldehyde and ketone compounds (Tsantis et al. 2020). They may adjust the ligation features by changing their denticity and basicity, and they are stable. Since the middle of the nineteenth century, and even before the report of the general production of the Schiff base ligands, its metal complexes have been recognized (More et al. 2020). A thorough review and comparison of the existing literature on these kinds of compounds is necessary after extensive study on the physicochemical characteristics and molecular structure of complexes containing Schiff bases has produced some intriguing new findings. As Schiff bases have the capacity to form stable complexes with metal ions, therefore these plays an important role in the area of coordination chemistry, particularly in the synthesis of Schiff base complexes (Ceramella et al. 2022). Significant attention has recently been observed in the chemistry of Schiff base transition metal complexes due to their diverse coordination behavior and importance to understanding molecular processes (Dalia et al. 2018). In terms of its structure and coordination chemistry, metal complexes are the focus of considerable attention. By combining them with various ligands, they exhibit a variety of optical, chemical, and magnetic properties (Divya, Pinto, and Pinto 2017). In transition metal coordination chemistry, Schiff base metal complexes are regarded as among the most significant stereochemical models because of their readily available preparation methods and diverse structural composition (Sheikh et al. 2016). Metal complexes, which often contain oxygen, nitrogen, or sulphur as ligand atoms, have grown in importance because Schiff bases can interact with various metal centers using multiple coordination sites and enable the productive synthesis of metal complexes (Raczuk et al. 2022). The

DOI: 10.1201/9781003291459-6

preparation of their solid complexes makes use of the Schiff base's strong chelation affinity for transition metal ions. Literature shows that such complexes are bioactive substances and have anti-inflammatory (Salih Abood, Ramadhan, and Hamza 2020), anti-microbial, anti-cancer, antioxidant, DNA binding and cleavage (Arunadevi et al. 2019), anti-malarial, anti-tubercular (Sr et al. 2020), analgesic, anthelmintic, and anti-diabetic actions (Uddin, Ahmed, and Alam 2020).

Inflammation is the primary physiological defensive mechanism against infection, burns, poisonous substances, allergies, and other unpleasant stimuli. Many of these chronic diseases may have an etiologic aspect that is an untreated and persistent inflammation (Paulpriya, Tresina, and Mohan 2016). Despite being a protective mechanism, the many processes and substances associated in an inflammatory reaction can cause, keep, or worsen a number of diseases. Anti-inflammatory medications now in use are linked to several serious negative effects (Intahphuak, Khonsung, and Panthong 2010). Therefore, it is essential to produce effective anti-inflammatory medications with fewer harmful or side effects. Schiff bases and their transition metal complexes are now frequently used in the development of numerous medications due to their vast range of biological activity (Leung et al. 2015). It is found that antioxidants with 1,1-diphenyl-picrylhydrazyl (DPPH) radical scavenging activity are gaining more and more attention these days. Interesting anti-cancer, anti-aging, and anti-inflammatory properties have been reported for them. Therefore, substances having antioxidant characteristics should be anticipated to provide defence against inflammation and rheumatoid arthritis. Therefore, substances having antioxidant characteristics should be anticipated to provide defense against inflammation and rheumatoid arthritis and to lead to potentially effective drugs (Pontiki, Hadjipavlou-Litina, and Chaviara 2008). Thus, this chapter outlines the uses of Schiff bases and their metal complexes as anti-inflammatory drugs.

6.2 ANTI-INFLAMMATORY ACTION OF SCHIFF BASE AND SCHIFF BASE DERIVED METAL COMPLEXES

Nitthinchandra and co-workers described the synthesis of a series of sydnone Schiff bases 3-[1(-4-isobutylphenyl)ethyl]-4-(3-aryl-4-sydnonylidene)amino-5-mercapto-1,2,4-triazoles (1a-1c) yield aminomethyl-3-[1-4-isobutylphenyl)ethyl]-4-(3-aryl-4-sydnonylidene)amino-1,2,4-triazole-5-thiones (2a-2e) (Scheme 6.1) and tested for their anti-inflammation and analgesic activities. Elemental analyses, mass spectra, IR, NMR, and NMR data were used to analyse the geometries of both the compounds. Both the compounds were screened for anti-inflammation activity against indomethacin as standard drug and results reveal that compound 1c showed better activity than 1a and 1b. It was conclusively shown that sydnone's electron-releasing group would increase its anti-inflammatory action. The compounds **2a** and **2b** showed significant anti-inflammatory activity as compared to the standard drug, that is, Indomethacin. Derivatives of compound **2** will aid enhanced biological activity due to the presence of piperidine and morpholine moieties in their structure (Nithinchandra et al. 2012).

A series of novel Schiff bases (3a-3f) was synthesized through the condensation reaction of 4-aminoantipyrine and different substituted benzaldehydes. The synthesized compounds were screened for anti-inflammatory activity. The results of the study on the structure-activity relationship showed that the Schiff base analogues of 4-aminoantipyrine's antioxidant activity were significantly influenced by the position and type of the substitution on the benzylidene phenyl ring. The most effective antioxidant and anti-inflammatory was discovered to be compound 3f. Using the Griess method, the anti-inflammatory efficacy of 3f was assessed concerning its ability to reduce the formation of NO in LPS-pre-treated RAW 264.7 cells. Lipopolysaccharide (LPS), an endotoxin generated by the cell wall of Gram-negative bacteria, can trigger the invention of inflammatory modulators such as PGE2, NO, and interleukins through a number of signaling pathways. The outcomes showed that 50 g/mL of 3f prevented COX-2 mRNA levels from increasing in response to LPS (Alam, Choi, and Lee 2012).

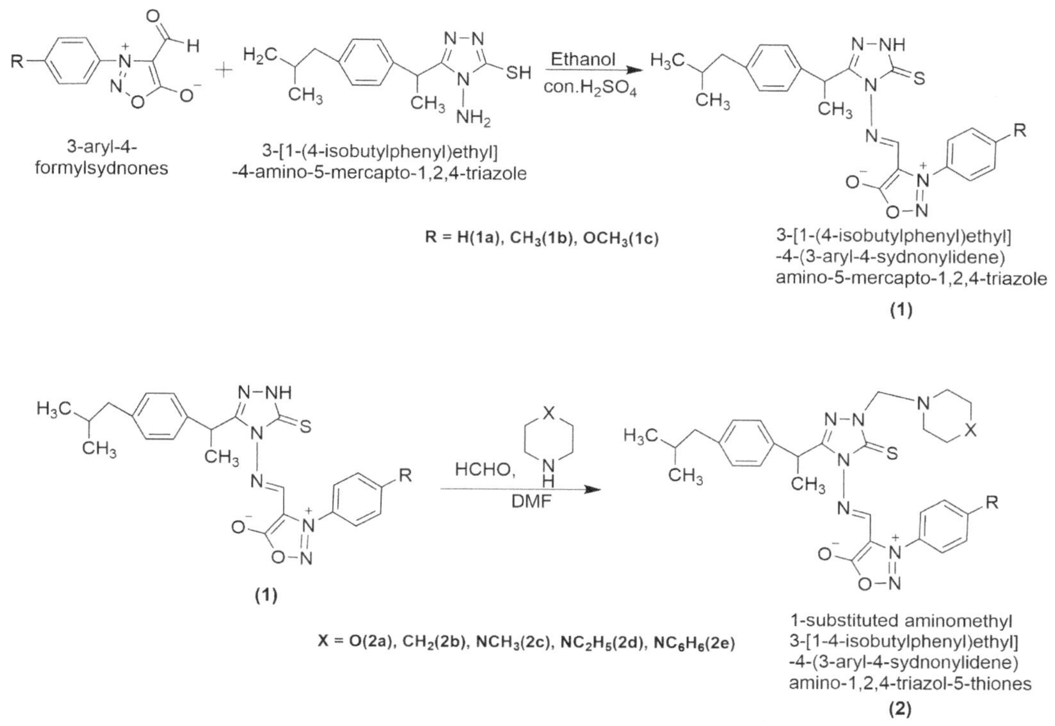

SCHEME 6.1 Synthesis of Sydone moiety containing Schiff bases.

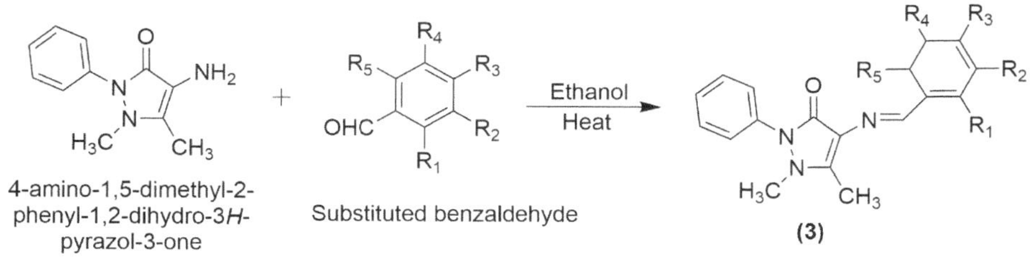

SCHEME 6.2 Synthesis of 4-amino-1,5-dimethyl-2-phenylpyrazol-3-one analogues (3a-3e), 3a- R_1, R_2, R_3, R_4, R_5 is H; 3b- R_1, R_2, R_4, R_5 is H and R_3-Cl; 3c- R_1, R_2, R_4, R_5 is H and R_3-NMe_2; 3d- R_1, R_2, R_4, R_5 is H and R_3-OH; 3e- R_1, R_2, R_4, R_5 is H and R_3-OMe.

Schiff base synthesized from 2-[(2,6-dichloroanilino) phenyl] acetic acid (diclofenac acid) that yield a series of S-substituted phenacyl 1,3,4-oxadiazoles and Schiff bases 4, 5 and their substituted compounds (4a-4k and 5a-5h) (Scheme 6.3) were investigated for anti-inflammation activity by Bhandari et al. The carrageenan-induced rat paw edema process was utilized to investigate the anti-inflammatory effects of the compounds. The ability of these substances to reduce inflammation was examined in vivo. The compounds were tested for their analgesic activity and to see if they may cause ulceration by ulcerogenic and histopathological investigations. The compounds that demonstrated considerable activity (similar to the standard medication diclofenac sodium) underwent these tests. In the carrageenan-induced rat paw edema paradigm, eight novel compounds out of the total of 18 were discovered to have substantial anti-inflammation action. They were also discovered to have

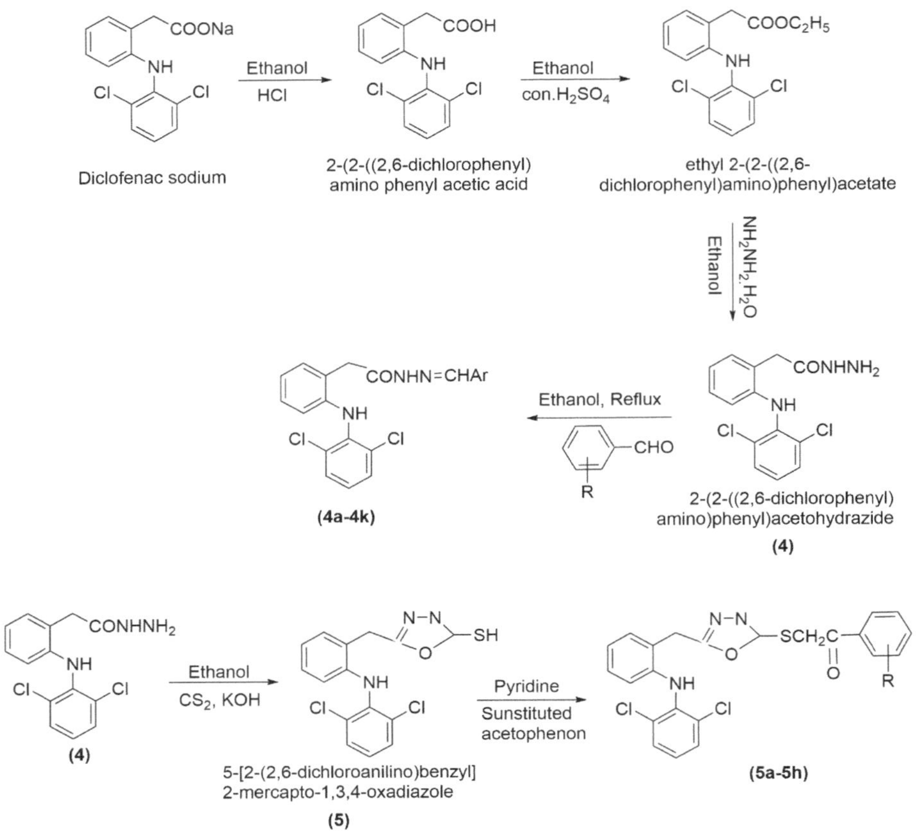

SCHEME 6.3 Synthesis of nonulcerogenic derivatives of diclofenac acid derived Schiff bases.

significant analgesic effect in the acetic acid-induced writhing model without any ulcerogenicity. Compounds N-(4-Bromobenzylidenyl)-[2-(2,6-dichloroanilino)benzylcarbazide] (**4K**) and 5-[2-(2,6-Dichloroaniline)benzyl]-S-(3-methoxyphenacyl)-1,3,4-oxadiazole-2-thiol (**5b**) showed the most pronounced and persistent anti-inflammatory action among all the produced compounds. It was concluded that the compounds not only maintained the anti-inflammatory description of diclofenac acid but have also assisted in improving the anti-inflammatory action and are free of the more dangerous gastrointestinal toxicities from the in-depth investigation of histopathological studies (Bhandari et al. 2008).

Pontiki et al. reported the synthesis of Schiff base which derived from 2-thiophene-carboxaldehyde and dipropylenetriamine and novel Cu(II) dien based metal complexes [Cu(dien)Y_2], [Cu(dien) XXY2], [Cu(1,3-propanediamine) X] (X=Cl_2, Br_2). Compounds were characterized using various spectroscopic techniques. Structures of tested compounds are shown in Scheme 6.4. As a template for acute inflammation, the rat carrageenin-induced paw edema assay was used to evaluate the monobasic Schiff base and complexes' anti-inflammatory efficacy. The medicine indomethacin was used as a reference. Carrageenin-induced edema has been characterized as developing in two phases. Histamine and serotonin act as mediators during the initial stage of the inflammatory response. The second stage is regulated by prostaglandins and kinins, most likely. The investigated compounds exhibited significant scavenging action and reduce carrageenin-induced rat paw edema upto 52.0–82.6%. The experimental analysis confirmed the highest potency (82.6%)

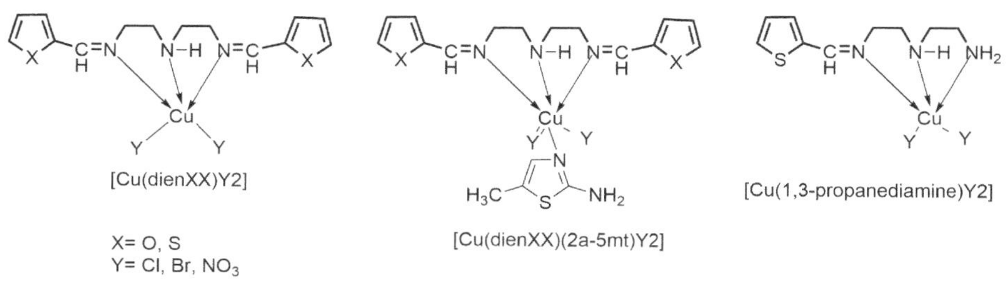

SCHEME 6.4 Chemical structures used in anti-inflammation study.

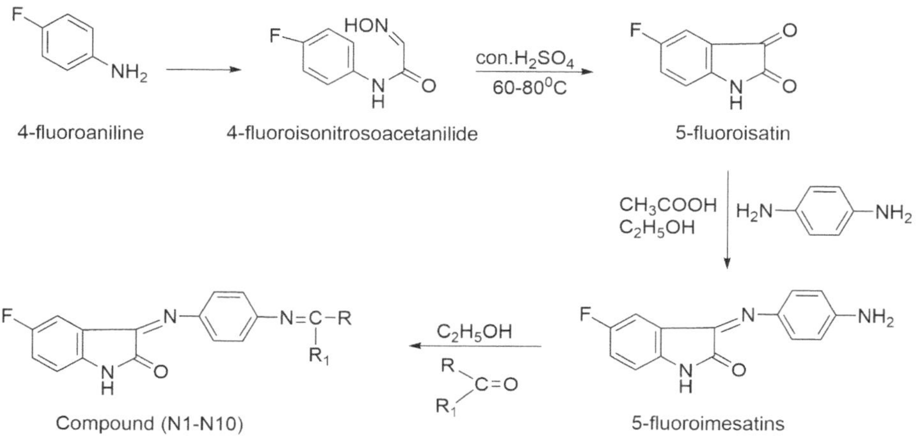

SCHEME 6.5 Synthesis of 3-(4-amino)phenylimino)-5-fluoroindolin-2-one derivatives (N1-N10).

of [Cu(dien)OO] Br_2 complex in the *in-vivo* experiment. The findings confirmed that, in comparison to starting material, [Cu(dienXXY_2] demonstrated enhanced anti-inflammatory activity (Pontiki and Chaviara 2008).

The synthesis and spectrum analysis of a number of novel derivatives of 3-(4-(benzylideneamino) phenylimino) 4-fluoroindolin-2-one (N1-N10) (Scheme 6.5) were carried out using IR, mass and ^{1}H NMR techniques by Nirmal and team. The compounds had their anti-inflammatory potency assessed. The results showed that compounds N2, N3, and N8 among the investigated title compounds all exhibited anti-inflammatory effectiveness that was comparable to that of the reference standard, diclofenac sodium. This initial investigation revealed that alkyl groups at the N-4 position had stronger anti-inflammatory effect than aryl groups (Nirmal et al. 2010).

Two novel Schiff bases derived from the condensation of dapson and 8-formyl-7-hydroxy-4-methylcoumarin (6) and 5-formyl-6-hydroxycoumarin (7), respectively, and their Co(II), Ni(II), and Cu(II) complexes (8 and 9) were synthesized by Manjunatha et al. and analyzed on the basis of elemental and spectroscopic techniques. All the synthesized compounds were tested for their action as anti-inflammation. For this purpose, Formalin induced edema in the right hind paws of rats was utilized. Diclofenac sodium was used as the standard drug. Rats in the control group gradually increased the size of their edematous paws (Formalin treated). However, the synthesized compounds significantly reduced the edema paw volume in the test groups. The maximum anti-inflammatory impact of Schiff base 6 developed at 1 h (71.4%), and the anti-inflammatory effect began to appear at 1–2 h, gradually rose, and attained a maxima of 71.4%, 70%, and 65.5% at 5 h, respectively.

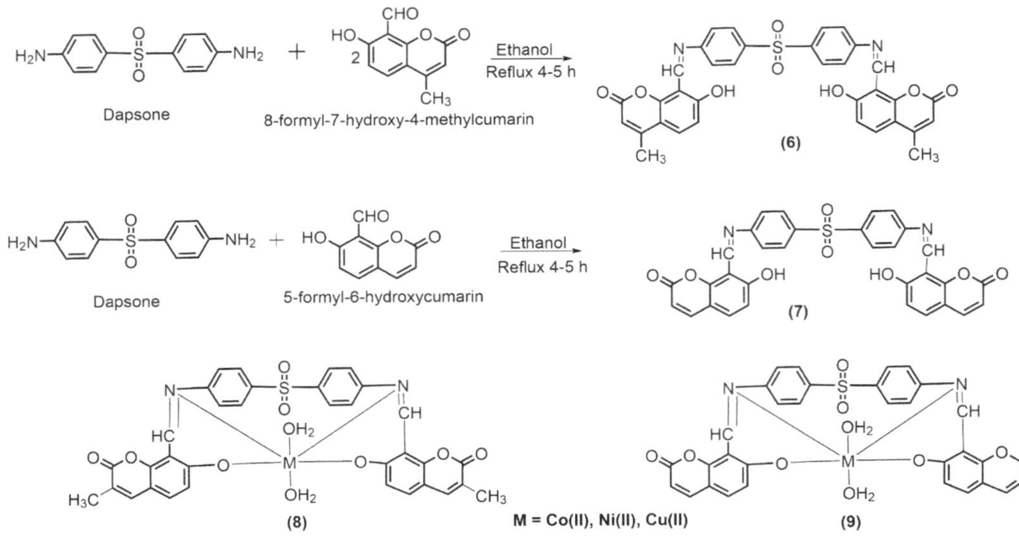

SCHEME 6.6 Synthesis of dapson derived Schiff base and metal complexes.

Diclofenac sodium's anti-inflammatory impact gradually grew and peaked at 75% after two hours. The findings of the inflammation tests revealed that the Schiff bases play a key role in inhibiting inflammation, which increases when metal ions are introduced (Manjunatha, Naik, and Kulkarni 2011) (Scheme 6.6).

Hariprasath et al. synthesized sulphadiazine compounds using aromatic aldehydes such as para diethyl amino benzaldehyde and para-dimethyl amino benzyldehyde, Schiff bases (A & B) of the sulphadiazine. By reacting with methyl iodide (MeI), the produced Schiff bases were changed into their cationic amphiphilic bases. The cationic Schiff bases were treated with metals such as copper chloride ($CuCl_2$), zinc chloride ($ZnCl_2$), and cadmium chloride to produce metal complexes ($CdCl_2$) (Scheme 6.7). Elemental analysis, IR, and H1 NMR were used to characterize synthesized compounds. The ability of Schiff base as well as metal complexes to reduce inflammation was tested against standard drug diclofenac sodium standard drug. Paw edema method was used to perform the experiment. This anti-inflammatory function may be explained by how cationic lipids interact with the PKC pathway to suppress pro-inflammatory mediators.

Additionally, cationic lipids naturally have anti-inflammatory properties. Studies show that anti-inflammatory efficacy requires the release of cationic lipids into the cytoplasm of macrophages. In comparison to the reference drug diclofenac sodium (10mg/kg), Cu(II) metal complexes (20mg/kg) of A1 and B1 showed superior anti-inflammatory efficacy in carrageenan-induced edema technique. Thus, it was established that copper complexes are a special class of anti-inflammatory medications that may be more beneficial therapeutically. These findings show that cationic lipids are potential new anti-inflammatory compounds (Hariprasath et al. 2014).

Shilpa and co-workers reported the synthesis of Schiff base ligand N'-(2-hydroxybenzylidene)-3-((4-(trifluoromethyl)phenyl)amino)benzohydrazide and its Cu(II), Co(II), Ni(II) and Zn(II) complexes (Scheme 6.8). The formation of synthesized compounds was analyzed by using various spectroscopic (IR, UV-Visible, ^{1}NMR) and elemental techniques. The anti-inflammation activity of Schiff base and metal complexes was carried out against standard drug flufenamic acid. Among Cu(II), Co(II), Ni(II) and Zn(II) only Cu(II) and Zn(II) complexes were used for the analysis. A mucolytic enzyme called Hyaluronidase was used to test the anti-inflammatory activity of the Schiff base ligand, the Cu(II) and Zn(II) complexes at concentrations of 10 micrograms, 50 micrograms, and

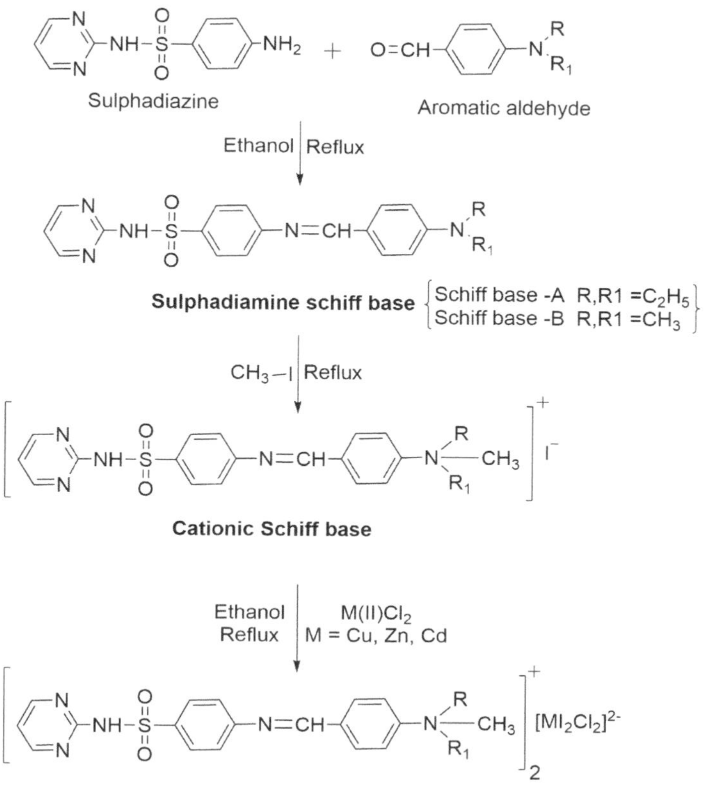

SCHEME 6.7 Synthesis of cationic Schiff bases and their metal complexes.

100 micrograms, respectively, in addition to the standard drug flufenamic acid. It has been observed that at a concentration of 10 micrograms, the ligand and Zn(II) combination shows greater activity than the normal medication. This was the finding of the study that was conducted. The Zn(II) complex has a higher level of activity than the ligand at doses of 50 and 100 micrograms. Based on the findings of this investigation, metal complexes seem to possess superior levels of biological activity compared to free ligand and the reference medication. The influence that metal ions have on the processes that normally occur in cells is the primary reason for the rise in the biological activity of metal complexes. The overtone idea and chelation treatment both provide an explanation for the enhanced activity of such metal complexes (Gopinath et al. 2021).

The curiosity in Schiff base complexes has grown as bioinorganic chemistry. It is now understood that most of these complexes may act as models for biologically significant species. In the above section, we focused mainly on the anti-inflammatory action of the Schiff bases and their metal complexes. A summary of other applications is also mentioned in Table 6.1.

6.3 CONCLUSION AND FUTURE PERSPECTIVE

Schiff base metal complexes have a long history in medical chemistry. Carbonyl compounds (Aldehydes/ketones) and primary amines are affordable, readily available starting ingredients that make it simple to build Schiff bases. Schiff bases can be produced as stable products using this straightforward, synthetically accessible reaction with high yields and purities and little to no product preparation. Azomethine (-C=N-) linkage, a typical property of Schiff bases, is an effective

TABLE 6.1
Pharmacological Applications of Schiff Base and Their Metal Complexes

S. No.	Ligand	Metal Ions	Activities	Ref no.
1.	1,3- propanediamine derived Schiff bases	Cu(II)	Anti-inflammation, Antioxidant	(Pontiki, Hadjipavlou-Litina, and Chaviara 2008)
2.	2-imino-4-thiobiuret and methoxybezaldehyde derived Schiff bases	VO(II)	Anti-inflammation	(Shukla and Mishra 2019)
3.	Histidine and 3-phenylpropane derived Schiff base	Cu(II)	Anti-inflammation	(Abood, Ramadhan, and Hamza 2020)
4.	Mixed ligand derived from acetaminophen and diclofenac potassium salt	Co(II), Cu(II), Ni(II), Zn(II)	Anti-inflammation, Anti-microbial	(Authors 2021)
5.	Carbohydrazide and salicyldehyde derived Schiff base	Co(II), Cu(II), Ni(II), Zn(II)	Anti-inflammation, Anti-microbial	(Gopinath et al. 2021)
6.	Derivatives of 2-hydroxybenzoic acidbenzylidene hydrazide	-	Anti-inflammation, Antioxidant	(Anusuya et al. 2020)
7.	Sulphadiazine derived Schiff Base	Cu(II), Zn(II), Cd(II)	Anti-inflammation, Anti-depresent	(K.Hariprasath, I. Sudheer Babu 2014)

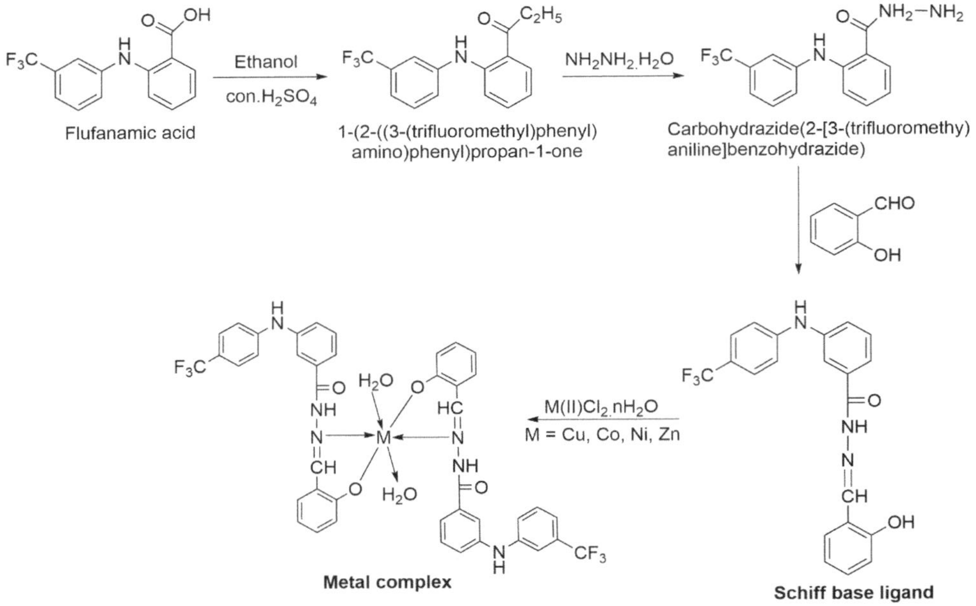

SCHEME 6.8 Synthesis of flufenamic acid derived Schiff base and metal complexes.

way to connect structurally varied, physiologically active scaffolds. This enables scientists to amass huge libraries of hybrid molecules with fascinating biological properties and a wide range of structural diversity. Their numerous uses in areas including medical, catalysis, electronics, industry, and material science have drawn more and more attention in recent years. As a result, creating drugs with metal complexes and Schiff base scaffolds is becoming more and more popular. A summary of anti-inflammatory action of Schiff bases and their metal complexes has been gathered in this chapter. The articles under evaluation discuss different aspects of Schiff bases and their metal complexes' anti-inflammatory applications. There are already a number of anti-inflammation medications in medical use, but the emergence of inflammation resistance calls for the creation of new and potent anti-inflammatory drugs. On the basis of finding, it can be concluded that Schiff base and their metal complexes are the most popular anti-inflammatory drugs, which also significantly influence medical treatments.

REFERENCES

Abood, Huda S., Usama H. Ramadhan, and Hussam Hamza. 2020. "Synthesis and Anti-Inflammatory Activity Study of Schiff Bases Complexes." *Biochemical and Cellular Archives* 20 (2): 5627–5631.

Alam, Mohammad Sayed, Jung-Hyun Choi, and Dong-Ung Lee. 2012. "Synthesis of Novel Schiff Base Analogues of 4-Amino-1,5-Dimethyl-2-Phenylpyrazol-3-One and Their Evaluation for Antioxidant and Anti-Inflammatory Activity." *Bioorganic & Medicinal Chemistry* 20 (13): 4103–4108. doi:https://doi.org/10.1016/j.bmc.2012.04.058.

Anusuya, V., G. Sujatha, G. B. Broheshnu, and G. L. Balaji. 2020. "Anti Oxidant, Anti Inflammation and Anti Diabetic Activity of Novel Schiff Base Derived from 2-Hydroxy Benzoic Acid (3-Hydroxy Benzylidine) Hydrazide and 2-Hydroxy Benzoic Acid (3-Hydroxy – 4-Methoxy Benzylidine)-Hydrazide." *European Journal of Molecular and Clinical Medicine* 7 (4): 2523–2532. www.embase.com/search/results?subaction=viewrecord&id=L2010204950&from=export.

Arunadevi, Alagarraj, Jeyaraman Porkodi, Lakshmanan Ramgeetha, and Natarajan Raman. 2019. "Biological Evaluation, Molecular Docking and DNA Interaction Studies of Coordination Compounds Gleaned from a Pyrazolone Incorporated Ligand." *Nucleosides, Nucleotides \& Nucleic Acids* 38 (9): 656–679. doi:10.1080/15257770.2019.1597975.

Bhandari, Shashikant V., Kailash G. Bothara, Mayuresh K. Raut, Ajit A. Patil, Aniket P. Sarkate, and Vinod J. Mokale. 2008. "Design, Synthesis and Evaluation of Antiinflammatory, Analgesic and Ulcerogenicity Studies of Novel S-Substituted Phenacyl-1,3,4-Oxadiazole-2-Thiol and Schiff Bases of Diclofenac Acid as Nonulcerogenic Derivatives." *Bioorganic & Medicinal Chemistry* 16 (4): 1822–1831. doi:https://doi.org/10.1016/j.bmc.2007.11.014.

Ceramella, Jessica, Domenico Iacopetta, Alessia Catalano, Francesca Cirillo, Rosamaria Lappano, and Maria Stefania Sinicropi. 2022. "A Review on the Antimicrobial Activity of Schiff Bases: Data Collection and Recent Studies." *Antibiotics* 11 (2). doi:10.3390/antibiotics11020191.

Dalia, Sadia, Farhana Afsan, Md Hossain, Md Nuruzzaman Khan, Choudhury Zakaria, Md Kudrat-E Zahan, and Md Ali. 2018. "A Short Review on Chemistry of Schiff Base Metal Complexes and Their Catalytic Application." *International Journal of Chemical Studies* 6: 2859–2866.

Divya, Kumble, Geetha M. Pinto, and Asha F. Pinto. 2017. "Application of Metal Complexes of Schiff Bases As an Antimicrobial Drug: A Review of Recent Works." *International Journal of Current Pharmaceutical Research* 9 (3): 27. doi:10.22159/ijcpr.2017.v9i3.19966.

Gayakwad, D. R., S. R. Sarda, S. U. Tekale, R. B. Nawale, Rajani, J. V. Bharad, R. P. Pawar. 2020. "Synthesis, Characterization and Biological Screening for Antifungal, Antimalarial and Antitubercular Activities of Novel Bis-Imines and Their Metal Complexes." 11 (February): 14–21.

Gopinath, Shilpa Kondareddy, Malathesh Pari, Archanamedehal Rudrannagari, Ishwari Boodihaal Kattebasaveshwara, and Shivaprasad Kengunte Halappa. 2021. "Synthesis, Characterization and Electrochemical Sensor Based upon Novel Schiff Base Metal Complexes Derived from the Non-Steroidal Anti-Inflammatory Drug, Flufenamic Acid for the Determination of Uric Acid and Their Biological Applications." *Biointerface Research in Applied Chemistry* 11 (4): 11390–11403. doi: 10.33263/BRIAC114.1139011403.

Hariprasath, K., I. Sudheer Babu, P. Venkatesh, and U. Upendra Rao. 2014. "Antidepressant and Anti-Inflammatory Activities of Cationic Amphiphilic Complexes of Sulphadiazine." *Global Journal Inc* 14 (7).

Intahphuak, S., P. Khonsung, and A. Panthong. 2010. "Anti-Inflammatory, Analgesic, and Antipyretic Activities of Virgin Coconut Oil." *Pharmaceutical Biology* 48 (2): 151–157. doi:10.3109/13880200903062614.

Leung, Chung Hang, Sheng Lin, Hai Jing Zhong, and Dik Lung Ma. 2015. "Metal Complexes as Potential Modulators of Inflammatory and Autoimmune Responses." *Chemical Science* 6 (2): 871–884. doi:10.1039/c4sc03094j.

Manjunatha, M., Vinod H. Naik, and Ajaykumar D. Kulkarni. 2011. "Activities, and Spectroscopic Studies of Co (II), Ni (II), and Cu (II) Complexes of Biologically Potential Coumarin Schiff Bases." 64 (24): 4264–4275.

Md Saddam Hossain, Pijush Kanti Roy, C. M. Zakaria and Md Kudrat-E-Zahan. 2018. "Selected Schiff base coordination complexes and their microbial application: A review." International Journal of Chemical Studies 6 (1): 19–31.

Nirmal, R., K. Meenakshi, P. Shanmugapandiyan, and C. R. Prakash. 2010. "Synthesis Pharmacological Evaluation of Novel Schiff Base Analogues of 3-(4-Amino) Phenylimino) 5-Fluoroindolin-2-One." *Journal of Young Pharmacists* 2 (2): 162–168. doi:https://doi.org/10.4103/0975-1483.63162.

Nithinchandra, B. Kalluraya, S. Aamir, and A. R. Shabaraya. 2012. "Regioselective Reaction: Synthesis, Characterization and Pharmacological Activity of Some New Mannich and Schiff Bases Containing Sydnone." *European Journal of Medicinal Chemistry* 54: 597–604. doi:https://doi.org/10.1016/j.ejmech.2012.06.011.

Obaleye, J. A., A. A. Aliyu, A. O. Rajee and K. E. Bello. 2021. "Synthesis, Characterization, In-Vitro Anti-Inflammatory and Antimicrobial Screening of Metal(II) Mixed Diclofenac and Acetaminophen Complexes." *Bulletin of the Chemical Society of Ethiopia* 35 (1): 77–86.

Paulpriya, K., Dr. Tresina P., and Veerabahu Mohan. 2016. "Evaluation of Anti-Inflammatory Activity of Aerial Part Extract of Daphniphyllum Neilgherrense (Wt.) Rosenth." *International Journal of Pharmaceutical Sciences Review and Research* 37: 98–100.

Pontiki, E., D. Hadjipavlou-Litina, and A. T. Chaviara. 2008. "Evaluation of Anti-Inflammatory and Antioxidant Activities of Copper (II) Schiff Mono-Base and Copper(II) Schiff Base Coordination Compounds of Dien with Heterocyclic Aldehydes and 2-Amino-5-Methyl-Thiazole." *Journal of Enzyme Inhibition and Medicinal Chemistry* 23 (6): 1011–1017. doi:10.1080/14756360701841251.

Pontiki, E. and A. T. Chaviara. 2008. "Evaluation of Anti-Inflammatory and Antioxidant Activities of Copper (II) Schiff Mono-Base and Copper (II) Schiff Base Coordination Compounds of Dien with Heterocyclic Aldehydes and 2-Amino-5-" 6366 (Ii). doi:10.1080/14756360701841251.

Raczuk, Edyta, Barbara Dmochowska, Justyna Samaszko-Fiertek, and Janusz Madaj. 2022. "Different Schiff Bases—Structure, Importance and Classification." *Molecules* 27 (3). doi:10.3390/molecules27030787.

Salih Abood, Huda, Usama Ramadhan, and Hussam Hamza. 2020. "Synthesis And Anti-Inflammatory Activity Study of Schiff Bases Complexes." *Biochemical and Cellular Archives* 20: 5627–5631.

Sheikh, Rayees Ahmad, Mohmmad Younus Wani, Sheikh Shreaz, and Athar Adil Hashmi. 2016. "Synthesis, Characterization and Biological Screening of Some Schiff Base Macrocyclic Ligand Based Transition Metal Complexes as Antifungal Agents." *Arabian Journal of Chemistry* 9: S743–S751. doi:https://doi.org/10.1016/j.arabjc.2011.08.003.

Shukla, Shraddha and A. P. Mishra. 2019. "Metal Complexes Used as Anti-Inflammatory Agents: Synthesis, Characterization and Anti-Inflammatory Action of VO(II)-Complexes." *Arabian Journal of Chemistry* 12 (7): 1715–1721. doi:10.1016/j.arabjc.2014.08.020.

Tsantis, Sokratis T., Demetrios I. Tzimopoulos, Malgorzata Holynska, and Spyros P Perlepes. 2020. "Oligonuclear Actinoid Complexes with Schiff Bases as Ligands—Older Achievements and Recent Progress." *International Journal of Molecular Sciences* 21 (2). doi:10.3390/ijms21020555.

Uddin, Mohammad Nasir, Sayeda Ahmed, and S. M. Alam. 2020. "REVIEW: Biomedical Applications of Schiff Base Metal Complexes." *Journal of Coordination Chemistry* 73: 3109–3149. doi:10.1080/00958972.2020.1854745.

7 Role of Schiff Base Metal Complexes in the Fight Against Tumors and Cancer

*Namita Misra**
Silver Oak University, Ahmedabad, Gujarat, India
*Corresponding author
Email: namitamis@gmail.com

CONTENTS

7.1 INTRODUCTION

Cancer consists of uncontrolled growth and the rapid proliferation of a group of abnormal cells originating from a single cell [1]. These cells have the capability to abolish the normal body cells. Cancer can occur at any stage of human life and can initiate in any part of the body, which is made up of more than 30 trillion cells. Normally the body cells grow, split, and die in an orderly manner, whereas cancerous cells continuously proliferate and form a mass or lump that resemble overgrowth or swelling called a tumor. Tumors grow constantly in a disorderly fashion. Tumors are classified into two types: benign and malignant. Benign tumors are generally not dangerous whereas malignant tumors are life-threatening, grow very fast, and move to different body parts. These tumors are known as cancer. Cancer cells are of variable size: larger/smaller or abnormal shape than normal cells [2–3]. Cancer is caused within the cells by mutations in the DNA. Mutation of the cells inside DNA force a cell to stop its normal function and allows the cell to convert into a cancerous cell [4].

In 2015, a report by the World Health Organization reported one out of six deaths occur due to cancer which indicates around 20 million deaths in 2030 [5]. It has been estimated that in 2021 there were more than 1.9 million newly diagnosed cancer patients and 608,570 deaths by cancer in the United States [6]. Based on the present scenario, the National Cancer Registry Programme 2020

DOI: 10.1201/9781003291459-7

Report, issued by the ICMR and NCDIR, India, it is estimated that up to 2025 cancer cases will be more than 15.7 lakhs in the country [7]. Chemotherapy is the main treatment of this global life-threatening disease though it has severe side effects. A chemotherapeutic drug destroys the speedily dividing cells, consequently reducing the disease from spreading to other parts.

Various formulations have been used for the treatment of cancer. Bioinorganic transition metal compounds are the most commonly used anticancer drugs. Among them platinum containing chemotherapeutic compounds are used in almost 50% of cancer treatment cases. The oldest compound among them is cisplatin (Figure 7.1) discovered by Barnett Rosenberg in 1960 [8]. Because of several side effects and resistance towards different cancer, oncologists have focused on developing new kinds of complexes with less toxicity [9–11]. In the last few years Schiff bases metal complexes were noticed to possess substantial anticancer properties and inhibit cancer cell growth [12–16].

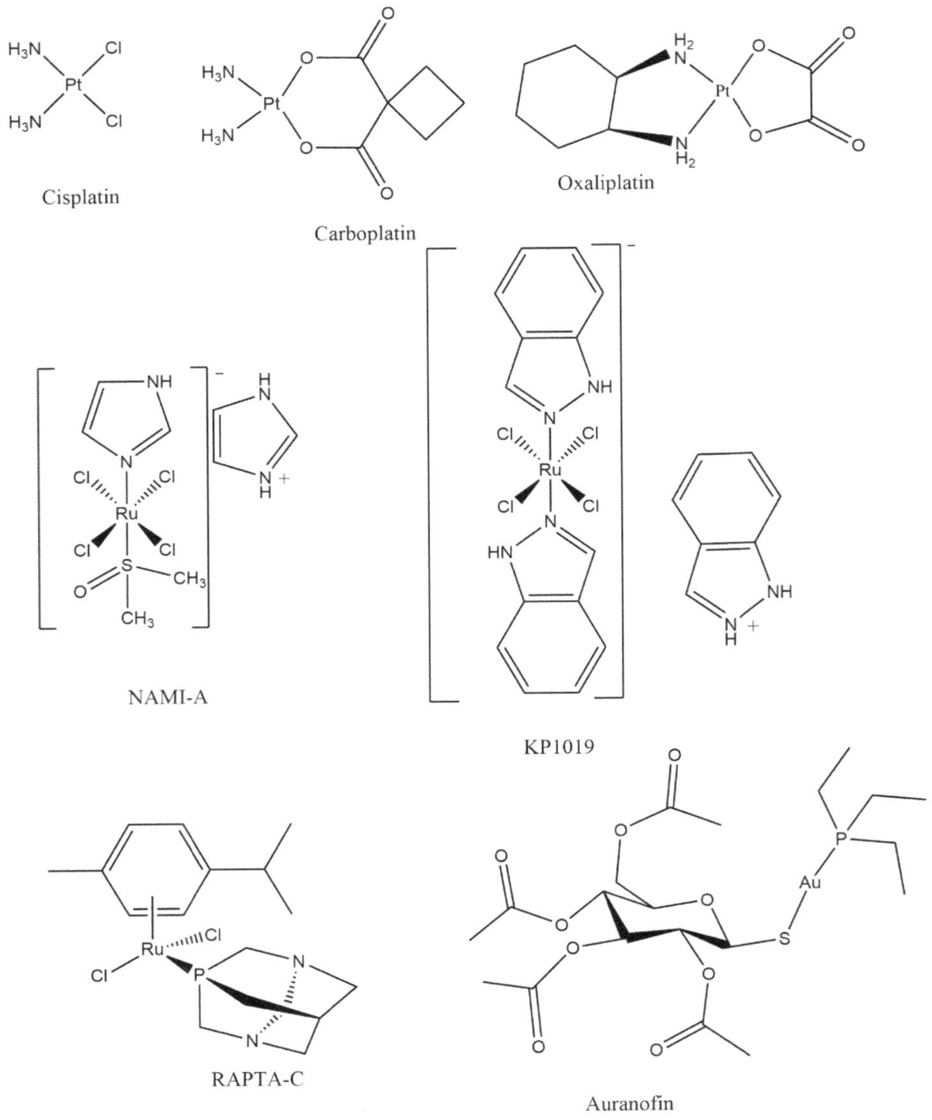

FIGURE 7.1 Metal containing anticancer drugs.

Schiff base ligands are the privileged class of compound and their activity is enhanced when they form complexes with different metals [17–19]. We begin this chapter introducing the role of transition metals in cancer and their Schiff base complexes and then describe different classical and non-classical transition metal complexes with their *in vitro/in vivo* activity reported against different cancer cell lines.

7.2 ROLE OF TRANSITION METAL IN CANCER TREATMENT

Introduction of a metal containing drugs into the body allows the drug to react with many substances in the biological media. Protein and DNA are the main targets for these metallo drugs. Cisplatin was the first metal containing compound which entered into clinical use for cancer treatment [20]. Cisplatin is a Pt^{+2} metal containing the square planer complex, cis-diamminedichloroplatinum (II) [21]. Earlier, on the discovery of cisplatin, chemists limited their research only up to natural product chemistry and organic chemistry for exploring new anticancer agents. The discovery of cisplatin energized inorganic chemists to synthesize different platinum and non-platinum metal complexes analogous that would become a useful anticancer drug (Figure 7.2).

Almost 50% of all chemotherapy-receiving patients are treated with cisplatin or its analogues carboplatin or oxaliplatin. Despite the success rate of cisplatin and its analogues, these anticancer compounds have two main disadvantages, that is, resistance and side effects which include neuropathy, ototoxicity and nephrotoxicity [22–23], which researchers continued to explore other transition metal complexes with improved efficacy and fewer or no side effects.

The selection of a ligand as well as metal is the greatest substantial feature in the designing of a metal-containing cytotoxic drug to find the optimum conjugation of in vivo kinetic and thermodynamic stabilities [24, 25]. Transition metals are d-block elements and are contained within III–XII groups of the periodic table and studied as an alternative to platinum [26]. Several other metal complexes including ruthenium, gold, palladium, rhodium, chromium, copper, cobalt, zinc, and

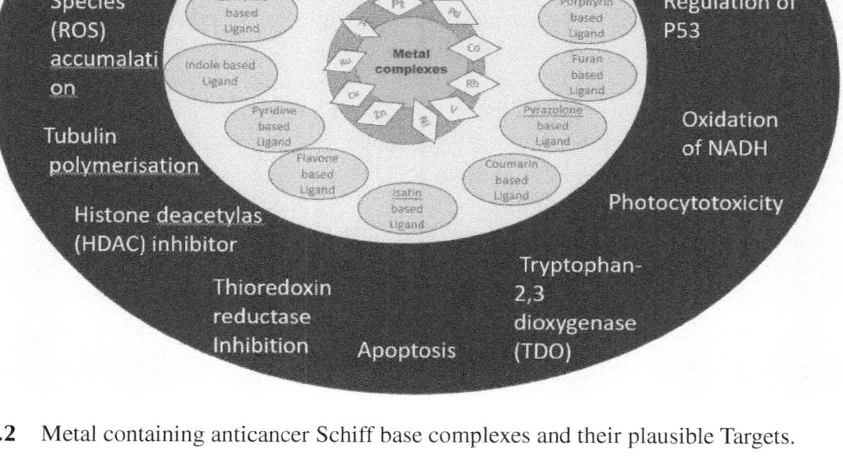

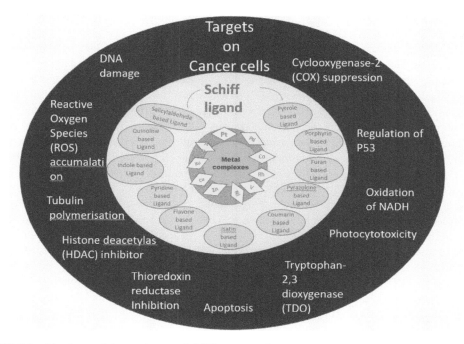

FIGURE 7.2 Metal containing anticancer Schiff base complexes and their plausible Targets.

nickel metal center have been synthesized and evaluated against different cell lines [27–30]. They possess unique properties that involve partially filled d or f shell, a tendency to undergo redox reaction, possess different oxidation states, and permit functionalization of groups. Transition metals act as a positively charged ion in aqueous medium and bind to negatively charged biomolecules [31].

7.3 TRANSITION METAL COMPLEXES OF MULTIFUNCTIONAL SCHIFF BASES

Schiff base ligands were synthesized from the condensation of the primary amines and carbonyl compounds. Schiff bases are a privileged compound group, having the functionality (>C = N-R), called azomethine, where the atom or group attached to nitrogen cannot be hydrogen. The C = N group possesses important biological significance due to loan pair electrons present on the nitrogen atom [32–34].

It is well established that metals have been useful in treating various diseases in humans. Because of the convenient versatile preparation and structural variability of Schiff ligands and the unique properties of transition metal complexes in cancer therapy, Schiff based metal complexes gained the attention of researchers to study and investigate their anticancer properties [35–36]. Schiff bases have a high ability to reduce cancerous cells and this property can be enhanced by complexation with the transition metal.

Schiff bases derived from different sources act as a chelating agent with the different donor atoms like O, N, S, and so on. A myriad of transition metal complexes of multidentates Schiff base with different hetero atoms have been studied. In this chapter, anticancer studies carried out on different transition metal complexes containing Schiff base ligand including platinum are discussed. In addition to transition metal, Schiff base complexes of main group metals are also studied for their anticancer activities, as discussed here.

7.3.1 Platinum and Palladium Complexes as an Anticancer Drug

Cisplatin and its different analogues were used as medication in cancer patients, either singly or in combination with additional drugs, despite their toxicity and resistance. Cisplatin continues the gold standard to which classical platinum and non-platinum drugs are compared. Different platinum analogues were synthesized from the Schiff base ligand and evaluated against different cancer cell lines *invitro* or *invivo*. Li and co-workers [37] reported ten new Pt(II) complexes **4 a-j** by Schiff base ligand having reduced amino pyridine. Their cytotoxicity was investigated against HeLa, MCF-7 and A549 human cancerous cell lines with MTT assay. Schiff bases ligand **3** were synthesized with 2-Amino pyridine **2** and different salicylaldehyde **1** in methanol, followed by their reduction with sodium borohydride. Complexes were synthesized by adding K_2PtCl_4 to the solution of pyridine Schiff base (Figure 7.3). Findings suggested that nearly all the complexes showed better activity than standard drugs against tested carcinoma cell lines (Table 7.1). Complex **4j** displayed the greater cytotoxic activity against HeLa and MCF-7 cell line versus cisplatin, but somewhat less activity against A549 than cisplatin. Complex **4b** has also shown better activity against HeLa cell line than standard drug. The electron-withdrawing group present at the C-4 position of the aromatic ring was important to increase the antitumor activity. In a previous report Li et al. [38] synthesized few new Pt(II) complexes **7–11** formulated as [Pt(L)], where L = symmetric Schiff base tetradentate ligands **5–6**, prepared from salicylaldehyde derivative and ethylenediamine or 1,2-cyclohexanediamine (Figure 7.4). All the complexes were evaluated for DNA interaction capability. Sterically hindered complexes [Pt(L1)] **7–9** showed the greatest interaction with DNA. Large substituents on the ligands provided less susceptibility to deactivation by sulphur conatining proteins and helping to overcome resistance mechanisms. Li et al. [39] also reported stable Pt(II) complexes **14a-f** with Schiff bases ligands of reduced amino acid esters (Figure 7.5). Cytotoxic activities of metal(II) complexes were checked against HeLa, BGC-823, HL-60, and HepG2 carcinoma cell lines using MTT assay. All the

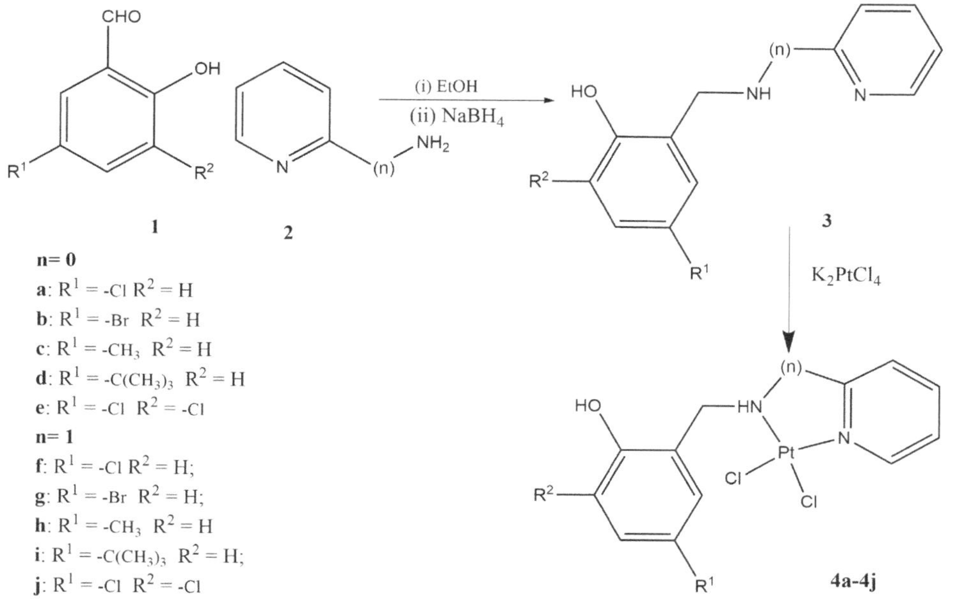

FIGURE 7.3 Synthesis of reduced amino pyridine functionalized Schiff base Pt-complex.

compounds showed selectivity against cancerous cell lines but the complexes **14e** and **14f** revealed improved cytotoxicity over the standard drug against BGC-823 and HepG2 cell lines, respectively (Table 7.1).

Jean et al. [40] reported noval Pt(II) organometallic complex synthesis, by the condensation of lithiated salicylaldimines **15** with the dimer [PtCl2(η2-coe)]2 where coe = *cis*-cyclooctene. Platinum complexes contained the Schiff base ligand, *cis*-cyclooctene, and chloride. The ligand was synthesized by the condensation of salicylaldehyde with different amines (Figure 7.6). The salicylaldiminato Pt(II) complexes were studied for cytotoxic activity against MB231 cell line and RCC cell lines. To increase the lipophilicity, complexes **16 a-c** were synthesized from long chain aliphatic amines and complexes **16d-g** from different aromatic amines appended with electron-donating group **16e** as well as electron-withdrawing groups **16f**, **16g**, to study their effect on cytotoxic activities. Complex **16g** is unstable and has not been studied for anticancer activities. The longest aliphatic chain complex **16c** appended with an octyl group was the most active in MB231 cell line for inducing apoptosis whereas complexes **16d-f** were the most promising against RCC cell lines. Unfortunately, complexes also displayed considerable cytotoxic activity towards non-cancerous cell lines. Complex **16d** showed square planar structure and the imine nitrogen was *trans* to the cyclooctene's alkene group and deprotonated hydroxyl group remained trans to the chloride ligand.

Similar to above, six bis Schiff base (salicylaldiminato)Pt(II) complexes **17a-f** and six mono Schiff base Pt(II) complexes **16h-m** were prepared by Patterson et al. [41]. Anticancer activities were evaluated against different human cancerous cell lines LN405, LN18, and Hs683 (Figure 7.6). To reduce the limitations of cisplatin, the lipophilicity of the metal complexes **17a-c** increased by appending the aliphatic chain of different lengths. Aromatic ring containing complexes **17d-f** were also tested and it was found that the presence of an aromatic ring did now show any impact on cytotoxicity. Complex **17e** had electron-donating group at para position whereas **17f** had electron-withdrawing group at para position. Complexes **17b** and **17c** were proved most effective against LN18 cells. Mono Schiff base organometallic complexes **16h-m** were also prepared and it was found complexes **16h-i** and **16l-m** exhibited better results than the standard drug.

TABLE 7.1
Anticancer Activity of Platinum and Palladium Metal Complexes

	Cell lines (IC_{50} = μM)		
Complex	**Hela**	**A549**	**MCF-7**
4a	21.3	51.3	37.5
4b	20.8	20.1	36.1
4c	22.4	22.9	44.0
4d	21.4	40.5	32.6
4e	32.8	28.7	31.8
4f	29.2	49.5	24.3
4g	41.3	16.5	25.6
4h	43.8	20.0	31.3
4i	44.5	17.0	30.9
4j	21.9	12.4	15.1
Cisplatin	34.3	9.7	28.4

	Cell lines (IC_{50} = μM)			
Complex	**Hela**	**BGC-823**	**HL-60**	**HepG2**
14a	10.28	9.78	22.09	18.45
14b	9.92	6.36	4.08	21.46
14c	41.75	15.99	24.44	24.13
14d	43.12	26.96	21.68	28.00
14e	47.41	20.59	40.56	30.57
14f	38.91	15.62	17.32	23.80
Cisplatin	4.41	6.48	2.29	20.60

	Cell lines (IC_{50} = μM)					
Complex	**HeLa**	**MCF-7**	**MCF-12A**	**Caco-2**	**Hep-G2**	**PC-3**
25	66.62	>100	22.95	>100	>100	25.1
26	86.4	>100	53.5	>100	>100	>100
27	45.5	64.5	24.9	15.81	13.0	20.9
29	84.86	23.3	>100	16.63	0.3	29.5

	Cell lines (IC_{50} = μM)
Complex	**T-lymphoblastic leukemia.**
30	2.5 μg cm^{-3}
31	2.9 μg cm^{-3}

	Cell lines (IC_{50} = μM)		
Complex	**Hep-G2**	**HT-29**	**HaCaT**
32	7.0±0.3	6.5±0.3	>40
33	6.5±0.2	6.0±0.2	>40
34	5.8±0.2	5.5±0.1	>40
35	8.2±0.4	7.0±0.2	>40

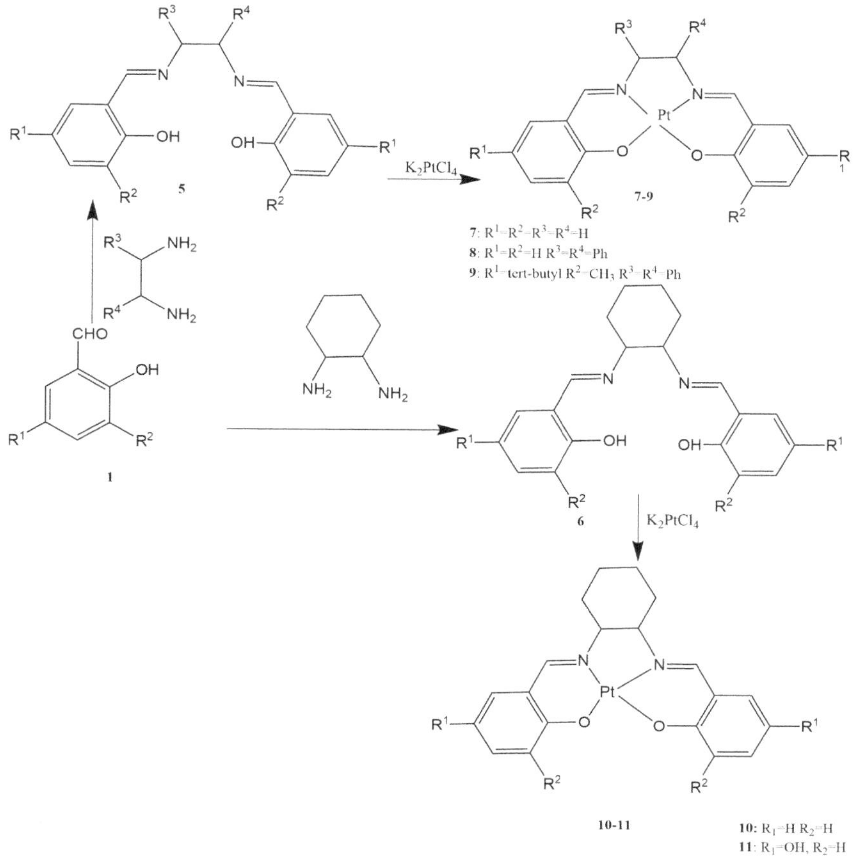

FIGURE 7.4 Synthesis of symmetric Schiff base Pt-complex.

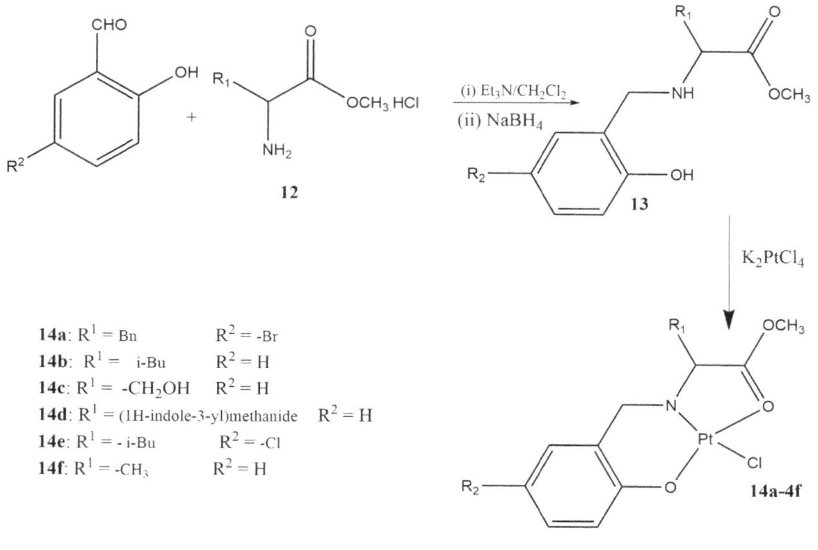

FIGURE 7.5 Synthesis of amino acid esters Schiff base Pt (II) complex.

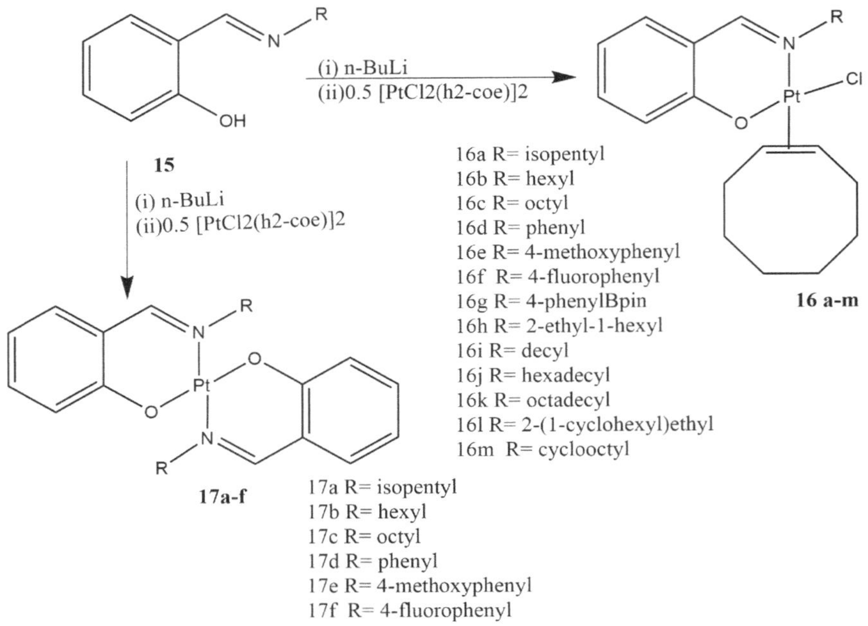

FIGURE 7.6 Synthesis of mono and bis Schiff base Pt-complex.

Over the last two decades chemists have shown interest in palladium compounds as plausible potential anticancer agents. Due to the structural and thermodynamic similarity of palladium (II) complexes with the platinum complexes, they are the most preferred target as an anticancer drug [42–44]. High kinetic lability is the only issue with the palladium complexes. The rapid hydrolysis of palladium-ligand bonds resulted in a highly reactive species which is incapable of reaching the target cells. An approach used to decrease this constraint was the use of bulky monodentate or polydentate ligands which could strongly bind to the palladium metal. Many palladium (II) complexes were synthesized and evaluated as anticancer agents. Padeliporfin (TOOKAD1) is a first approved palladium (II) drug for clinical use.

It is exciting that Pd(II) complexes exhibited cytotoxicity equal to or greater than the standard first-, second- or third-generation drugs cisplatin, carboplatin and oxaliplatin (Figure 7.1). However, most of the trans Pd(II) complexes revealed better anticancer activity than the platinum drug, and also cis-palladium isomers. Mbugua et al. [45] reported pyrrole appended Pd(II) and Pt(II) Schiff base square planar geometry complexes similar to cisplatin, but have bulky groups around the metal center, and consequently steric shielding of the metal. Bulky group hindered the ligand substitution which in turn reduces the hydrolysis rate and increases the efficacy to DNA binding. Five different metal complexes of Pd and Pt were formed by the metallation of three different ligands **22–24** with Pd(cod)Cl_2, Pd(cod)ClMe, and Pt$(DMSO)_2Cl_2$, respectively. These ligands were obtained by the reactions of 1H-Pyrrole-2-carbaldehyde **18** with different amines (benzyl amine **19**, 2-furanmethyl amine **20**, 2-picolyamine **21**) (Figure 7.7). Metal bonded to the Schiff base through the imine-N only as a monodentate,without the hetero atoms of rings. The main objective is to use the bulky group around the metal to block the axial position thereby reducing the ligand substitution. It was believed that complex **25** and **26** have trans geometry due to steric factors and complexes **27** and **28** have cis geometry. Trans analogues could exhibit better or equivalent anticancer properties and are needed to be explored further. Anticancer activity was investigated by MTT assay and apopercentage assay against different cancerous and non-cancerous cell lines HeLa, MCF-7, Caco-2, PC-3, Hep-G2 and MCF-12A respectively (Table 7.1). The complexes showed different results for dissimilar cell

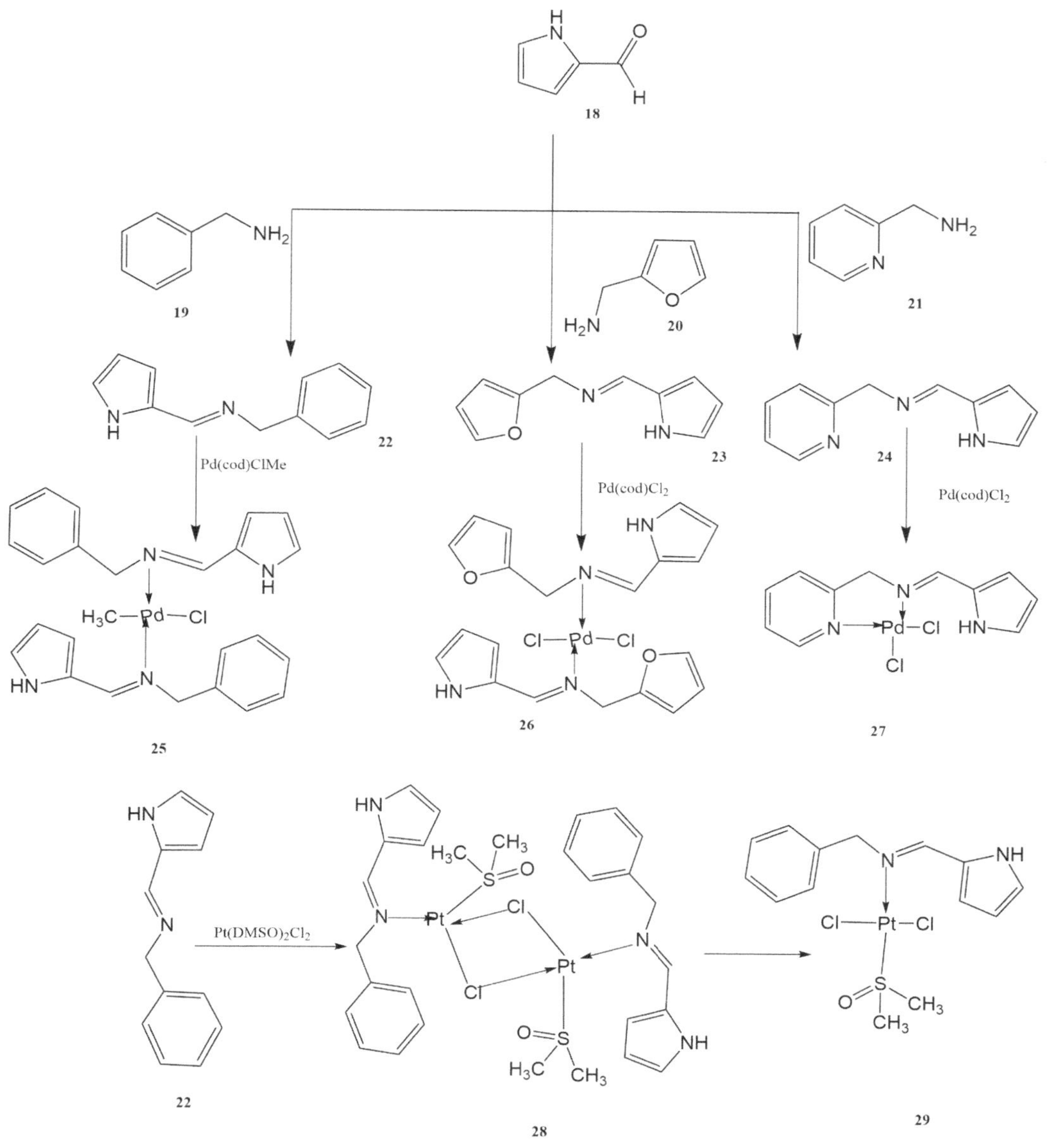

FIGURE 7.7 Synthesis of Pyrrole based Pd & Pt-complex.

lines. Cis geometry complexes **27** and **29** showed improved cytotoxic activity versus trans-geometry complexes **25–26**. Complex **27** reduced cell viability by more than 80% in all the tested cell lines whereas **29** showed selective toxicity and reduced the cell viability by 60% or more but had no effect on viability of the MCF- 12A cell line (non-cancerous). Complexes **25** and **26** showed selective toxicity. Complex **26** was highly toxic to HeLa, PC-3, Hep-G2, and MCF-12A cell line but did not notably decrease the viability of MCF-7 and Caco-2 cells. Complex **27** (cis geometry) showed high toxicity to all the six cell lines. It was concluded that it is important to balance the shielding extent and availability of the metal center for the ligand. The complex **29** displayed enhanced and selective toxicity to cancerous cells and displayed strong DNA intercalation ability. Complex **29** has been proposed as a workable drug candidate for the therapy of cancer and requires upper-level testing.

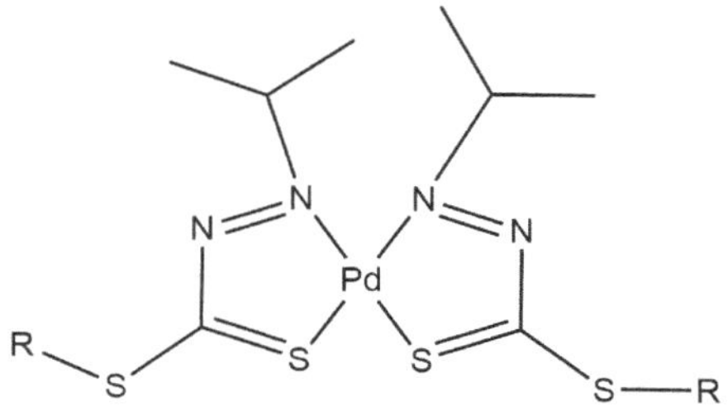

30-R=methyl
31-R=Benzyl

30-31

FIGURE 7.8 Structure of based Pd (II) complex.

M. A. Ali et al. [46] has prepared two Pt(II) and Pd(II) complexes formulated as $[M(NS)_2]$ of acetone-based Schiff ligand of S-substituted-dithiocarbazate (Figure 7.8). All the complexes possessed distorted square planar *cis* structure and were screened for cytotoxicity against T-lymphoblastic leukemia cell. Acetone based Schiff ligand of *S*-methyldithiocarbazate (Hasme) and their [Pt(asme)] complex exhibited much less activity, while the *S*-benzyl ligand (Hasbz) and their [Pt(asbz)] complex was inactive. Palladium(II) complexes **30**, **31** showed strong cytotoxicity against this cancer in comparison to the standard anticancer drug, Tamoxifen (Table 7.1).

In the year 2019, Prabhakaran et al. [47] reported coumarin appended three tetranuclear palladacyclic compounds **32–34** and a mononuclear palladium(II) complexes **35** (Figure 7.9). All the complexes were synthesized from 3-acetyl-coumarin appended substituted thiosemicarbazone ligands with potassium tetrachloropalladate and tested for their cytotoxicity using MTT assay against HepG2, HT-29, and HaCaT cell lines. In the complexes **32–34**, ligand coordinated in CNS manner by C-Csingle bond rotation of thiosemicarbazone moiety at C-3 position followed by para metallation at C4 position whereas in the complex **35,** ligand coordinated with the chromene oxygen, sulphur atom, and azomethine nitrogen. The complex **34** showed the better cytotoxicity and induced apoptotic cell death. The order of activity was **34>33>32>35>ligand>cisplatin** for both the tested cancerous cell line HepG2 and HT-29 (Table 7.1). All the complexes were noticed to be non-toxic in the HaCaT non-cancerous cell lines. All the complexes as well as their respective ligand displayed higher activity than the cisplatin. Complex **34** showed better activity among all the Pd(II) complexes because of the presence of an ethyl group at the terminal nitrogen. Introduction of palladium(II) metal complexes enhanced the anticancer activity.

7.3.2 Gold Complexes as an Anticancer Drug

Earlier reported work on Auranofin, a gold compound (Figure 7.1), and its analogues promotes the researcher to use gold metal for treating cancer. The gold (III) ion possess d8 electronic configuration, isoelectronic with Pt(II) which indicates it as a possible anticancer agents similar to cisplatin [48–50]. Bian and co-workers [51] synthesized stereoisomeric gold (III) complexes (Figure 7.10) and tested their antitumor activity against hepatocellular carcinoma which is mostly associated with liver cancer. Inhibition of overexpression of cellular protein thioredoxin reductase is an important clinical target for the treatment of the liver cancer. Thioredoxin reductase transfers the electron or scavenging of reactive oxygen species. Schiff base ligand used in this study was synthesized by the reaction of p-fluorobenzaldehyde **35** and 1,2-bis(2-hydroxylphenyl)-1,2-diaminoethane **36** in 1:2 molar ratio. Biological activity was tested against different human liver cancerous cells SMMC-7721, HepG2, and Hep3B by using MTT assay. All the complexes possess square planer geometry and four-fold coordinated via the phenolic oxygen and azomethine nitrogen. Structural study showed the three staggered conformations: two gauche structures **37b** and **38b** and one with an

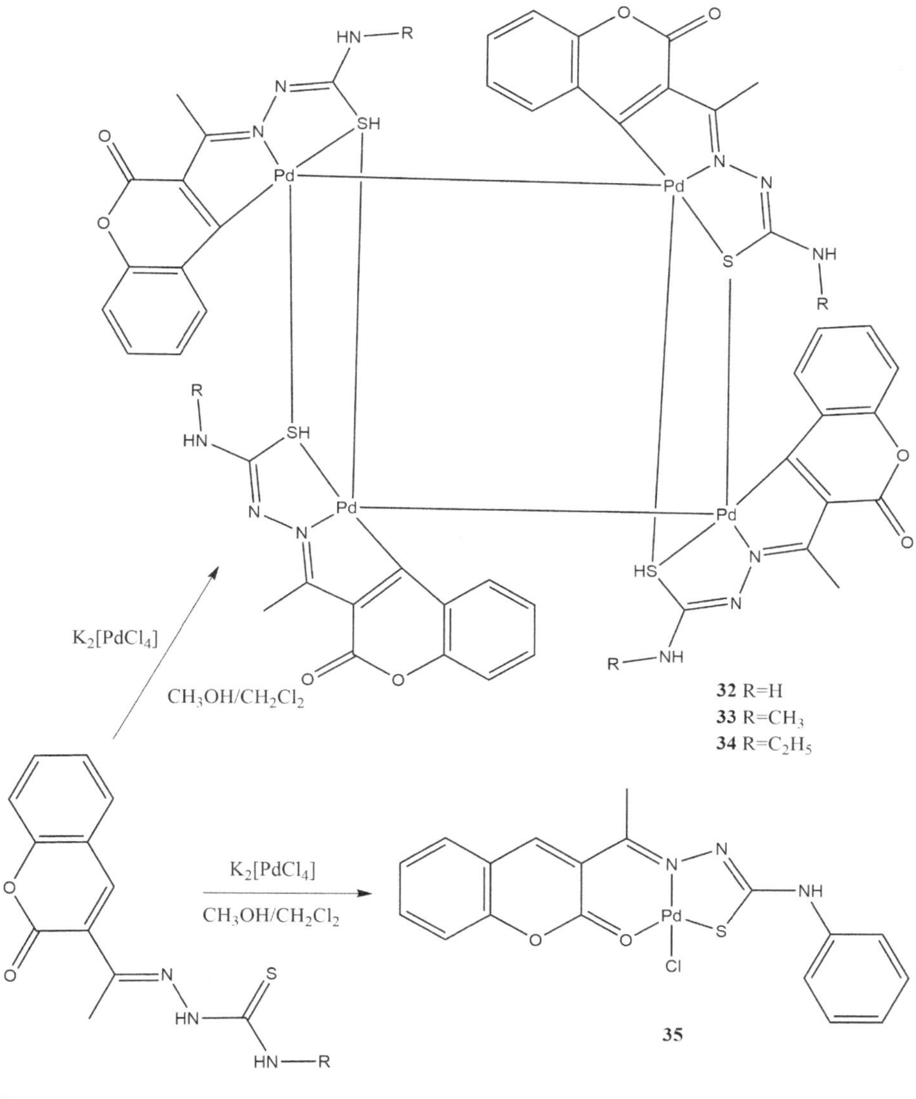

FIGURE 7.9 Synthesis of coumarin appended Schiff base Pd-complex.

anti-conformation **39b**. Complex **40b** was stereoisomeric mixture of R,R- and S,S. Among all the gold (III) complexes, complex **37b** is the most potential TrxR inhibitor and was tested in vivo. TrxR elevates reactive oxygen species in HepG2 cells, activating the endoplasmic reticulum stress. Excessive presence of reactive oxygen species leads to mitochondrial dysfunction and ultimately causes apoptosis. Cytotoxic activity of complex **37b** showed higher IC_{50} values in MCF-7 (Table 7.2). Compound **37b** was able to inhibit hepatocellular carcinoma and tumor growth in mice models. Thereafter, **37b** may signify as a novel Au (III) thioredoxin reductase inhibitor and a possible candidate for hepatocarcinogenesis therapy. Further research is required on complex **37b** to improve its carcinogenic property.

Sankarganesh et al. [52] synthesized an Au(III) complex with the formula $[AuL_2]Cl_3$ **42** from morpholine and pyrimidine containing Schiff base ligand **41**. Ligand **41** was prepared from an 2-hydrazino-4-(trifluoromethyl)pyrimidine and 4-Morpholin-4-yl-benzaldehyde (Figure 7.11). Square

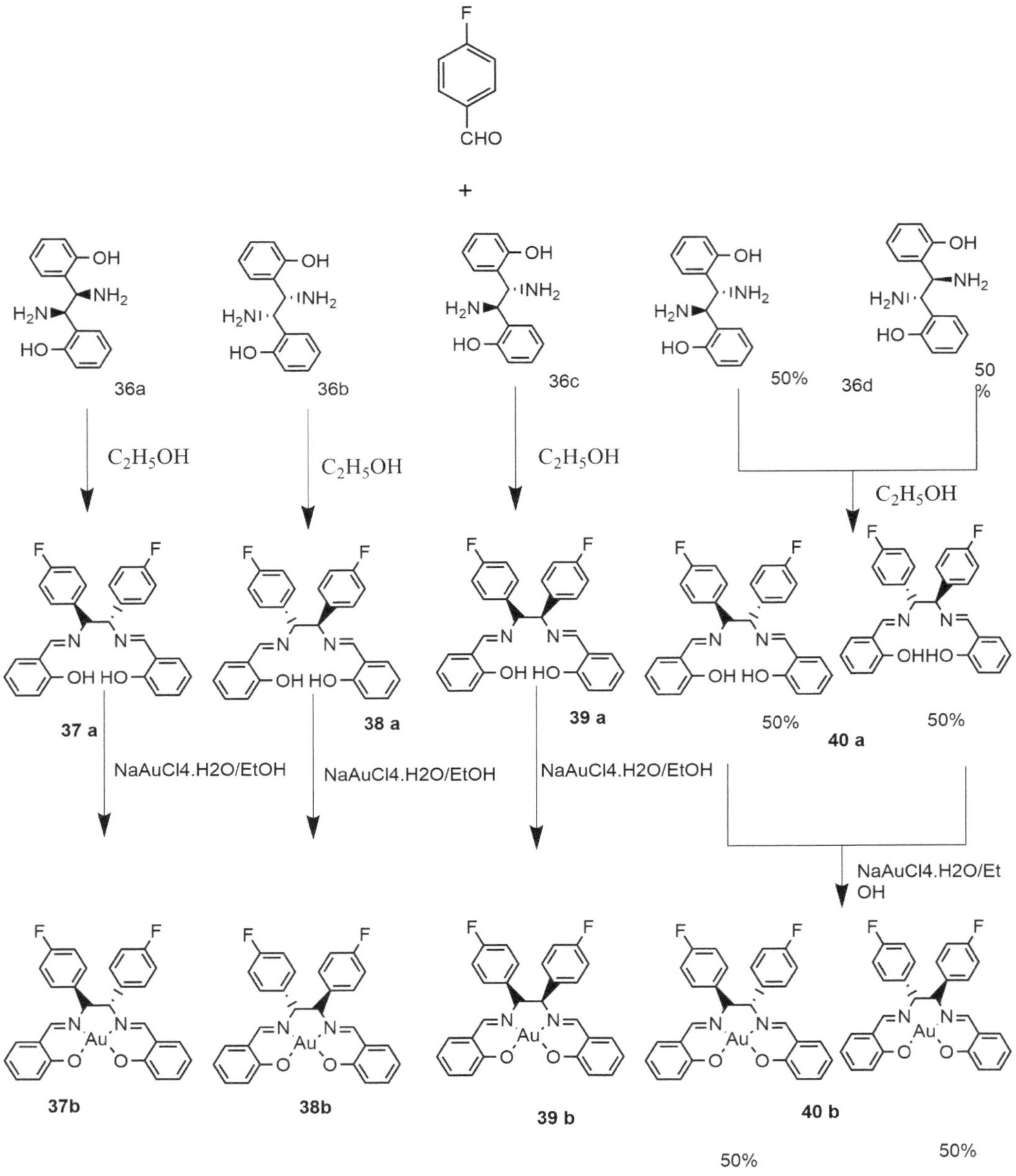

FIGURE 7.10 Synthesis of stereoisomeric Schiff base Gold-complex.

planar geometry was suggested for the Au(III) complex. *In vitro* cytotoxicity was studied by MTT assay against cancerous cell line A549, HeLa, HepG2, MCF-7, and non-cancerous NHDF cell lines. Gold(III) complex **42** displayed low half-maximal inhibitory concentration values against cancerous cell lines in comparison to its ligand **41**. Furthermore, the Au(III) complex possessed thrice low cytotoxicity in the tested cancerous cell lines and ten-fold less toxic activity on non-cancerous cell lines as compared to standard drugs. Findings revealed that complex **42** targeted the carcinoma cell lines without affecting the non-cancerous cell line and was promoted to explore the *in vivo* activity.

TABLE 7.2
Anticancer Activity of Gold Complexes

Complex	Cell lines (IC_{50} = µM)		
	SMMC-7721	HepG2	Hep3B
37a	>20	>20	>20
37b	10.05 ± 1.19	5.40 ± 0.99	10.89 ± 1.26
38b	10.17 ± 1.88	7.50 ± 0.77	12.01 ± 1.82
39b	8.25 ± 1.29	10.29 ± 1.14	9.03 ± 1.26
40b	9.45 ± 1.89	6.49 ± 0.89	9.48 ± 1.86
Auranofin	2.24 ± 0.24	1.74 ± 0.32	1.93 ± 0.21
Cisplatin	2.76 ± 0.41	1.42 ± 0.21	1.32 ± 0.14

Complex	Cell lines (IC_{50} = µM)				
	HeLa	HepG2	MCF-7	A549	NHDF
41	76.26 ± 3.81	78.25 ± 3.91	75.82 ± 3.79	77.19 ± 3.86	100.48 ± 5.02
42	32.00 ± 1.60	22.68 ± 1.13	20.60 ± 1.03	33.19 ± 1.66	109.65± 5.48.14
Cisplatin	7.26 ± 0.36	6.94 ± 0.35	6.93 ± 0.35	9.62 ± 0.48	10.28 ± 0.51

Complex	Cell lines (IC_{50} = µM)	
	OVCAR-3	HOP-62
46	09.40 ± 0.17	07.25 ± 0.21
47	05.27 ± 0.11	08.16 ± 0.43
48	09.11 ± 0.12	07.55 ± 0.43
Cisplatin	05.89 ± 0.12	03.91 ± 0.20

In vivo study was conducted on Swiss albino mice, carrying the tumor Ehrlich Ascites Carcinoma. In vivo analyses suggested Au(III) complex as a possible anticancer drug.

In 2021 five gold (I) complexes were prepared by Babgi et al. [53] and were checked for their anticancer activity against the cell lines OVCAR-3 and HOP-62. Complexes **44** and **45** were prepared by reaction of p-ethynylbenzaldehyde **43** with $AuCl(PR_3)$ and base potassium tert-butoxide(excess) in $MeOH\text{-}CHCl_3$ solvent mixture (Figure 7.12). These gold(I) alkynyl aldehyde complexes **44**, **45** on further reaction with p-aminophenol or o-aminophenol gave Schiff bases gold(I) complexes **46–48**. Schiff base metal complexes **46–48** displayed much higher cytotoxicity (IC_{50} = 5.27 and 9.40 µM) against the OVCAR-3 and HOP-62 cancer cell lines than the complexes **44** and **45** (IC_{50} = 12.45 to 15.86 µM) (Table 7.2). Compound **47** possessed hydroxyl group at the para position showed cytotoxicity equivalent to that of standard drug cisplatin against OVCAR-3 cell line. These results delivered the possibility for exploring gold(I)-based Schiff base complex as an anticancer drug.

7.3.3 Ruthenium Complexes as an Anticancer Drug

Ruthenium complexes emerged as a promising alternative to platinum complexes. Ruthenium complexes have shown higher selectivity to cancerous cells and low toxicity to normal cells in comparison to platinum and possess multiple oxidation states. Ruthenium can mimic iron when bound to DNA or proteins. Various Ru(II) and Ru(III) complexes were evaluated for their *in vitro* anticancer property. Complexes of Ru(II) prefer S- and N-donar ligands, whereas Ru(III) has considerable affinity to O- and N-donor ligands. Ru(III) complexes,

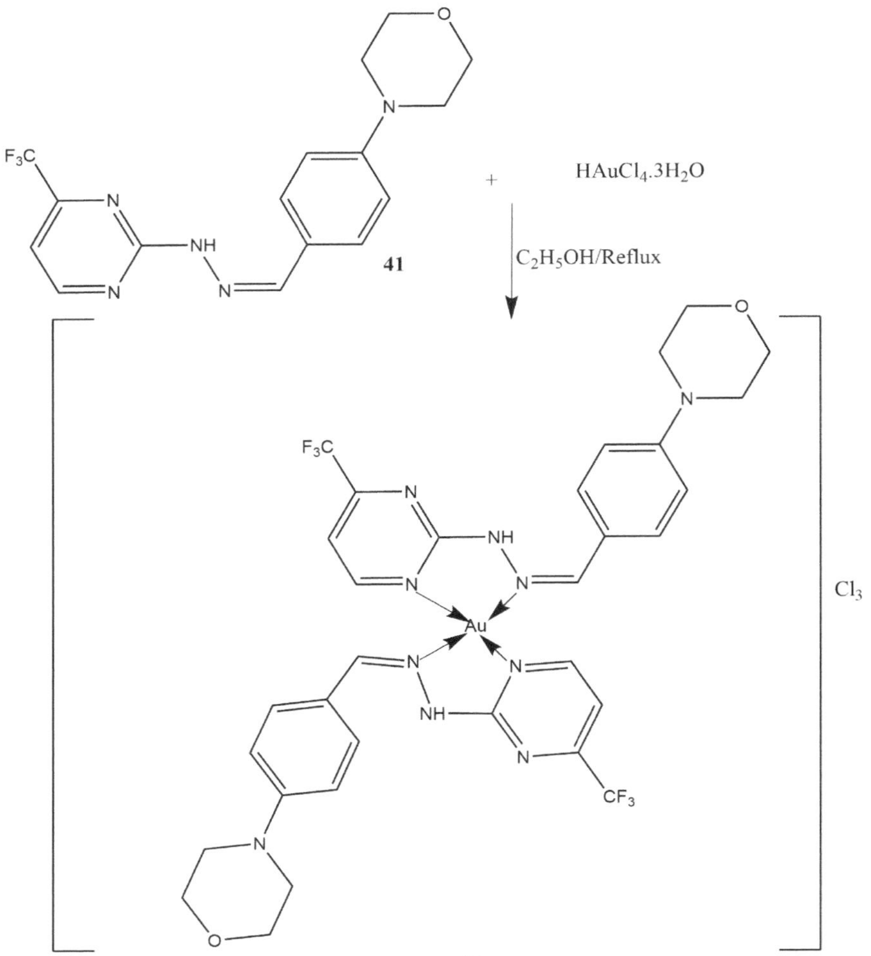

FIGURE 7.11 Synthesis of pyrimidine and morpholine Schiff base Gold (III) complex.

Ruthenium-DMSO complex (NAMI-A) and Keppler-type complex (KP1019), have completed I and II phase clinical trials (Figure 7.1). [54–58] Alsalme and co-workers [59] synthesized two complexes of platinum **50** and ruthenium **51** from tridentate (-ONO-) Schiff base ligand and evaluated anticancer activity for HepG2 cell line. Schiff base ligand **49** used was derived from 2,3 dihydroxybenzaldehyde and α-amino acid L-alanine (Figure 7.13). Ru(II) complexes exhibited superior cytotoxicity compared to platinum complex due to improved generation of reactive oxygen species by showing better affinity for proteins. IC_{50} values of platinum complex **50** and ruthenium complex **51** on HepG2 human cancer cell was 30 ± 0.7 and 20 ± 0.2 μM, respectively (Table 7.3). Ejidike and Ajibade [60] synthesized mononuclear complexes of Ru(III) represented as $[Ru(LL)Cl_2(H_2O)]$ where LL = tridentate monobasic Sanion: DAE **52**, DEE **53**, MBE **54** and HME **55** (Figure 7.14) and valuated their *in vitro* anticancer activities against three different cell lines, TK-10, UACC-62 and MCF-7, using sulforhodamine B assay. The drug parthenolide was the positive control in the study. Octahedral geometries were proposed for all the complexes. It was noted that variation in the ligand was accountable for the variation in bioactivity. Table 7.3 showed that complex **52** $[Ru(DAE)Cl_2(H2O)]$

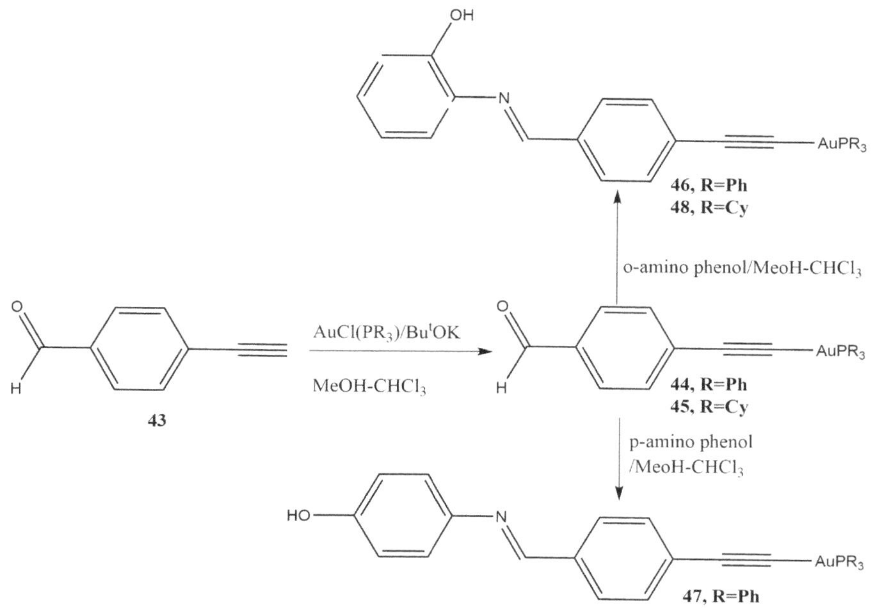

FIGURE 7.12 Synthesis of Gold (I) complex.

TABLE 7.3
Anticancer activity of Ruthenium, Iridium and Rhodium Metal Complexes

	Cell Lines ($IC_{50} =$ μM)		
Complex	**HepG2**		
50	30 ± 0.7		
51	20 ± 0.2		
	Cell lines ($IC_{50} =$ μM)		
Complex	**MCF-7**	**UACC-62**	**TK-10**
52	3.57	6.44	9.06
53	3.43	5.14	13.10
55	4.88	6.31	41.09
57	3.63	6.63	10.34
58	3.99	6.27	14.47
59	3.79	4.88	11.85
Parthenolide	0.44	0.89	0.50
	Cell lines ($IC_{50} =$ μM)		
Complex	**MCF-7**		
65	12.86		
66	7.24		
67	0.90		
68	21.19		
Cisplatin	12.33		

(continued)

TABLE 7.3 (Continued)
Anticancer activity of Ruthenium, Iridium and Rhodium Metal Complexes

	Cell lines (IC_{50} = μM)		
Complex	**BGC-823**	**SGC-7901**	**HeLa**
70	16.14	21.60	18.39
71	13.83	15.86	16.04
72	9.21	10.38	13.09

	Cell lines (IC_{50} = μM)		
Complex	**A549**	**BGC-823**	**MDA-MB-231**
73	29.7±2.22	18.75±0,42	50.62±1.66
74	7.97±0.19	3.60±0.06	12.47±0.69
75	37.65±1.78	29.39±0.42	59.21±1.45

	Cell lines (IC_{50} = μM)				
Complex	**MCF-7**	**A549**	**SW620**	**HeLa**	**WI38**
76	4.09 ± 0.78	0.68 ± 0.88	1.99 ± 0.56	1.66 ± 0.48	2.51 ± 98.88
77	4.59 ± 0.42	1.05 ± 0.79	3.2 6 ± 0.17	2.23 ± 0.82	2.74 ± 0.48
78	4.70 ± 0.16	1.38 ± 0.41	3.00 ± 0.20	2.26 ± 1.09	4.01 ± 2.08
79	8.13 ± 0.95	26.85 ± 24.9	5.24 ± 0.86	5.80 ± 4.34	31.62 ± 69.51

	Cell lines (IC_{50} = μM)	
Complex	**HeLa**	**MCF-7**
80	178.21	184.15
81	53.12	79.08
82	67.54	93.18
85a	22.62 ± 0.58	26.32 ± 0.42
85b	08.25 ± 0.76	17.42 ± 0.85
85c	07.95 ± 0.45	07.85 ± 1.16
85d	09.45 ± 0.62	19.6 ± 1.26
85e	25.12 ± 0.72	28.4 ± 0.18
85f	25.42 ± 0.64	29.22 ± 1.22
85g	09.76 ± 0.48	16.2 ± 1.22
85h	24.26 ± 0.78	27.10 ± 1.60
85i	10.76 ± 0.79	20.1 ± 0.50
85j	08.00 ± 0.88	07.96 ± 1.26
85k	09.46 ± 0.98	17.10 ± 0.54
85l	08.66 ± 0.43	16.20 ± 0.66
85m	09.46 ± 0.48	14.30 ± 0.74
85n	10.66 ± 0.88	18.10 ± 1.90
85o	07.76 ± 0.88	07.10 ± 1.28
85p	09.86 ± 0.48	16.22 ± 1.37
85q	08.56 ± 0.82	10.20 ± 0.80
85r	23.86 ± 0.72	27.30 ± 1.26
Doxorubicin	5.2 ± 0.28	6.8 ± 0.76
Cisplatin	9.42±0.52	16.2 ± 0.48

TABLE 7.3 (Continued)
Anticancer activity of Ruthenium, Iridium and Rhodium Metal Complexes

Complex	Cell lines (IC_{50} = μM)
	K562
86	87.93 ± 4.22
87	59.81 ± 2.53
88	12.46 ± 0.47
89	53.25 ± 2.59
90	11.8 ± 0.49
91	65.83 ± 2.76

Complex	Cell lines (IC_{50} = μM)		
	MCF-10A	MCF-7	MCF-7CR
92	>200	>200	>200
93	28 ± 7	130 ± 28	123 ± 35
94	34 ± 14	>200	>200
95	10 ± 6	65 ± 25	>200
96	>200	>200	>200
97	125 ± 21	42 ± 13	60 ± 31
Cisplatin	18.5 ± 1.2	36.2 ± 3.4	>200

Complex	Cell lines (IC_{50} = μM)				
	MCF-7	A549	A549cisR	HuH-7	LoVo
98	2.63 ± 0.3	1.39 ± 0.15	3.39 ± 0.47	1.71 ± 0.2	2.04 ± 0.1
99	2.91 ± 0.35	1.41 ± 0.23	5.70 ± 0.33	1.98 ± 0.30	2.09 ± 0.77
Cisplatin	4.24 ± 0.28	2.68 ± 0.16	17.24 ± 1.52	3.04 ± 0.29	3.45 ± 0.27

Complex	Cell lines(IC_{50} = μM)		
	HeLa	MDA-MB-231	HepG2
100	11.4 ± 0.3	9.1 ± 0.2	8.9 ± 0.2
101	13.5 ± 0.5	9.7 ± 0.3	9.3 ± 0.4
102	8.9 ± 0.4	6.6 ± 0.1	6.4 ± 0.2
103	7.3 ± 0.4	5.7 ± 0.4	6.1 ± 0.2
104	8.9 ± 0.3	8.4 ± 0.2	7.5 ± 0.3
105	4.2 ± 0.2	2.9 ± 0.3	5.9 ± 0.2

Complex	Cell lines (IC_{50} = μM)		
	MCF-7	A459	NIH 3T3
106	35.7 ± 1.2	30.3 ± 1.2	240.1 ± 0.4
107	50.5 ± 0.5	48.4 ± 2.8	235.5 ± 1.2
108	48.4 ± 1.2	40.1 ± 0.5	245.1 ± 0.8
109	18.5 ± 0.6	11.9 ± 0.7	225.3 ± 0.5
110	40.2 ± 0.3	45.0 ± 1.1	230.5 ± 2.3
111	35.8 ± 1.5	31.4 ± 3.5	220.1 ± 1.4
Cisplatin	12.6 ± 0.8	18.5 ± 1.4	175.1 ± 1.8

Complex	Cell lines (IC_{50} = μM)	
	MCF-7	HepG2
112	74.9 ± 2.7	115.5 ± 4.1
113	>200	>200

(continued)

TABLE 7.3 (Continued)
Anticancer activity of Ruthenium, Iridium and Rhodium Metal Complexes

	Cell lines (IC_{50} = μM)			
Complex	**HT29**	**T47D**	**A2780**	**A2780cisRa**
117	2.18 ± 0.39	5.48 ± 0.17	6.61 ± 0.12	6.42 ± 0.13
118	0.98 ± 0.02	2.27 ± 0.04	1.87 ± 0.04	1.77 ± 0.04
119	7.76 ± 0.04	6.41 ± 0.23	7.12 ± 0.14	4.67 ± 0.07
120	>50	15 ± 1	69 ± 7	36 ± 2
121	2.40 ± 0.13	4.37 ± 0.11	7.40 ± 0.04	7.46 ± 0.12
122	9.16 ± 0.98	6.67 ± 0.29	6.09 ± 0.18	4.43 ± 0.12
123	2.40 ± 0.07	2.34 ± 0.22	1.82 ± 0.02	2.07 ± 0.06
124	>50	8.97 ± 0.24	8.05 ± 0.09	5.27 ± 0.16
125	5.37 ± 0.02	22 ± 1	6.64 ± 0.08	4.36 ± 0.08
Cisplatin	9.5 ± 0.2	38 ± 2	1.54 ± 0.07	15 ± 1

	Cell lines (IC_{50} = μM)	
Complex	**ARPE-19**	**HT-29**
132	85.31 ± 14.86	56.95 ± 11.76
133	93.45 ± 11.34	89.42 ± 18.33
134	83.03 ± 14.76	82.32 ± 15.55
135	>100	96.93 ± 5.31
136	97.39 ± 4.53	46.17 ± 12.78
137	>100	83.74 ± 28.17
138	>100	93.16 ± 11.84
139	>100	88.09 ± 20.63
Cisplatin	6.41 ± 0.95	0.25 ± 0.11

was found to be the most active against MCF-7, TK-10 and UACC-62 cell lines, respectively (Table 7.3). All the complexes displayed great inhibition against different cell lines. To enhance the activity they synthesized some more complexes of Ru(III) **56–59** formulated as [Ru(L)$Cl_2(H_2O)$] where L = tridentate NNO donor ligands and checked against the three different selected cell lines used previously [61] (Figure 7.14). Schiff base ligand was synthesized by different aldehydes, 2,4- dihydroxyacetophenone and bridging ligand 1,2-Ethylenediamine. These tridentate ligands on further reaction with $RuCl_3.3H_2O$ gave different Ru(III) complexes. Complexes **57**, **58**, and **59** were found to induce more effective apoptosis towards MCF-7 cell line and the order was **57>59>58** based on the substituent's nature (Table 7.3). Prakash and co-workers [62] synthesized eight hexa coordinated Ru(III) isothiosemicarbazone compounds **61–68** having molecular formula [RuCl(EPh3)L_{1-4}] where E = P or As and reported their *in vitro* anticancer activity against cell line MCF-7 using MTT assay. These Ru(III) complexes were synthesized from the rection between bis(salicylaldehyde)-*S*-methylisothiosemicarbazone or bis(2-hydroxynaphthaldehyde)-*S*-methylisothiosemicarbazone ligands **60** with $RuCl_3(EPh_3)_3$] (Figure 7.15). The anticancer activity of triphenylarsine complexes **65–68** were reported in Table 7.3. The standard drug used in the study was cisplatin. Complexes **66** & **67** displayed higher activity than cisplatin whereas **65** showed a similar activity profile to cisplatin and complex **68** exhibited much less cytotoxicity in comparison to standard drugs. Cell inhibition increased with the increase in the concentration of the complexes. Zhang and co-workers [63] synthesized three new binuclear Ru (II) complexes **70–72** formulated as [$(bpy)_2Ru(BL_{1-3})$$Ru(bpy)_2](PF_6)_4$ appended with 1,10-phenanthroline containing dinucleating bridging ligand and 2, 2′-bipyridine (Figure 7.16). The anticancer activity of all the metal (II) complexes were

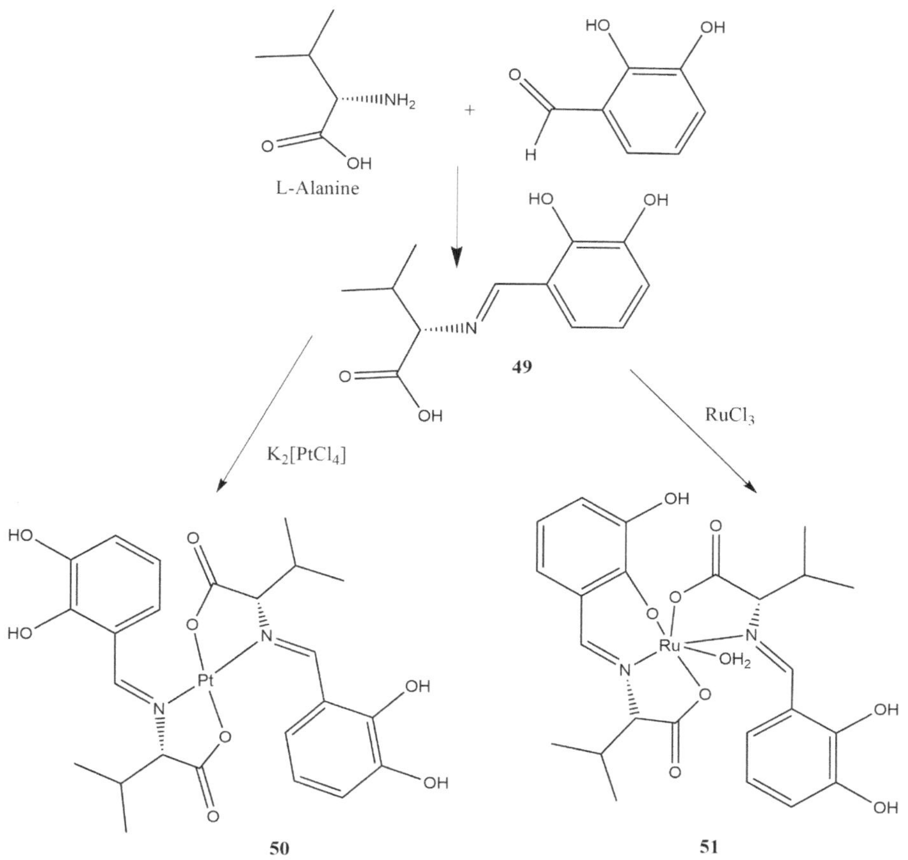

FIGURE 7.13 Synthesis of Ruthenium (I) complex.

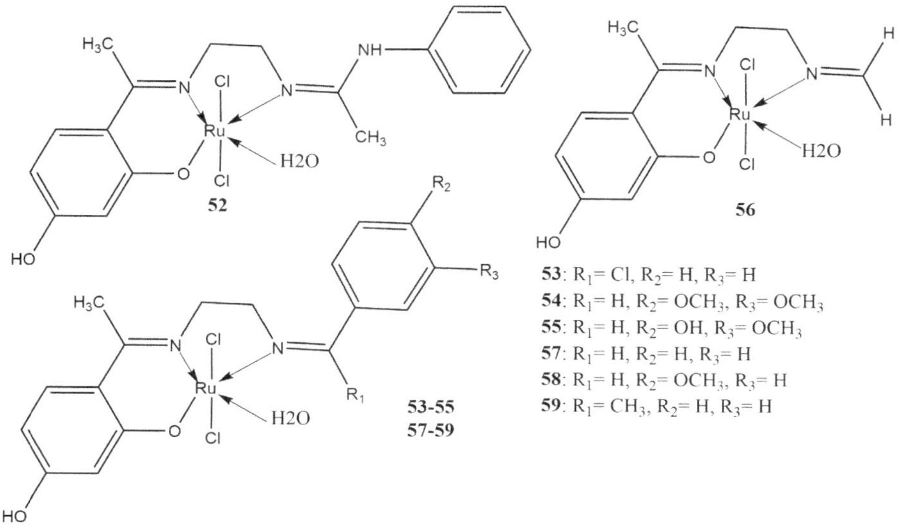

FIGURE 7.14 Structure of Ruthenium (III) complex.

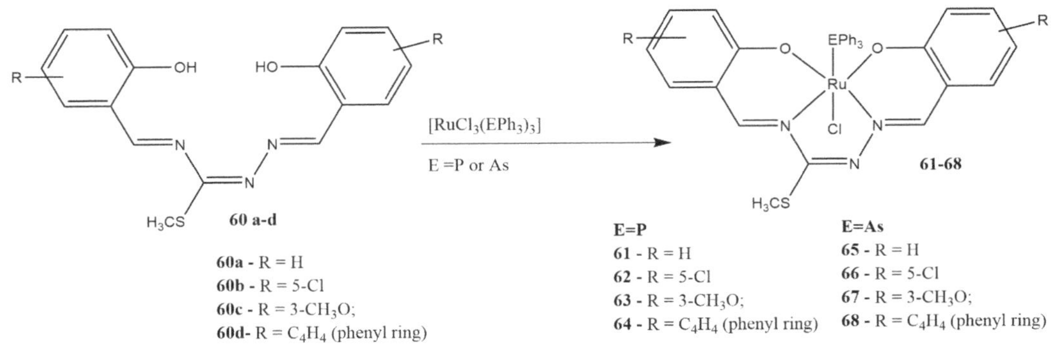

FIGURE 7.15 Synthesis of Ruthenium (III) complex.

reported against different cell lines SGC-7901, HeLa, and BGC-823 using the MTT assay. The Schiff base ligands **69**$_{a\text{-}c}$ were polyaromatic bridging ligands, synthesized by condensation of 5-amino-1,10-phenanthroline with dialdehyde followed by their reduction. The ligand had two phenanthroline nuclei, linked by a flexible alcoxyphenyl group. All the complexes were phen based and differ in the methylene group's number in the bridging chain. The activity of the complexes enhances with an increment of the methylene group's number in the flexible alkyl chain. The order of activity was **70** <**71** <**72** for different cell lines used in the study (Table 7.3). In another study they reported [64] the synthesis of three different mononuclear Ru (II) complexes formulated as [Ru(bpy)$_2$salH]PF$_6$ **73**, [Ru(dmb)$_2$salH]PF$_6$ **74** and [Ru(phen)2salH]PF$_6$ **75** where salH represented the salicylaldehyde, bpy represented 2,2′-bipyridine, dmb represented 4,4′-dimethyl-2,2′ bipyridine, and phen was 1,10-phenanthroline, respectively (Figure 7.16). All the complexes were tested for the *in vitro* and *in vivo* anticancer activity against the cell lines BGC-823, A549, and MDA-MB-231 utilizing MTT assay. Complex **74** showed better activity among all the complexes and was tested for *in vivo* activity. The IC_{50} values of complex **74** was 7.97 μM (A549), 3.60 μM (BGC-823), and 12.47 μM (MDA-MB-231) against the three tested cell lines respectively (Table 7.3). Complex **74** suppresses the BGC-823 growth *in vivo*. Kahrovi and co-workers [65] reported binuclear Ru(II) octahedral coordinated complexes **76-79** represented as $[Ru_2L_2Cl_2(Et_2NH)(H_2O)]$·nH2O, in which L = Schiff base ligand, synthesized from 2-aminopyridine and 5-substituted salicylaldehyde (Figure 7.17). Complexes were checked against cell line HeLa, A549, SW620, MCF-7 and control cell line WI-38. All the complexes have been active in the very low range 0.1-100 μM. All the studied compounds showed exceptionally low IC_{50} value (1.9-5.2 μM) against SW620. Complex **76** displayed extremely low IC_{50} of 0.68 μM against the cell line A549. Complex **79** containing NO_2 substituent displayed lower toxicity to WI38 whereas IC_{50} values for the other three cell lines MCF-7, SW620, and HeLa was slightly less than the compared cell line (Table 7.3). Subbaiyana et al. [66] designed three new Ru(III)complexes (Figure 7.17) from benzothiazole appended Schiff base ligand. Ligand was synthesized from salicylaldehyde and 2-aminobenzothiazole. Cytotoxic activity was validated against the MCF-7 and HeLa cancer cell lines. Results were shown in the order **80>82>81**, indicating an effect of substituent PPh_3 compared to $AsPh_3$ (Table 7.3) which could be ascribed to the lipophilicity of PPh_3 in the metal complexes, helped to cross the cytoplasmic membrane. Malik and co-workers [67] synthesized two Ru(III) complexes of mixed ligand (Schiff ligand and co-ligand) formulated as [Ru(Cl)$_2$(SB)(phen)] **83** and [Ru(Cl)$_2$(SB)(bpy)] **84**, in which phen = 1,10-phenanthroline, bpy = bipyridine and SB = Schiff base ligand. Ligand was synthesized from salicylaldyhyde and 1*H*-indol-3ethyl amine (Figure 7.17). Both the complexes checked for bioactivity against lung cancer cell line H1299. It was concluded for both the complexes that cytotoxicity also increased with the concentration and there was

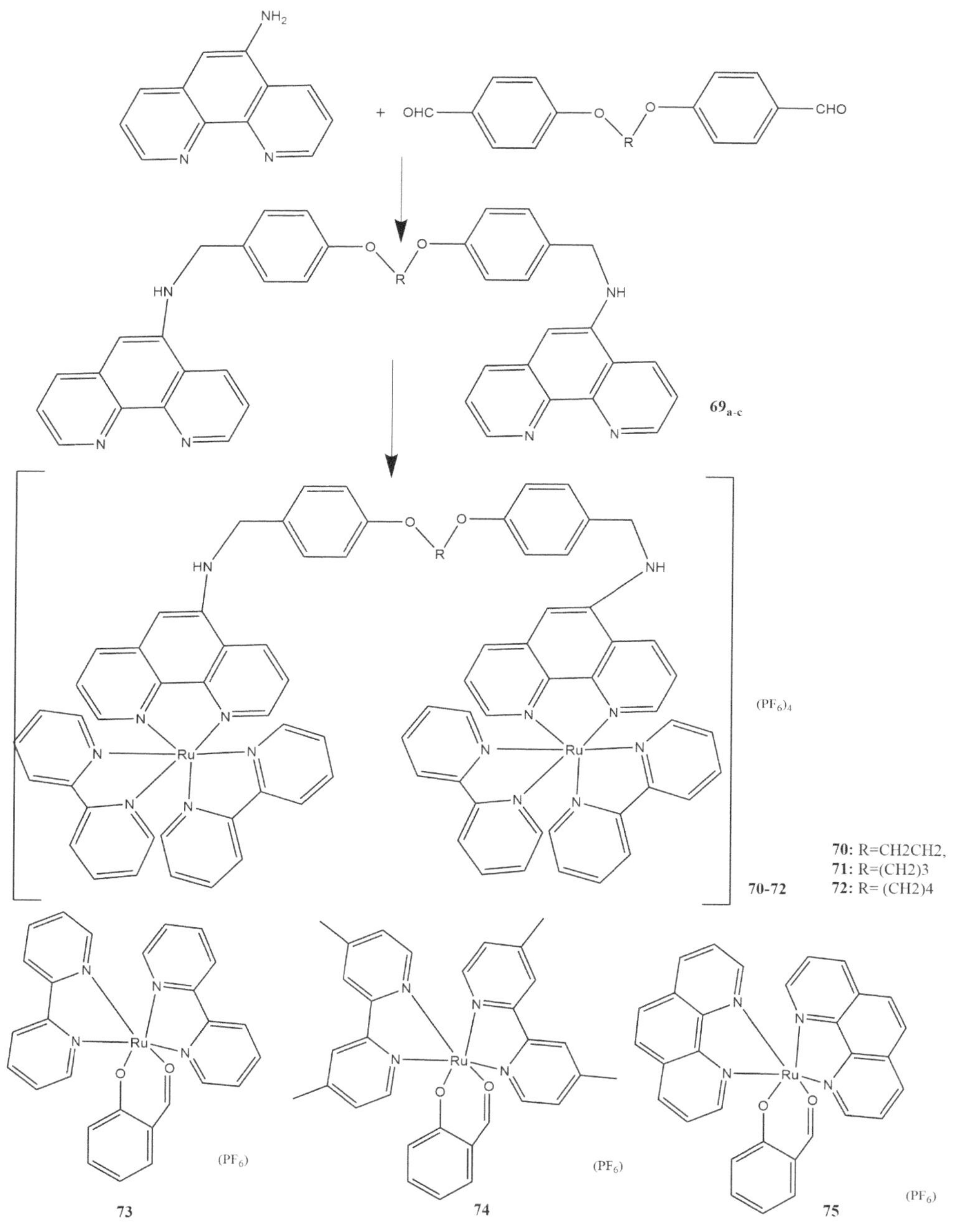

FIGURE 7.16 binuclear and mononuclear Ru (II) complexes.

a decrease in cell viability. The result revealed the cytotoxicity with an IC_{50} value of **83** was 10–12.5 µg/ml while the IC_{50} value for **84** was 15–20 µg/ml compared to cisplatin (6.0–7.5 µg/ml) which confirmed the effect of steric hindrance as well.

Jadhav and co-workers [68] reported a list of half-sandwich Ru–arene pyridinylmethylene complexes **85a-r** (Figure 7.17) from three component Pyridine-2-carboxaldehyde, 2-aminopyridine and Ru–arene dimer. Anticancer activity was evaluated using MTT bioassay against cell lines MCF-7 and HeLa. Cisplatin and doxorubicin were used as a standard drug at concentrations of 1–100 µM

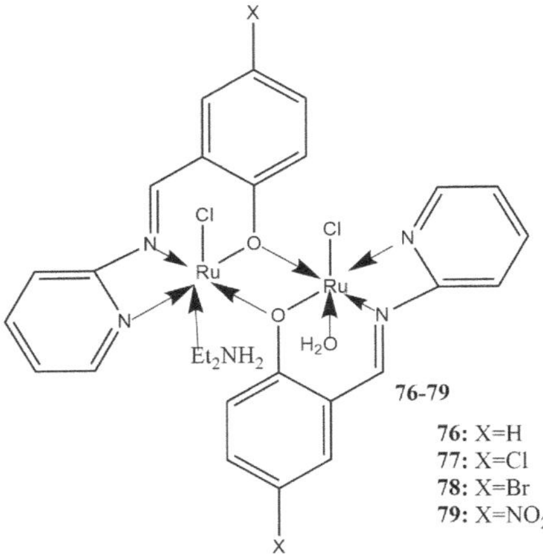

76-79

76: X=H
77: X=Cl
78: X=Br
79: X=NO_2

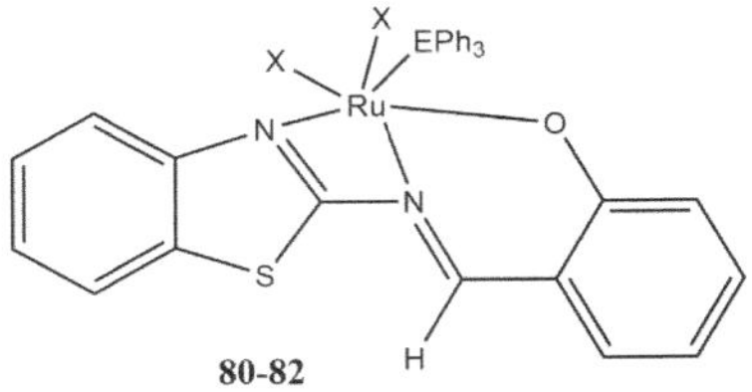

80-82

80: X=Cl; E= P
81: X=Br; E= P
82: X= Cl; E=As

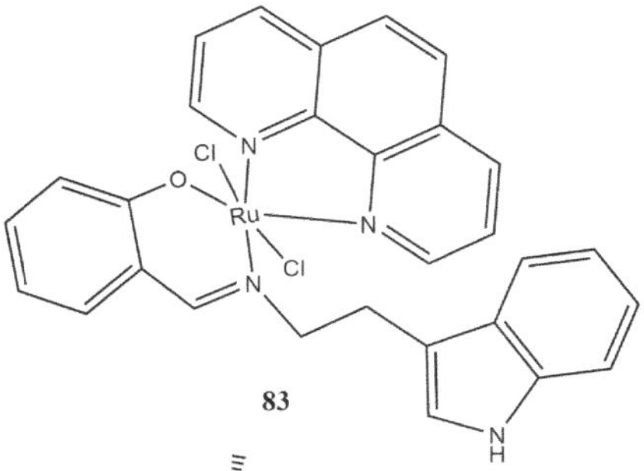

83

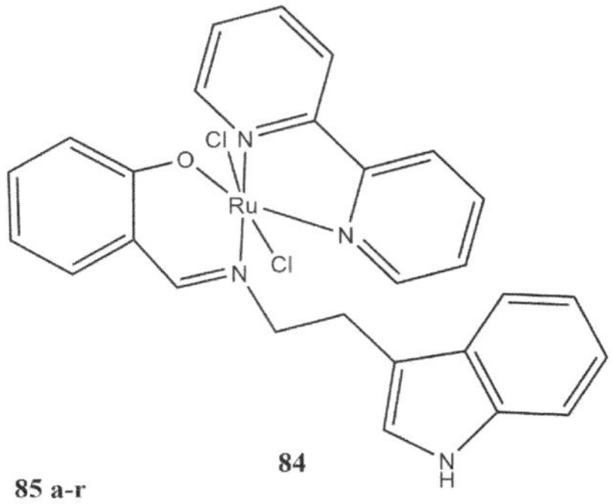

84

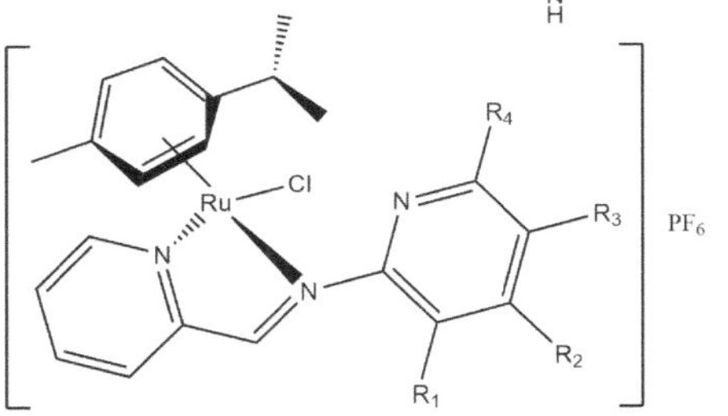

85 a-r

85 a-r
a.R1 = R2 = R3 = R4 = H
b.R1 = Br, R2 = CH3, R3 = R4 = H
c.R1 = R2 = Br, R3 = R4 = H
d. R1 = R3 = R4 = H, R2 = Br
e.R1 = CH3, R2 = R3 = R4 = H
f.R1= R2 = R4 = H, R3 = CH3
g.R1 = OH, R2 = R3 = R4 = H
h.R1 =R3 = R4 = H, R2 = CH3
i.R1 = R3 = R4 = H, R2 = Cl
j. R1 = R2 = Br, R3 = H, R4 = CH3
k. R1 = R2 = Cl, R3 = R4 = H
l. R1 = R3 = R4 = H, R2 = I
m. R1 = NO2, R2 = Br, R3 = CH3, R4 = H
n. R1 = R4= H, R2 = Br, R3 = CH3,
o. R1 = R2 = NO2, R3 = R4 = H
p. R1 = R3 = R4 = H, R2 = NO2
q. R1 = NO2, R2 = Br, R3 = R4 = H
r. R1 = R2 = R3 = H, R4 = CH3

FIGURE 7.17 Structure of binuclear Ru(II) (76-79), Ru(III) complexes (80-84) and half-sandwich Ru–arene complex (85).

for 24 h. Ru–arene complexes possess amphiphilic properties. The hydrophobic arene ring stabilizes by hydrophilic metal center and improves the cell penetration. Arene donated electrons into the empty d orbitals of Ru metal and in exchange $4d_6$ metal orbitals donated electron into unoccupied arene orbitals. 18-electron configuration of the Ru–arene complex was also stabilized by arene. Bioactivity results displayed that most of the complexes displayed a greater activity than cisplatin and somewhat lesser than the drug doxorubicin. IC_{50} values for ruthenium pyridinyl complexes were in the range of 7–25 μM in MCF and 7-29 μM in HeLa cell lines, respectively (Table 7.3). Pitchaimani and co-workers [69] synthesized Ru(II)–arene complexes **86–91** with N-monodentate ligand (4-aminoantipyrine) and N,N' and N,O -chelating bidentate ligands. The anticancer property

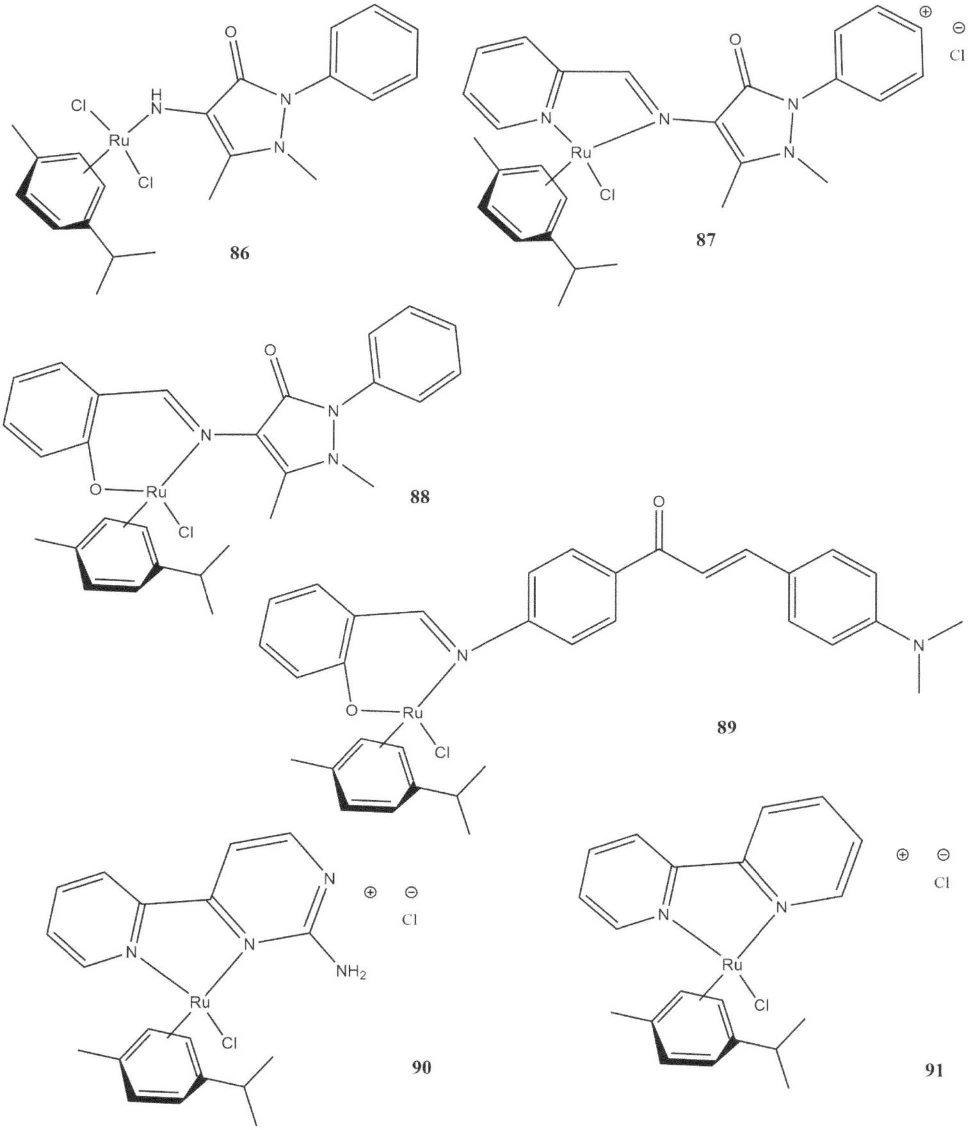

FIGURE 7.18 Structure of Ru(II)-Arene complexes.

of these complexes (Figure 7.18) was assessed against cell line K562 using MTT assay. It was found that complex **88** (IC_{50} = 12.46 μM) and **90** (IC_{50} = 11.8 μM) showed strong activity (Table 7.3) whereas **86** & **91** did not shown inhibition of the growth of leukemia cells. The result concluded that hydrophilic functionality of N,O' and N,N'-bidendate ligand and ionic nature like positive charge might be appropriate to explore antileukemic agent. Pettinari et al. [70] synthesized pyrazolone-appended Ru(II)–arene Schiff ligand complexes **92–97**. The anticancer activity of the complexes (Figure 7.19) was evaluated against cancerous cell lines MCF-7 and MCF-7CR and non-cancerous cell line MCF-10A. All The *p*-cymene substituted complexes **93**, **95**, and **97** were the highly active in exactly stimulating apoptosis in breast cancer cell lines. Complex **97** showed promising selectivity, as it possessed equivalent activity against the MCF-7CR cell line also, as well as lesser

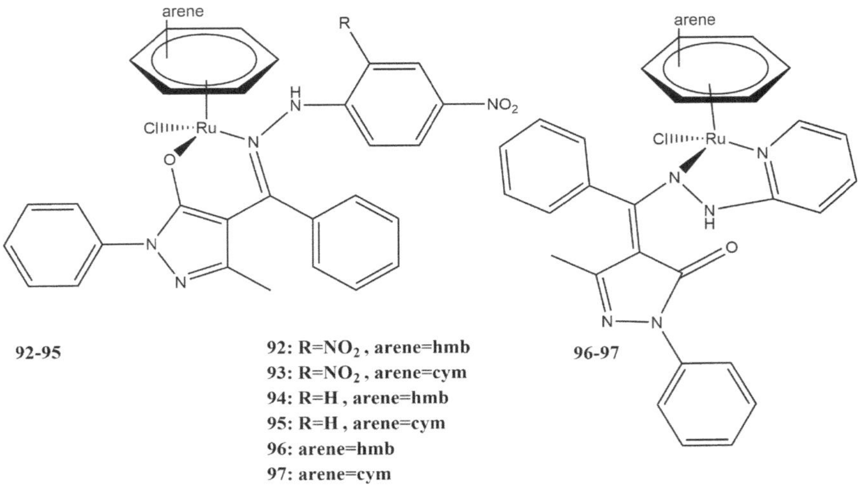

FIGURE 7.19 Structure of pyrazolone-appended Ru(II)-Arene complexes.

amount of toxicity to MCF-10A cell line (Table 7.3). Subarkhan and co-workers [71] reported tetranuclear Ru(II)–arene complexes **98** and **99** (Figure 7.20) and reported their anticancer activity against different cancer cell lines MCF-7, A549, A549cisR, HuH-7, and LoVo cell line. Both the complexes displayed a better activity profile in comparison to the standard drug cisplatin for all the screened cell lines. Both the complexes **98** (3.39 ± 0.5 mM) and **99** (5.70 ± 0.3 mM) showed significantly lower IC_{50} value for cisplatin resistant lung cancer cell lines in comparison to cisplatin 17.24 ± 1.5 mM (Table 7.3). In another report, Subarkhan et al. [72] synthesized binuclear Ru(II)–arene complexes (**100–105**) (Figure 7.20) formulated as [(η6-arene)$_2$Ru$_2$(L)Cl$_2$] where L = benzil bis(benzoylhydrazone derivatives) and arene = *p*-cymene or benzene. All the complexes tested against HeLa, HepG2 and MDA-MB-231 cell lines. Complexes **102** and **103** showed better activity than the standard drug with lower IC_{50} values against the tested cell lines (Table 7.3). Kumar et al. [73] also synthesized six half-sandwich ruthenium(II) complexes **106–111** of the formula [Ru(η6-arene)(L)Cl] where arene = benzene or *p*-cymene and L = pyrene appended benzyl substituted ligands (Figure 7.21) and screened for the anticancer activities against different cell lines (A549, MCF-7 and NIH 3T3). Complex **109** having *p*-cymene group displayed a superior activity against A549 cell line in comparison to cisplatin (Table 7.3). The higher anticancer activity showed by the complex **109** may be because of substituents present at the arene group in addition to the presence of a lipophilic moiety pyrene. Gopalakrishnan and co-workers [74] synthesized two Ru(II)–*p*-cymene complexes (Figure 7.21) formulated as [Ru(η6-p-cymene)(L)Cl]PF_6 where L was bien in complex **112** and bpen in complex **113**. Complexes have tested for *in vitro* cytotoxic activity against HepG2 and MCF 7 cells using MTT assay. Complex **112** displayed higher activity against MCF-7 (74.9 ± 2.7) in comparison to HepG2 (115.5 ± 4.1) cancer cells, while **113** was found to be inactive (IC_{50} >200 μM) (Table 7.3).

7.3.4 Rhodium and Iridium Complexes as an Anticancer Drug

Like ruthenium, iridium and rhodium, members of neighbouring group nine also show variable oxidation states like M(I), M(III), M(IV) because of their ligand exchange property and slow kinetic activity [75–78]. Rhodium belongs to the platinum group. Other than platinum and ruthenium, rhodium and iridium complexes [79–81] have also been reported as active anticancer agents against numerous cancer cell lines and suggested as a possible alternate to routine metal-based

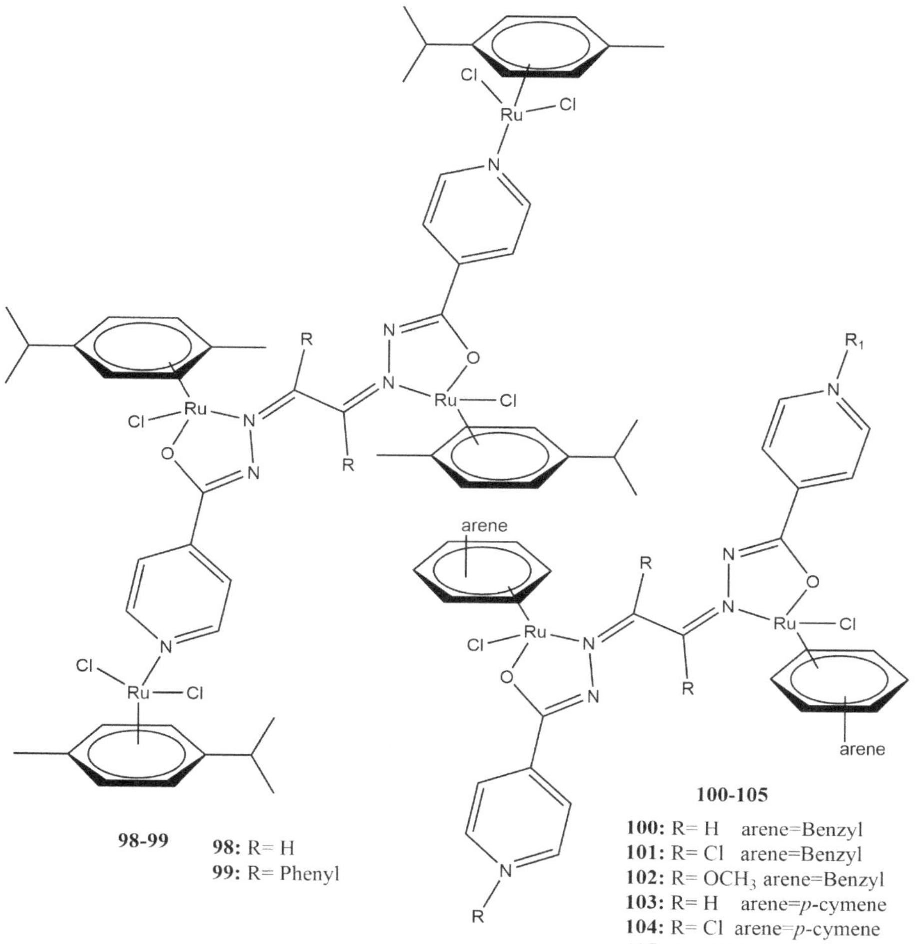

98-99

98: R= H
99: R= Phenyl

100-105

100: R= H arene=Benzyl
101: R= Cl arene=Benzyl
102: R= OCH_3 arene=Benzyl
103: R= H arene=*p*-cymene
104: R= Cl arene=*p*-cymene
105: R= OCH_3 arene=*p*-cymene

FIGURE 7.20 Structure of tetranuclear and binuclear Ru(II)-Arene complexes.

drugs. Half-sandwich complexes of these metals displayed promising anticancer activity. W. Aboura et al. [82] reported piano-stool N,N,S-Schiff base rhodium pentamethylcyclopentadienyl complexes **114–116** formulated as [(η5-C_5Me_5)Rh(L-OR)] where R = methyl, ethyl or isopropyl and L = triazole appended tridentate Schiff base ligand from the dinuclear precursor[(η5-C_5Me_5)$RhCl_2$]$_2$ (Figure 7.22). All the complexes were evaluated against cancerous cell lines A2780 and A2780cisR and non-cancerous HEK293 cell lines. Ethanolate complex **115** showed an IC_{50} value of 21 μM against the cancer cell line A2780. Similar to above, Yellol et al. [83] prepared half-sandwich Ru(II), Rh(III) and Ir(III) complexes **117–125** by using phenyl benzimidazole ligand (Figure 7.22). Ligand was prepared by the condensation of methyl 4-(alkylamino)-3-aminobenzoate or 4-(arylamino)-3-aminobenzoate with benzaldehyde, which resulted in the formation of benzimidazole ring. This phenyl benzimidazole ligand on further reaction with ruthenium(II) [$RuCl_2$(p-cymene)]$_2$ resulted in cyclometalated ruthenium complex. Similarly, phenyl benzimidazole ligand on reaction with pentamethylcyclopentadienyl chloride iridium(III) or rhodium(III) dimers resulted in corresponding cyclometalated metal complex. These complexes were tested for the cytotoxic activity against different T47D, HT29, A2780 and cisplatin resistant cancer cell lines A2780cisR as reported in

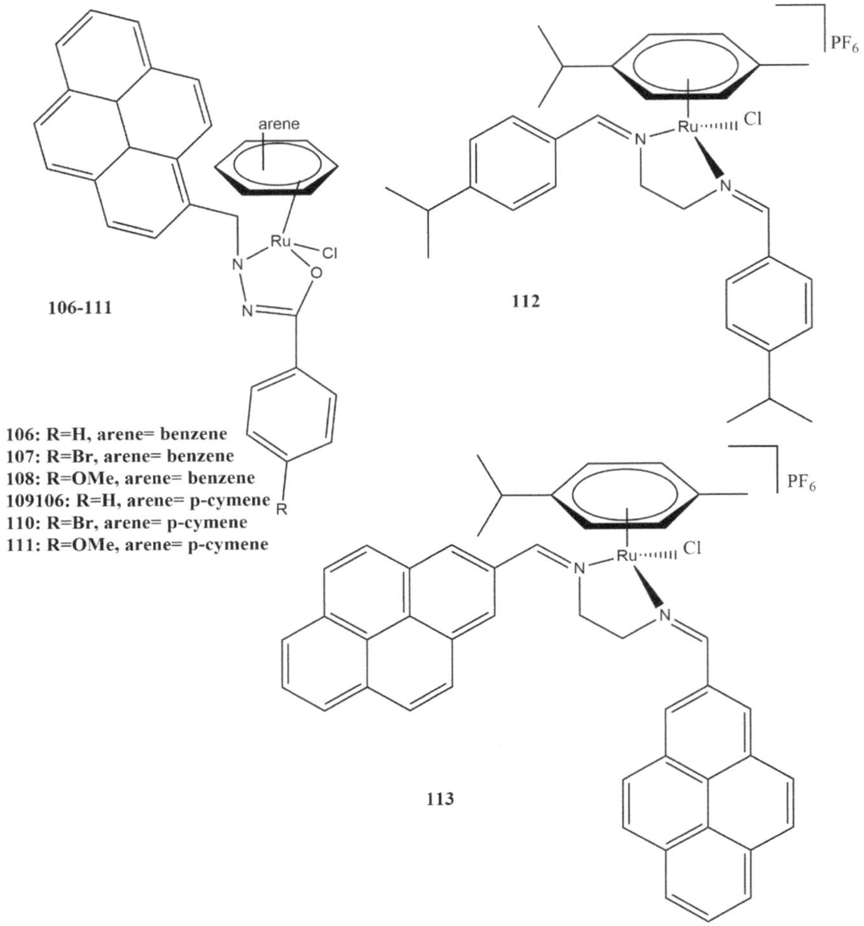

FIGURE 7.21 Structure of pyrene based Ru(II)-Arene (106-111) and Ru(II)-*p*-cymene complexes (112 and 113).

Table 7.3. All the complexes showed good anticancer activity against the tested cancer cell line. Most of the complexes were more active than the standard drug cisplatin towards colon cancer cell line HT29 and breast cancer cell line T47D (Table 7.3). Iridium (III)complexes **118** and **123** showed almost similar activity value to cisplatin.

In another study Mukhopadhyay et al. [84] synthesized six new mononuclear complexes of rhodium and iridium **126–131** and were tested for anticancer property against the A549 cancerous cell line with MTT assay. Different Schiff ligand were prepared via the reaction of 4-substituted benzaldehyde with *p- tert*-butylaniline. These Schiff base ligand on reaction with chloro-bridged dimeric rhodium(III) or iridium(III) pentamethylcyclopentadienyl chloridecomplexes, afford corresponding cyclometalated piano-stool metal complexes (Figure 7.23). Rh(III) complexes **126** and **127** displayed better activity than cisplatin. Rao et al. [85] synthesized mononuclear cationic half-sandwich Metal(III) complexes **132–139** of formula $[Cp^{*}M(L)Cl]^{+}$ where cp is pentamethylcyclopentadienyl ring in η5 manner, M = Rh(III) or Ir(III) and L = azine- Schiff ligand. First, 2-pyridylamidrazone was synthesized through the reaction of 2-cyanopyridine with hydrazine hydrate, which on further reaction with different aldehyde or ketone gave a corresponding azine-Schiff ligand. These ligands on

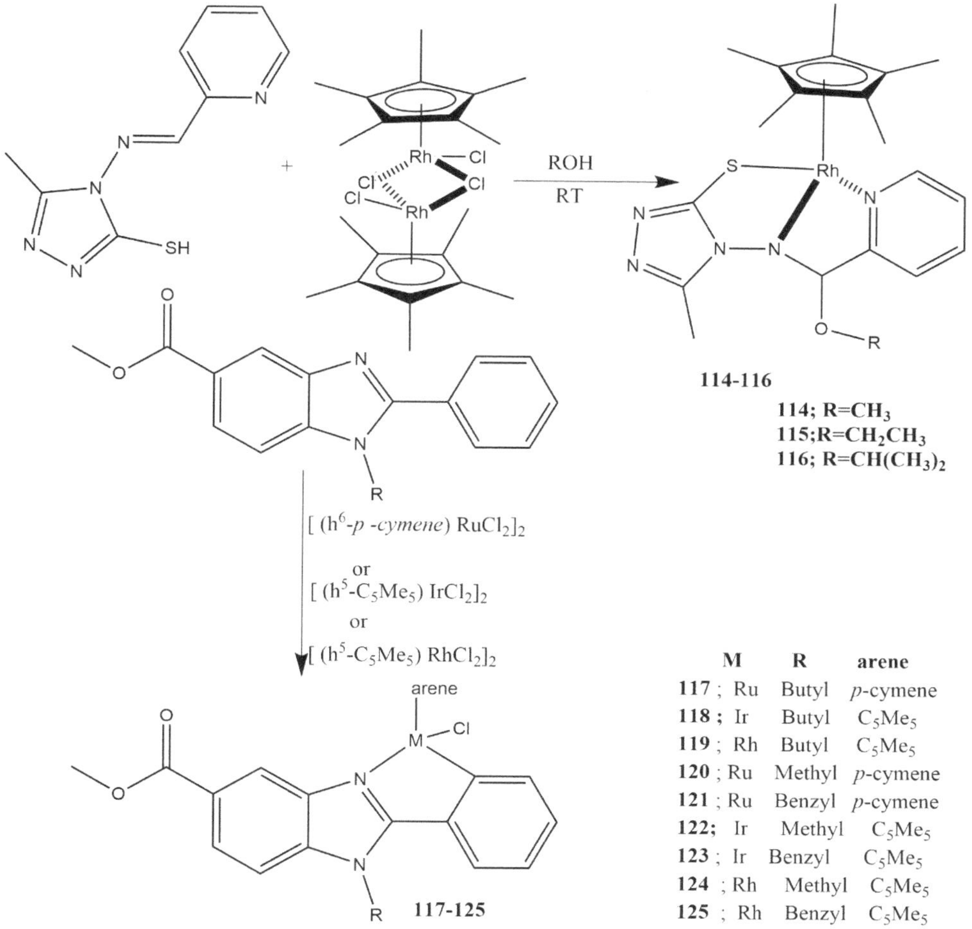

	M	R	arene
117	Ru	Butyl	*p*-cymene
118	Ir	Butyl	C_5Me_5
119	Rh	Butyl	C_5Me_5
120	Ru	Methyl	*p*-cymene
121	Ru	Benzyl	*p*-cymene
122	Ir	Methyl	C_5Me_5
123	Ir	Benzyl	C_5Me_5
124	Rh	Methyl	C_5Me_5
125	Rh	Benzyl	C_5Me_5

FIGURE 7.22 Synthesis of triazole based Rhodium piano-stool complexes (114–116) and benzimidazole Ru(II), Rh(III), Ir(III) complexes (117-125).

reaction with the metal precursor $[Cp^*MCl_2]_2$ (M = Rh/Ir) gave corresponding metal(III) complexes (Figure 7.23). The metal atom coordinated with the nitrogen atoms of azine and pyridine ring, displayed a typical three-legged piano-stool arrangement with the ligand with coordination sites occupied by one chloride group. Metal atom in all the complexes is situated in pseudo-octahedral geometry. All these complexes were evaluated for their anticancer activity against cancerous cell line HT-29 and non-cancer cell line ARPE-19 (Table 7.3). The complexes revealed moderate cytotoxicity to cancer cells. Complex **136** was the most potent among all the complexes.

7.3.5 Silver Complexes as an Anticancer Drug

Silver(I) complexes are known as antioxidant and antimicrobial agents. Researcher interest emerged in exploring these complexes due to their low toxicity and good potency [86–87]. Adeleke et al. [88] reported a list of silver(I)complexes and studied various biological activities. Quinolinyl Schiff base ligands on reaction with $AgNO_3$, $AgClO_4$ and $AgCF_3SO_3$ gave Metal(I) complexes. Quinolinyl Schiff ligands were obtained by condensation of 2-quinolinecarboxaldehyde with different aniline

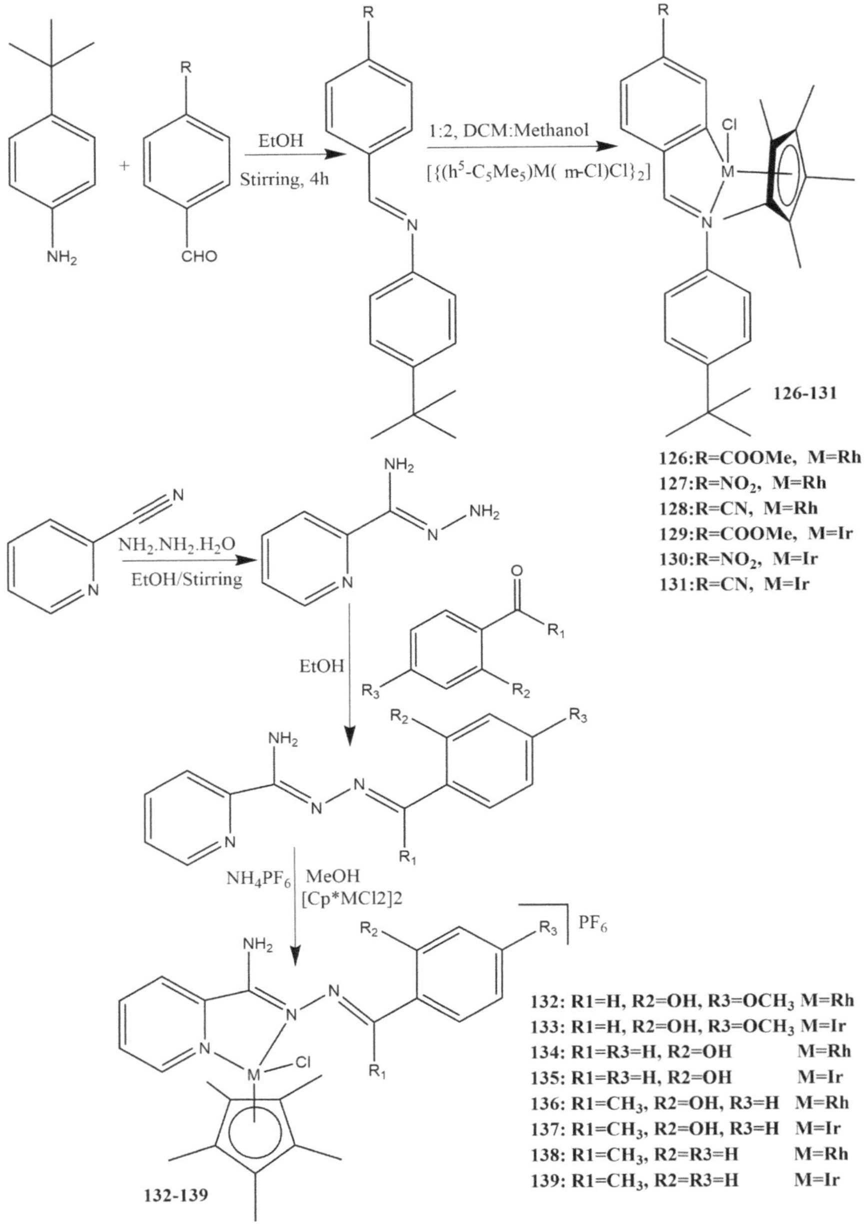

FIGURE 7.23 Synthesis of mononuclear Rhodium(III) and Iridium(III) complexes.

analogues (Figure 7.24). Cytotoxicity activity of three selected complexes **140**, **141** and **142** having perchlorate anion were examined against cell lines HeLa, MDA-MB231, and SH-SY5Y. Complexes **141** and **142** showed lower EC_{50} values of 22.80 ± 3.11 and 22.34 ± 4.86 μM, respectively, against HeLa cell line when compared with cisplatin. Gable et al. [89] synthesized a Schiff based morpholine containing ligand and its mononuclear complexes of different transition metals (M = Cd2+ , Zn^{2+}, Cu2+, Ag^{+}, Mn^{2+}, Ni2+, Co^{2+} and Fe^{3+}) (Figure 7.24). All the complexes were tested against different cancer cells PC-3, MCF-7, WI-38, and MDA-MB-231 using MTT assay. Ag(I) Complex

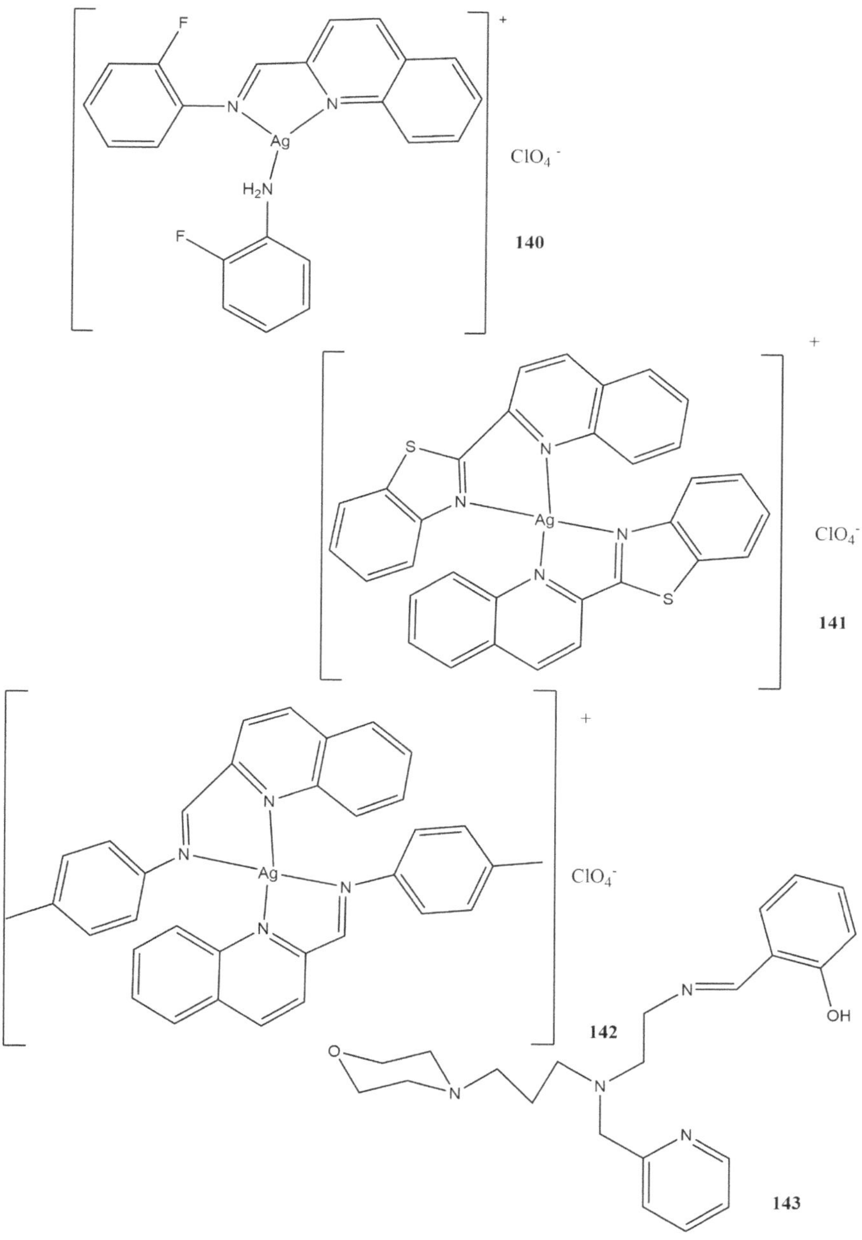

FIGURE 7.24 Synthesis of quinoline based Silver(I) complexes (140-142) and morpholine appended Schiff ligand (143).

[AgHL(NO3)] formed by the reaction of ligand **143** with the metal salt $AgNO_3$ and triethyl amine showed the highest anticancer activity among all the metal complexes against all the tested cell lines. Silver atom was coordinated to four atoms including the morpholine nitrogen, tertiary nitrogen, azomethine nitrogen, and nitrate group's oxygen atom. Findings suggested a need to explore silver (I) complexes for new anticancer agents.

7.3.6 First-Row Transition Metal Complexes as an Anticancer Drug

In the last two decades the interest of chemists has increased rapidly to explore the first-row transition metals such as iron, cobalt, copper, nickel, zinc, chromium, manganese and vanadium as a biological target [90–96]. These metals are present in the human body as an essential element and are responsible for various biological functions also stabilizing the structures of proteins and enzymes [97–98]. First-row transition metals appended with special heterocyclic structure like benzimidazole, thiazole, isoxazole, anthracene, pyridine, naphthalene, quinoline, 1,10 -phenanthroline and so on, providing a diversity of oxidation states and structures which expanded these metal complexes towards metallodrugs [99–106]. Several first-row d-block metal complexes have been reported as promising cancer targeted drugs in various studies. Zinc and copper are the second and third most important transition metals which participated in several biological functions in the human body [107–108]. Copper metal plays many significant roles in biological systems. The role of copper (II)complexes as anticancer compounds is well established and even referred to as best alternatives to cisplatin. Moreover, the investigation of several zinc(II) and copper (II) Schiff base complexes for new anticancer drug design and clinical applications were explored. Mei-Ju Niu and collaborators [109] synthesized mononuclear zinc complexes formulated as Zn(HL1)2 **144** and a binuclear zinc complex formulated as $[Zn_2(H_2L_2)(OAc)_2]_2$ **145** where HL_1 and H_2L_2 were Schiff base ligands. Schiff base ligands were obtained from the reaction of o-vanillin and (R)-(+)-2-amino-3-phenyl-1-propanol/2-amino-2-ethyl-1,3-propanediol, respectively. Metal complexes were prepared by treating $Zn(OAc)_2.2H_2O$ and Schiff ligand in 1:2 ratio (Figure 7.25). Metal complexes were checked against four different cell lines, namely A549, HL-60, HeLa, and K562 using MTT assays. Complex **144** revealed superior activity in comparison to complex **145**. Structural analysis of the complexes concluded that nuclearity and chirality have a large impact on their biological activities. In another study Kumar et al. [110] synthesized first series transition metal complexes from the dianionic tetradentate (N_2O_2) Schiff base ligand derived from vanillin, ethylenediamine, and acetoacetanilide (Figure 7.26). Metal complexes have been prepared by metal salt and ligand vanillin-(1,2-ethylenediimin)ethylacetoacetanilide in 1:1 molar ratio. Tetrahedral geometry is suggested for all the synthesized complexes **146** except for the Zn(II) complex that revealed the octahedral geometry **147**. All the complexes were evaluated for the invitro cytotoxic activity against DLA cell line. Copper complexes showed higher IC_{50} value (49 μl/ml). *In vivo* studies of copper complexes were carried out in the 24 Swiss albino mice at various concentrations and found efficient against Ehrlich Ascites Carcinoma (EAC)-induced ascites tumor in a dose-dependent manner.

Alturiqi et al. [111] reported octahedral geometry complexes of different transition metals Cr(III), Cu(II), Ni(II), Mn(II), Zn(II), and Ru(III) ions. All the complexes were synthesized from metal chloride salt and furan-based Schiff base ligand (H-MFMAQ) in the composition of 1:1. Tetradentate Schiff ligand was prepared from the condensation of 1-aminoquinolin-2(1H)-one and 5-hydroxymethylfuran-2-carbaldehyde. Different metal complexes were formulated as $[M(MFMAQ)Cl_2]$ for Cr(III) and Ru(III) **148–149** and [M(MFMAQ)Cl(H2O)]·*n*H2O for Co(II), Mn(II), Ni(II), Cu(II) and Zn(II) **150–154** (Figure 7.26). Cytotoxicity of all the metal complexes have been checked against A549 and MCF-7 cell line. Order of cytotoxic activity was Zn(II) complex<Cu(II) complex<Ni(II) complex< Co(II) complex< Mn(II) complex<Ligand< Ru(III) complex < Cr(III) complex. Anticancer activity of Cr(III) and Ru(III) complexes were comparable with cisplatin, while other complexes showed lower anticancer activities.

In another study isoxazole appended Schiff ligand (MIIMMP)-based mononuclear copper, nickel, and zinc complexes were synthesized [112] and examined for cytotoxic activity against carcinoma cell lines HeLa. Metal complexes were synthesized from metal salts and the Schiff base ligand prepared from 3-amino-5-methyl isoxazole and 5-methoxy salicylaldehyde. Stoichiometric proportion of the metal and ligand in the complexes was 1: 2 and formulated as $[M(L)_2(H_2O)2]$ **155–157** where M = Ni, Cu and Zn and L = MIIMMP. Metal coordinated with bidentate ligand via phenolic

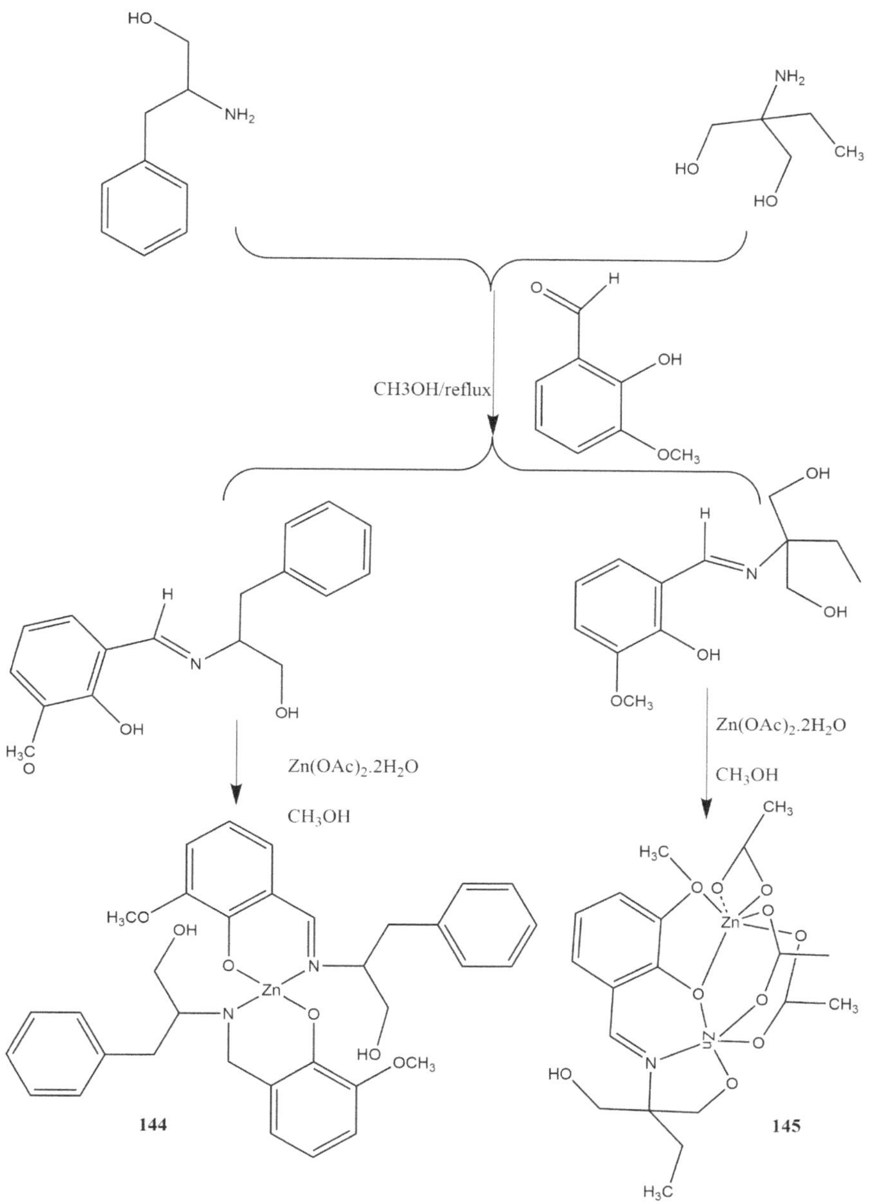

FIGURE 7.25 Synthesis of mononuclear zinc complexes (144) and binuclear zinc complex (145).

oxygen atom and nitrogen of azomethine respectively (Figure 7.26). The cytotoxic activity of 10 μg/mL of Ni, Cu, and Zn complexes are found to be 55.19%, 64.65% and 49.38%, respectively.

Cu(II) complexes having mixed ligand were synthesized by Rajeswari and co-workers [113]. Complexes were synthesized by adding copper(II) perchlorate with different Schiff base ligands and co-ligand 1,10-phenanthroline. The general formula of the complexes were [Cu(L)(phen)](ClO_4) **158–159** and [Cu(L)(phen)]$(ClO_4)_2$ **160–163** where L was a tridentate Schiff base ligand (2 nitrogen donor and 1 oxygen donor) for **158–159**, and (three nitrogen donor) for **160–163** (Figure 7.27). Effects of variable bulky group ligand pyridyl, (benz) imidazolyl or quinolyl groups

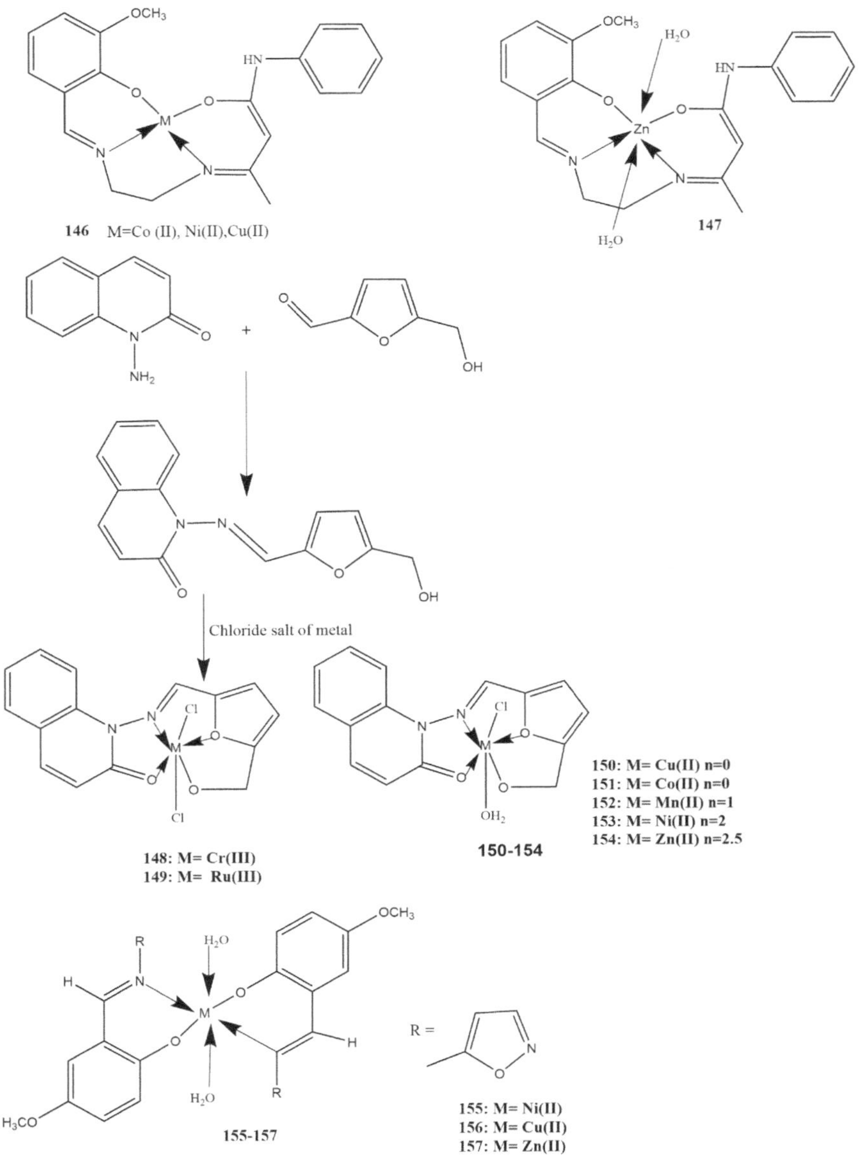

FIGURE 7.26 First-row transition Metal complexes.

and their hydrophobic properties on the coordination geometry were evaluated. Square pyramidal distorted trigonal bipyramidal (SPDTBP) structure was suggested for the complexes with 2NO ligand whereas complexes with 3N ligand showed trigonal bipyramidal distorted square pyramidal (TBDSP) structure. The metal complexes were tested with human cancer cell line HBL-100 with MTT assay. Cu(II) complex with mixed ligand displayed time dependent efficient cytotoxic activity in comparison to the corresponding chloride complex.

In an another study Jaividhya et al. reported [114] mixed ligand copper(II) complexes **164–168** formulated as $[Cu(L_{1\text{-}5})(phen)(ACN)]^{2+}$, where L was a anthracenyl group containing Schiff base bidentate ligand (L_1-L_5) and phen was 1,10-phenanthroline (Figure 7.27). The geometry of all

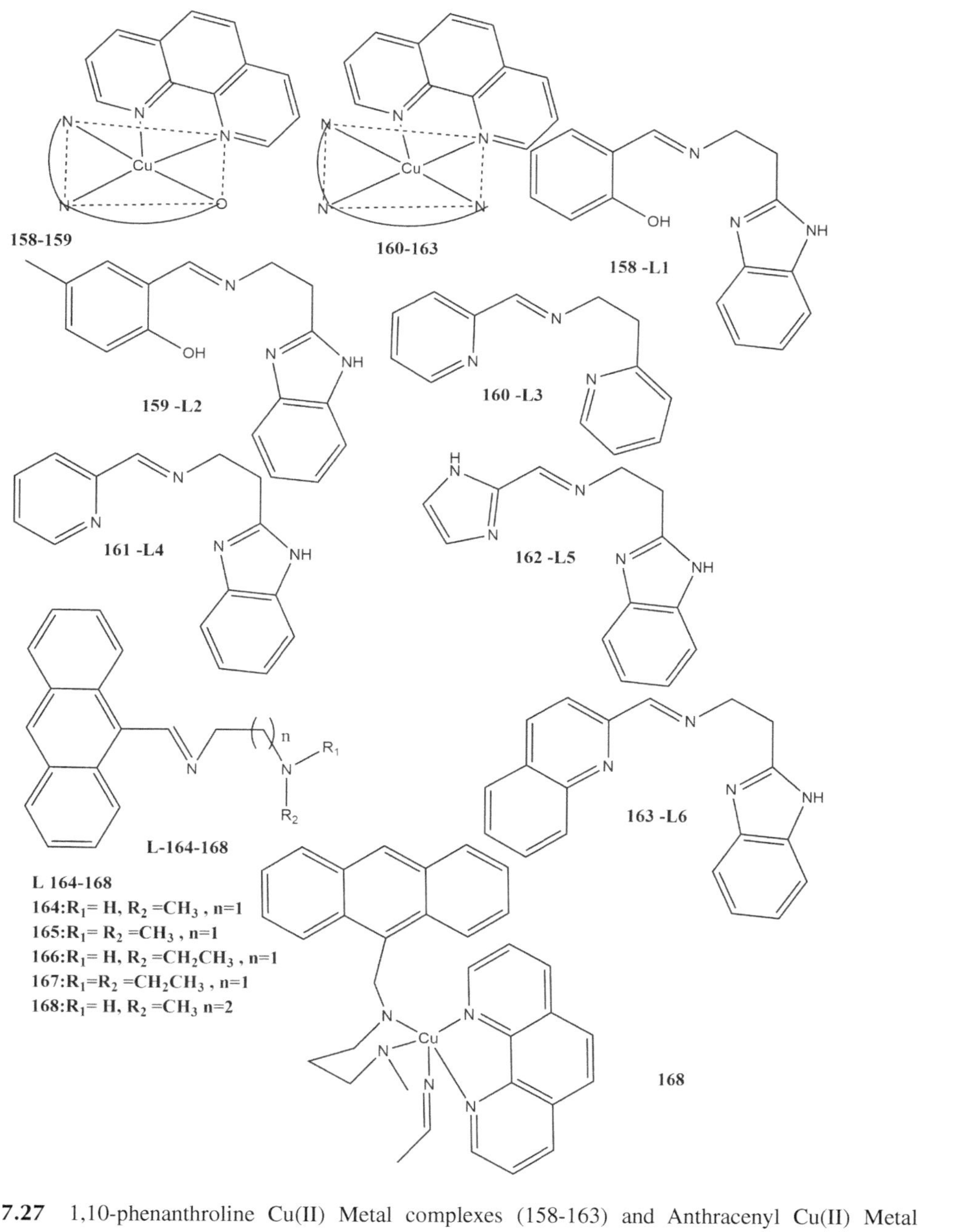

FIGURE 7.27 1,10-phenanthroline Cu(II) Metal complexes (158-163) and Anthracenyl Cu(II) Metal complexes (164-168).

the complexes was square pyramidal distorted trigonal bipyramidal. All the complexes exhibited higher cytotoxicity against MCF-7 cell lines. Cytotoxicity of the complexes has shown the order **168>167>166>165>164**. Surprisingly, complex **168** showed 15 times more effective anticancer activity than cisplatin. This study suggested to combine the aromatic groups like anthracene, benzimidazole, and so on, apart from phen co-ligand, in designing copper-based anticancer drugs. Jiao and co-workers [115] reported a new dinuclear copper(II) complex formulated as [Cu_2(dmapob) (dabt)(CH_3OH)(pic)]·DMF0.75·(CH_3OH)0.25 where H_3dmapob and dabt was a bridging and terminal ligand, respectively, pic was a picrate group also coordinated as a terminal ligand besides, a methanol molecule also coordinated in addition to three-fourth DMF, and one-fourth methanol as solvent molecules of crystallization. Dmapob coordinate with metal through nitrogen and oxygen of oxamido, and carboxyl oxygens. Cu(II) complex was prepared by the reaction of asymmetric N-[3-dimethylamino)propyl]-N′-(2-carboxylatophenyl)oxamide (dmapob) with Cu(pic)$_2$·6H_2O and 2,2'diamino-4,4'-bithiazole (dabt) in molar ratio 1:2:1. Five chelating rings were present around the metal ions, the three rings were five-membered and almost planar whereas two six-membered were folded. Cytotoxicity of the metal complex was estimated at SMMC-7721 and A549 cell line by using the sulforhodamine B assay and compared with the standard drug cisplatin. The IC_{50} values displayed by dinuclear complex was 12.5 ± 0.3 μg/mL for SMMC-7721 and 17.9 ± 0.8 μg/mL for A549 cell line.

Song et al. [116] synthesized two new Cu(II) complexes from benzimidazole heterocyclic group based Schiff base ligand formulated as [Cu(L)Cl]$_2$·CH_3OH **169–170** where L was the Schiff base synthesized from 2-methyl amine benzimidazole with substituted salicylaldehydde. Metal(II) was tetra coordinated, and showed distorted quadrilateral structures (Figure 7.28). Complex **169** showed better activity, IC_{50} = 16.9 ± 1.5 μmol L^{-1} and 16.5 ± 3.4 μmol L^{-1} against the cell line MCF-7 and COLO-205, respectively. Notash and co-workers [117] reported gabapentin drug appended Cu(II) complex **172**, obtained through the reaction of [Cu (OAc)$_2$]·4H_2O and a tridentate Schiff ligand **171** (Figure 7.28). Gabapentin (1-(aminomethyl) cyclohexaneacetic acid), a drug member of the group gamma aminobutyric acid (GABA), was used in the treatment of pain related to nerves. Ligand was prepared from 2-hydroxy-naphthaldehyde and gabapentin drug in equimolar ratio. Cu(II) Complex **172** was studied for *in vitro* cytotoxic activity against cancerous cell lines U-87 (IC_{50} = 28.4 μM), SK-OV3 (IC_{50} = 64.98 μM) and JURKAT (IC_{50} = 21.27 μM). Complex **172** exhibited high selective and dose-dependent cytotoxic activity. Prasad et al. [118] synthesized three new octahedral vanadium(IV) complexes **173-175** from a tridentate Schiff base benzimidazole ligand and bi ligand curcumin and napthalimide (Figure 7.28). Complexes **173–175** displayed exciting photocytotoxic activities against different cell lines MCF-7, HeLa, and HaCaT with MTT assays. Cisplatin and photofrin were utilized as positive controls. Complex **173** did not reveal cytotoxicity whereas complex **175** showed better activity against the cell line MCF-7 compared with the complex **174** and the free curcumin in the presence of light.

7.3.7 F-block Metal Complexes as an Anticancer Drug

In addition to d-block element Schiff base metal complexes, several f-block element metal complexes also showed promising anticancer activity [119–122]. Andiappan and co-workers [123] synthesized Erbium, Praseodymium, and Ytterbium metal Schiff base complexes and demonstrated their cytotoxicity against the HeLa and MCF-7 cell lines. Metal(III) complexes **177–179** were obtained from the reaction of metal salts with anthracene appended Schiff ligand **176**. 2,3-Diaminopyridine and anthracene-9-carbaldehyde on condensation gave Schiff base ligand **176** (Figure 7.29). Cytotoxicity activity of **177** and **178** complexes displayed significant cytotoxicity against the tested cancer cell lines and stimulated the apoptosis of MCF-7 and HeLa cells. These results suggested the possibility of these metal complexes as new antitumor drugs. Kumar et al. [124] developed two Schiff ligand and their mononuclear La(III) metal complexes formulated as [La(L_{1-2})2Cl_3]·7H_2O, **181–182** where L = bidentate Schiff bases ligand **180**. Ligand was obtained from the condensation of 2-aminobenzothiazole and heteroaryl ketone in equimolar ratio. Both the La(III) complex possessed capped octahedral structure. Ligand chelated with metal ions by N-atom of the azomethine and S-atom of the benzothiazole (Figure 7.30). Metal complex showed outstanding anticancer activity against the cell line PC-3. Prostate cancer is the second prominent cause of mortality in men. Ligand (**180a**) and its metal complex **181** were found more potent against the clinically tested cell lines

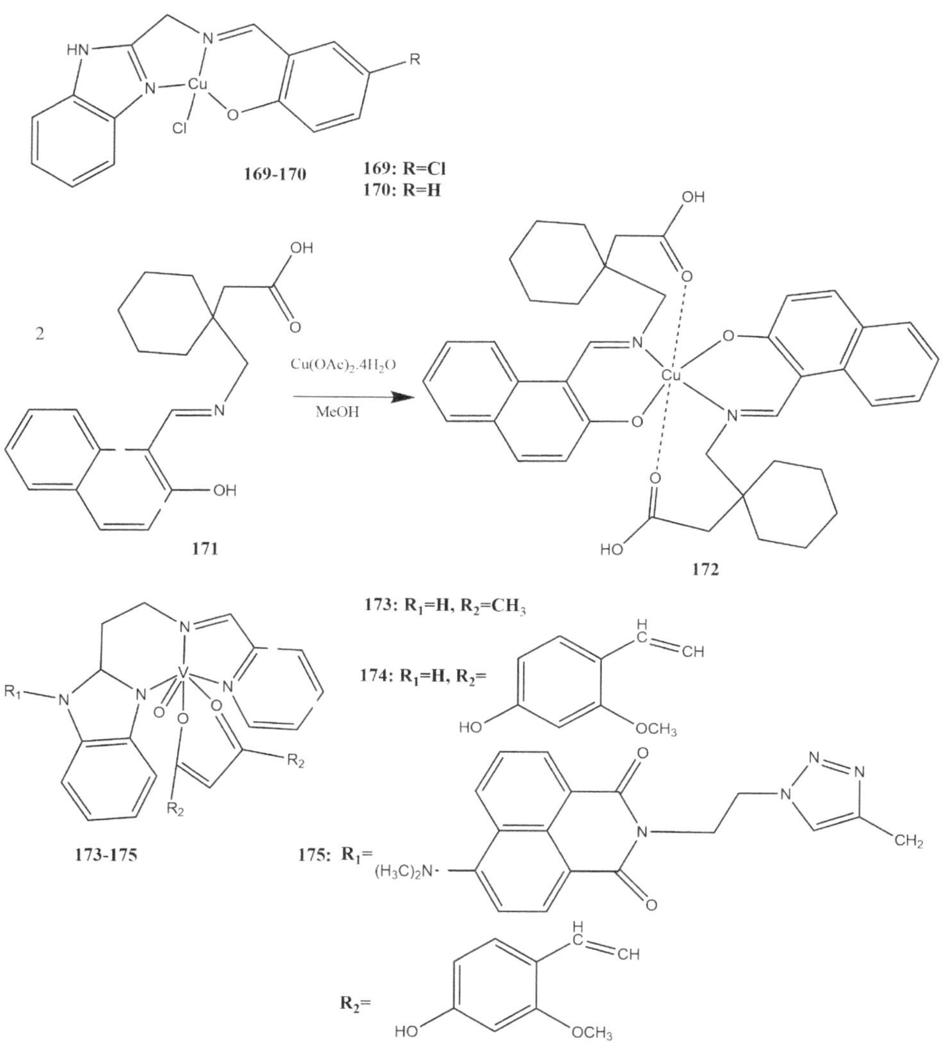

FIGURE 7.28 Cu(II) benzimidazole Schiff- base metal complexes (169-170), Gabapentin drug appended Cu (II) complex (172) and Vanadium (IV) complexes (173-175).

PC-3. The IC_{50} values of ligand **180a** was 102.1 μM and its La(III) complex **181** was 55.8 μM, respectively.

7.3.8 Tin Complexes as an Anticancer Drug

Among the main group complexes, organotin(IV) appeared as a most potent anticancer activities because of its high coordination ability. Several organotin complexes have been studied for anticancer activity [125–126]. Tian et al. [127] synthesized a new organotin binuclear complex **183** from triphenyl tin chloride and Schiff base ligand formulated as $_{Ph3}$Sn(HL)·Ph_2SnL where L = Schiff base ligand. Schiff ligand was synthesized from 3,5-dibromo-2-hydroxy benzaldehyde and potassium valinate (Figure 7.31). Structural analysis showed distorted trigonal bipyramidal structure with two dissimilar coordinations of carboxylate, *trans*-O_2SnC_3 and *trans*-O_2SnC_2N. Complex **183** was

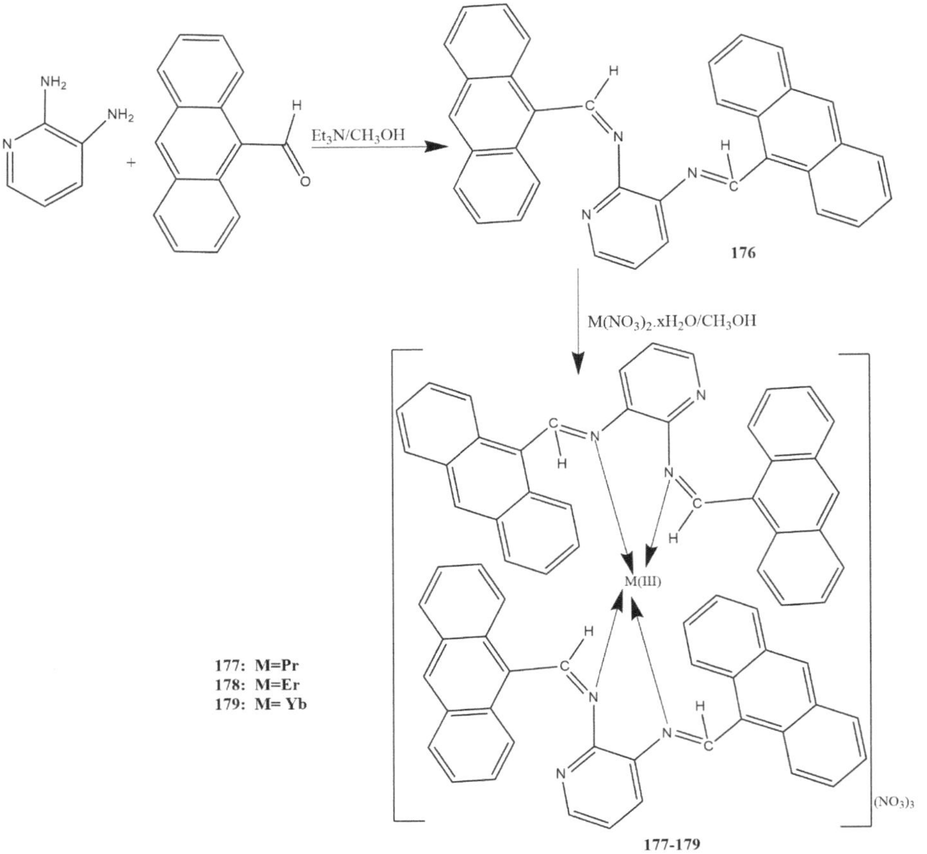

FIGURE 7.29 F-block Metal (III) complexes.

checked for cytotoxic activity against HeLa, COLO-205, and MCF-7 cell lines. *In vitro* cytotoxicity of organotin complex displayed better activity than the reference drug cisplatin. Hong et al. [128] reported synthesis of five new different organotin(IV) complexes **184–188** from substituted-4-carbohydrazone and di or trialkyltin(IV) precursors (Figure 7.31). Three complexes **184**, **185**, and **187** displayed distorted trigonal bipyramid structure whereas complex **188** displayed 72-membered macrocycle ring structure. An *in vitro* cytotoxic activity study against different cancer cell lines HT-29, A549, HL-60, Caco-2 and HCT-116 using MTT assay revealed that the R group attached with the metal atom affects the activity significantly. Complexes **185** and complex **187** have shown effective results, triggering them as potential candidates, especially in cisplatin resistant cells.

7.3.9 Coumarin Appended Metal Complexes as an Anticancer Drug

Coumarin are heterocyclic compounds containing oxygen and benzopyrone. They are well recognized drugs in various therapies [129–132]. Coumarin analogues are strong antitumor agents in numerous cancer types [133]. It is well established that various coumarin derivatives obstruct the cell growth in gastric carcinoma cell lines and they are considered selective active against various cancerous cell lines. In the year 2009 Creaven and co-workers [134] synthesized Cu(II) complexes **189a–k** of substituted 7-amino-4-methyl-coumarin derived bidentate Schiff base ligand derivatives (Figure 7.32) and explored their cytotoxicity against HT29 and MCF-7 cell line [135].

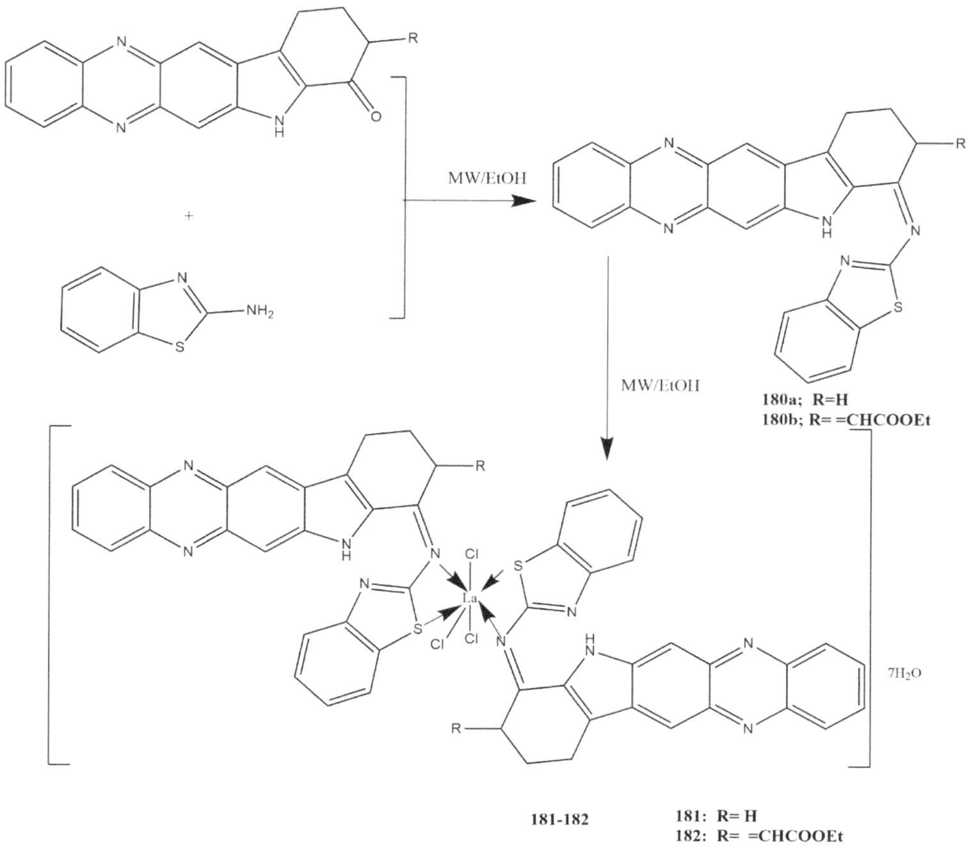

FIGURE 7.30 Synthesis of La (III) Metal complexes.

Schiff base ligands were synthesized by substituted salicylaldehydes and 7-amino-4-methyl-coumarin. These ligands' consequent reaction with copper(II) acetate gave the desired Cu(II) Schiff base coumarin derived complexes. The standard drugs used in the study were mitoxantrone and cisplatin. Cu(II) complexes **189** had not been cytotoxic to HT29 cells, whereas compounds **189i** and **189k** displayed IC_{50} values of 79.8 mM and 34.5 mM, respectively, when checked against the MCF-7 cells. Sahin and collaborators [136] presented coumarin-thiazole based Schiff base Pd(II) and Pt(II) complexes **190–191** formulated as ([Pd(L)$_2$] and [Pt(L)$_2$] where L = Schiff ligand. The anticancer activity of both the complexes was evaluated against different cancerous cell lines MCF-7, LS174T, and LNCAP using MTT assay. Schiff base ligand having thiazole rings was synthesized by the reaction of 2-amino-4-(3-coumarinyl)thiazole and 2-hydroxy-5-methylbenzaldehyde which gave metal complexes further reaction with metal chlorides (Na_2PdCl_4 and K_2PtCl_4) (0.5 mmol) in 2:1 molar ratio respectively (Figure 7.32). It was found that trans-Pd(II) complex possessed greater activity than the trans-Pt(II) complex against the screened cell lines. The cell viability percentage of the complexes reduces as a function of enhancing concentration. Anticancer activities of the complexes are both the structure and dose-dependent. The utmost dose is found to be 100 μM for both the complexes and ligand. Mestizo and co-workers [137] synthesized eight new metal complexes **192–199** from tetradentate bearing coumarin appended Schiff base ligand and metal acetate. The condensation reaction of 8-formyl-7- hydroxycoumarin with substituted 4,5- diamine benzene gave coumarin appended Schiff base ligand. Anticancer activities were checked against

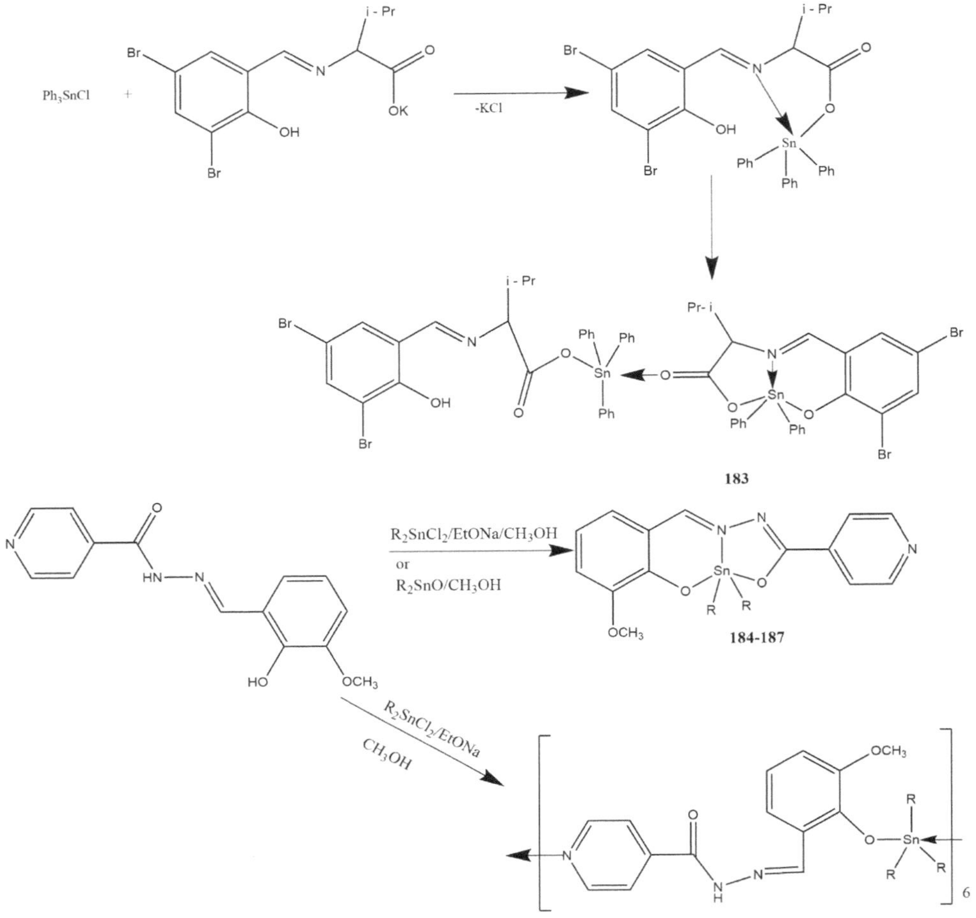

FIGURE 7.31 Synthesis of organotin(IV) complexes.

HeLa, HaCaT, and HFF-1 cancerous and non-cancerous human cell lines respectively. Cobalt complexes showed IC_{50} values, **193**(3.5 μM) and **197** (4.1 μM) against HeLa cell line that is greater than cisplatin. Mechanism of action of cobalt complexes suggested the activity results may be due to the accumulation of reactive oxygen species.

7.3.10 Isatin Appended Metal Complexes as an Anticancer Drug

Isatin is an indole alkaloid, formulated as 1H-indole-2, 3-dione, discovered in various plants. Isatin appended Schiff ligands are mostly synthesized by condensation of different aromatic primary amines or hydrazides with keto group of isatin [138–140]. Isatin-Schiff base ligand scaffolds coordinated with the various metal ions were reported to provide improved potency and mark specificity. Isatin-Schiff base ligands are an important therapeutic pharmacophore against different diseases.

Gomathi et al. [141] synthesized isatin based Schiff ligand and their metals complexes of Co(II), Ni(II) and Cu(II). Isatin appended ligand was synthesized by the reaction of N-phenylisatin and isonicotinohydrazide, which on reaction with metal(II) chloride gave corresponding metal complexes **200–202**, respectively (Figure 7.33). A stoichiometry ratio of metal and ligand was

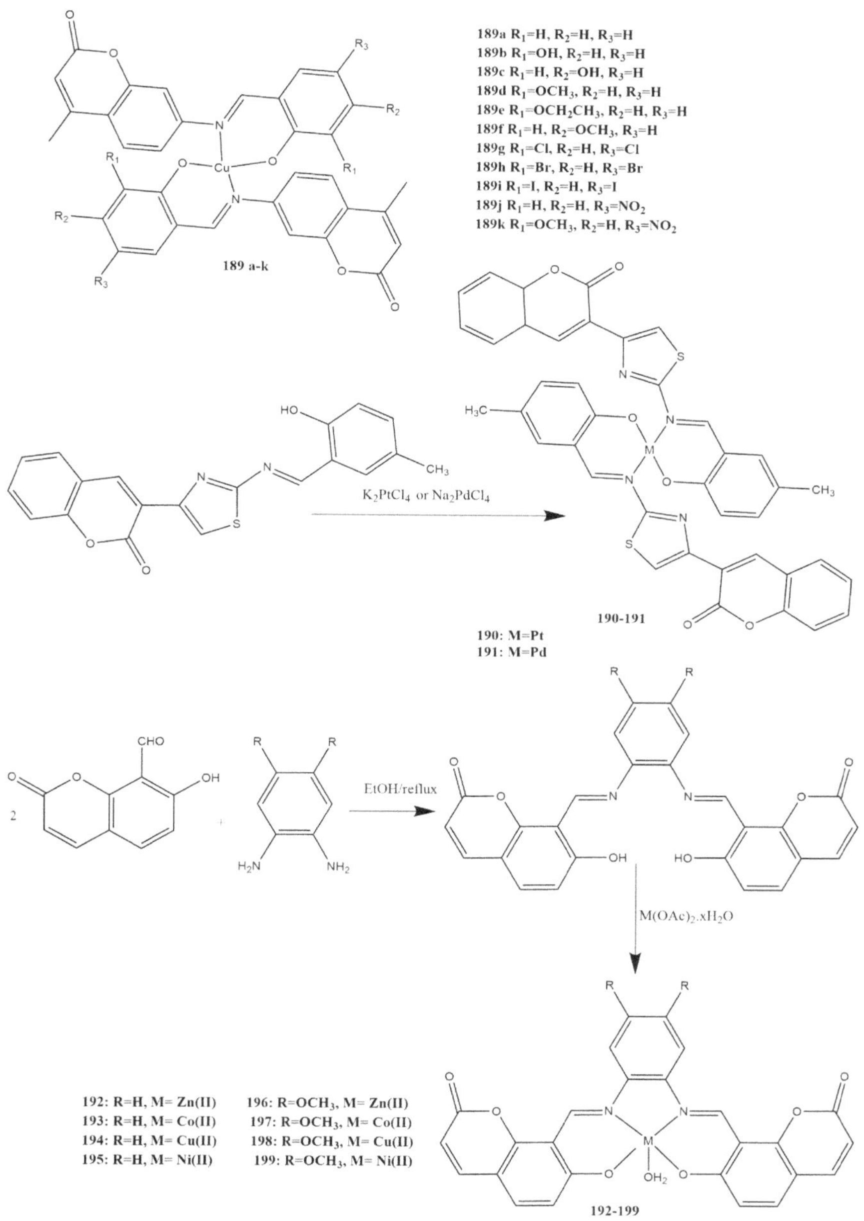

FIGURE 7.32 Coumarin appended Metal complexes.

1:2. All the complexes showed octahedral geometry. The standard drug used in the study was Mitomycin-C. Cu(II) complex **200** and Ni(II) complexes **202** showed significant cytotoxicity against AGS-human gastric cancer cell line. The Cu (II) complex **203** of isatin-Schiff base was synthesized by Bulatov et al. [142] and tested in numerous cancer cell lines (Figure 7.33). The impact of complex on the p53 protein was evaluated. Complex **203** exhibited cytotoxicity toward p53-positive as well as p53-negative MCF7 tumor cells, and promoted p53-dependent gene expression and stimulated apoptosis.

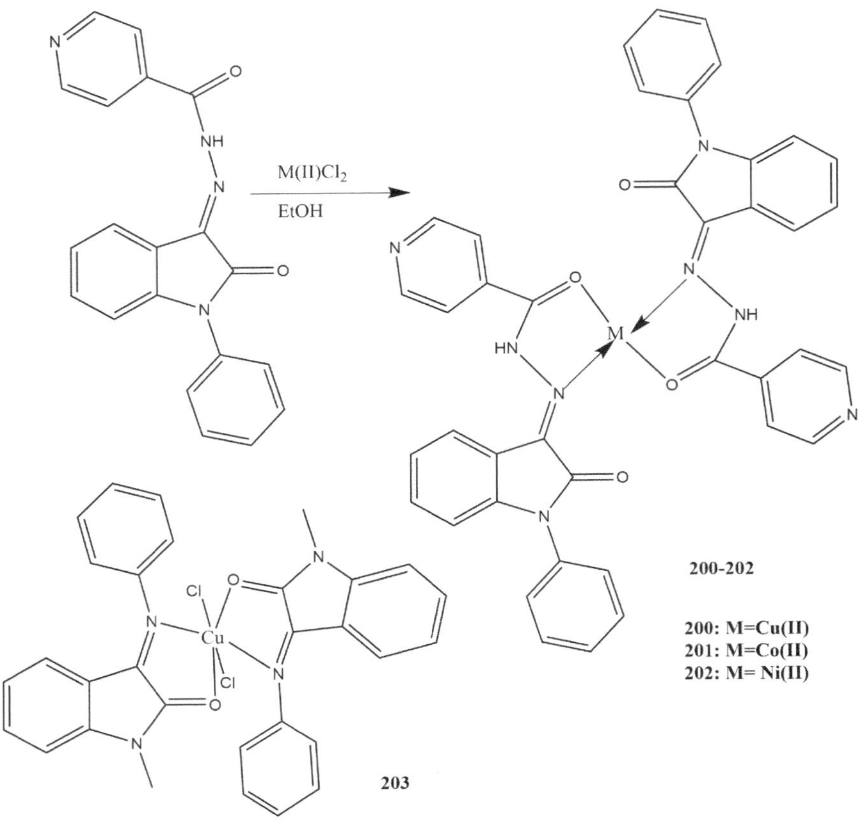

200-202

200: M=Cu(II)
201: M=Co(II)
202: M= Ni(II)

203

FIGURE 7.33 Isatin appended Metal complexes.

7.4 CONCLUSION

During the last two decades much interest has been drawn on developing anticancer mettalodrugs. The bioinorganic medicinal chemistry has a thriving area for cancer drug designing research. Coordination of various scaffolds to metal complexes resulted in a potential drug compound with target selectivity. An earlier used metal, platinum, was replaced by different transition metals like Rh, Ru, Ag, Au, Ir and Pd, F-block metals like La, Er, Pr, Yb, first-row transition metals like Ni, Co, Mn, Cu, and Zn, and main group metals like Sn, and so on, in order to develop a potential anticancer drug candidate. Metal complexes have been synthesized by reacting different metal salts with various Schiff base ligand. Schiff base ligand is a versatile precursor easily prepared with condensation of different aldehyde with amines. It was found that metal complexes displayed a captivating anticancer property against various cell lines even greater than their corresponding Schiff base ligands and in some cases also higher than the standard drugs inducing apoptosis or cell death. Different metal complexes also reported better activity towards standard drug resistance cells and low toxicity to non-cancerous cell lines, suggesting the need to explore different metal complexes further to find a new promising anticancer drug. Such complexes could yield a substitute to old metallo anticancer drugs and possibly defeat resistance toward the platinum drug.

ABBREVIATIONS

IC_{50} = Concentration needed to attain fifty percent of maximal inhibition
MTT = 3-(4,5-dimethylthiazol-2-yl)-2,5-diphenyl tetrazolium bromide
A2780 = Ovarian cancerous cells
A2780cisR = Cisplatin resistant Ovarian cancerous cells
A549 = Lung carcinoma cell line
A549cisR = Cisplatin Resistant human lung cancerous cells
AGS = gastric carcinoma cell line
ARPE-19 = Retinal pigment epithelia cell line (non-cancerous)
BGC-823 = Gastric cancer cell line
Caco-2 = Colon carcinoma cells
COLO-205 = Colorectal carcinoma cell line
DLA = Dalton's Lymphoma Ascites cell line.
HaCaT = Immortalized human keratinocyte cell line (non-cancerous)
HBL-100 = Breast milk epithelial cell line
HCT-116 = Colorectal cancer cells
HEK 293 = Non-cancerous embryonic cell line
Hep3B = Hepatocellular carcinoma cells
HFF-1 = foreskin fibroblast cell line
HL-60 = Acute promyelocytic leukaemia
HeLa = Cervical cancer cell line
HepG2 = Liver cancer cells
HOP-62= Lung cancer cell line
HS-683 = Oligodendroglioma cells
HT-29 = Colorectal cancerous cells
HuH-7 = Human hepatocyte carcinoma cell line
H1299 = Non-small cell lung cancerous cell line
JURKAT- = T lymphocyte cell line
K562 = Myelogenous leukemia cells
LS 174T = Colon cancer cell line
LN405 = Glioblastoma brain tumor cell lines
LN-18 = Glioblastoma brain tumor cell lines
LNCAP = Androgen-responsive prostate cell line
LoVo = Colon adenocarcinoma cell line
MCF-7CR = Cisplatin Resistant human breast cancer cell line
MCF 7 = Breast cancerous cell line
MDA-MB-231 = Metastatic triple negative breast cancer cell line
MCF-12A = Mammary epithelial cell line (non-cancerous)
MCF-10A = Non-cancerous breast epithelial cell line
NIH 3T3 = Fibroblasts cell line
NHDF = Normal human dermal fibroblasts cell line
OVCAR-3 = Ovarian cancer cell line
PC3 = Prostate cancer cell line
RCC = Renal Cell Carcinoma
SH-SY5Y = Neuroblastoma cell line
SK-OV3 = Ovarian adenocarcinoma cell line
SMMC-7721 = Hepatocellular carcinoma cells
SW620 = Human colon adenocarcinoma cell line

TK-10 = Renal carcinoma cell line
U-87 = Primary glioblastoma cell line
UACC-62 = Human melanoma cell line
WI38 = Human fetus lung cell line

REFERENCES

1. Zugazagoitia, J., C. Guedes, S. Ponce, I. Ferrer, S. Molina-Pinelo, and L. Paz-Ares. 2016. Current Challenges in Cancer Treatment. *Clinical Therapeutics* 38, no. 7: 1551–1566. https://doi.org /10.1016/j.clinthera.2016.03.0.
2. Esfahani, K., L. Roudaia, N. Buhlaiga, S.V. Del Rincon, N. Papneja, and W.H. Miller. 2020. A Review of Cancer Immunotherapy: From the Past to the Present, to the Future. *Curr Oncology* 27, no. 2 (April): 87–97. https://doi.org/10.3747/co.27.5223.
3. Hassanpour, S. H., M. Dehghani. 2017. Review of Cancer from the Perspective of Molecular, *Journal of Cancer Research and Practice* 4, no. 4: 127–129. https://doi.org/10.1016/j.jcrpr.2017.07.001.
4. Hanahan D, Weinberg RA. 2000. The Hallmarks of Cancer. *Cell.* 100: 57–70.
5. Hanusova, M. V., L. Skalova, V. Kralova, P. Matouskova. 2015. Potential Anti-cancer Drugs Commonly Used for Other Indications, *Curr. Cancer Drug Tar.* 15 no. 1: 35–52, https://doi.org/10.2174/15680 09615666141229152 81.
6. Siegel, L., K. D. Miller, H. E. Fuchs, A. Jemal. 2021. Cancer Statistics. *CA Cancer J Clin.* 71: 7–33. https://doi.org/10.3322/caac.21654.
7. ICMR and ICMR-NCDIR. The National Cancer Registry Programme Report 2020 www.ncdirindia.org/All_Reports/Report_2020/PB/Press_release.pdf.
8. Rosenberg, B. 1971. Some Biological Effects of Platinum Compounds: New Agents for the Control of Tumours. *Platinum Met Rev.* 15: 42–51.
9. Florea, A. M., D. Büsselberg. 2011. Cisplatin as an Anti-Tumor Drug: Cellular Mechanisms of Activity, Drug Resistance and Induced Side Effects. *Cancers* 3, no. 1: 1351–1371. https://doi.org/10.3390/cancers3011351.
10. Leisching, G., B. Loos, M. Botha, and A. M. Engelbrecht. 2015. Bcl-2 confers survival in cisplatin treated cervical cancer cells: circumventing cisplatin dose-dependent toxicity and resistance. *Journal of Translational Medicine*, 13, 328–35. https://doiorg.libsrv.wku.edu/10.1186/s12967-015-0689-4.
11. Zazuli, Z., S. Vijverberg, E. Slob, G. Liu, B. Carleton, J. Veltman, P. Baas, R. Masereeuw and A. H. Maitland-van der Zee. 2018. Genetic Variations and Cisplatin Nephrotoxicity: A Systematic Review. Frontiers in Pharmacology 9:1–17. http://dx.doi.org/10.3389/fphar.2018.01111.
12. Tadele, K.T., T. W. Tsega. 2019. Schiff Bases and Their Metal Complexes as Potential Anticancer Candidates: A Review of Recent Works. *Anticancer Agents Med. Chem.* 19, no. 15: 1786–1795. http://dx.doi.org/10.2174/1871520619666190227171716 4.
13. Amim, R.S., C. Pessoa, M.C.S. Lourenco, M.V.N. de-Souza J.A. Lessa, 2017. Synthesis, Antitubercular and Anticancer Activities of Pnitrophenylethylenediamine-derived Schiff Bases. *Med. Chem.* 13 no. 4: 391–397. http://dx.doi.org/10.2174/1573406412666161104123149.
14. Abou Melha, K. S.A., G. A. A. Al-Hazmi, M. S. Refat. 2017. Synthesis of Nano-metric Gold Complexes with New Schiff Bases Derived from 4-Aminoantipyrene, their Structures and Anticancer Activity. *Russian Journal of General Chemistry* 87 no. 12: 3043–3051. http://dx.doi.org/10.1134/S107036321 7120519.
15. Devi, J., M. Yadav, D. Kumar, L.S. Naik, D. K. Jindal. 2019. Some Divalent Metal(II) Complexes of Salicylaldehyde, Derived Schiff Bases: Synthesis, Spectroscopic Characterization, Antimicrobial and *In Vitro* Anticancer Studies. *Appl. Organometal. Chem. 33*, e4693. http://dx.doi.org/10.1002/aoc.4693.
16. Emam, S. M., I. E. T. El Sayed, M. I. Ayad, H. M. R. Hathout. 2017. Synthesis, Characterization and Anticancer Activity of New Schiff Bases Bearing Neocryptolepine. *J. Mol. Struct., 1146*: 600–619. http://dx.doi.org/10.1016/j.molstruc.2017.06.006.
17. Tandon, S. S., L. K.Thompson, M. E. Manuel, J. N. Bridson. 1994. *Inorg. Chem.* 33, no. 24: 5555–5570.
18. Collinson, S. R., D. E. Fenton,1996. Metal Complexes of Bibracchial Schiff Base Macrocycles. *Coordination Chemistry Reviews* 148: 19–40. https://doi.org/10.1016/0010-8545(95)01156-0.

19. Catalano, A., M. S. Sinicropi, D. Iacopetta, J. Ceramella, A. Mariconda, C. Rosano, E. Scali, C. Saturnino, P. Longo. 2021. A Review on the Advancements in the Field of Metal Complexes with Schiff Bases as Antiproliferative Agents. *Applied Sciences* 11, no. 13: 6027–6044. https://doi.org/10.3390/app11136027.
20. Rosenberg, B., L. Vancamp, J. Trosko, V. H. Mansour. 1969. Platinum Compounds: A New Class of Potent Antitumour Agents. *Nature* 222: 385–386. https://doi.org/10.1038/222385a0.
21. Carloni, P., W. Andreoni, J. Hutter, A. Curioni, P. Giannozzi, M. Parrinello. 1995. Structure and Bonding in Cisplatin and Other Pt(II) Complexes *Chemical Physics Letters* 234, no. 1–3 (March): 50–56. https://doi.org/10.1016/0009-2614(94)01488-H.
22. Sastry, J., S. J. Kellie. 2005. Severe Neurotoxicity, Ototoxicity and Nephrotoxicity Following High-dose Cisplatin and Amifostine, Pediatr. Hematol. Oncol. 22, no. 5: 441–445, https://doi.org/10.1080/08880010590964381.
23. Bhavsar, A. P., E. P. Gunaretnam, Y. Li, J. S. Hasbullah, B. C. Carleton, C. J. D. Ross, 2017. Pharmacogenetic Variants in TPMT Alter Cellular Responses to Cisplatin in Inner Ear Cell Lines. *Plos One* 12 no. 4: e0175711, 1–10. https://doiorg.libsrv.wku.edu/10.1371/journal.pone.0175711.
24. Marques, M. P. M. 2013. Platinum and Palladium Polyamine Complexes as Anticancer Agents: The Structural Factor. *International Scholarly Research Notices* 2013: Article ID 287353: 1–29, http://dx.doi.org/10.1155/2013/287353.
25. Arjmand, F., M. Muddassir, R. H. Khan. 2010. Chiral Preference of L-tryptophan Derived Metal-based Antitumor Agent of Late 3d-Metal Ions (Co (II), Cu(II) and Zn(II)) in Comparison to D- and DL-Tryptophan Analogues: Their *In Vitro* Reactivity towards CT DNA, 5'-GMP and 5'-TMP. *Eur. J. Med. Chem.* 45: 3549–3557. doi:10.1016/j.ejmech.2010.04.031.
26. Pattan S. R., S. B. Pawar, S. S. Vetal, U. D. Gharate, S. B. Bhawar. 2012. The Scope of Metal Complexes in Drug Design-a review. *Indian Drugs* 49 no. 11: 1–12. https://doi.org/10.53879/id.49.11.p0005.
27. Allardyce, C. S., P. J. Dyson. 2016. Metal-based Drugs That Break the Rules. *Dalton Transactions* 45, no. 8: 3201–3209.
28. Muhammad, N., Z. Guo. 2014. Metal-based Anticancer Chemotherapeutic Agents. *Current Opinion in Chemical Biology* 19: 144–153. https://doi.org/10.1016/j.cbpa.2014.02.003.
29. Van Rijt S. H., P. J. Sadler. 2009. Current Applications and Future Potential for Bioinorganic Chemistry in the Development of Anticancer Drugs. *Drug Discovery Today* 14, no. 23-24: 1089–1097. https://doi.org/10.1016/j.drudis.2009.09.003.
30. Paprocka, R., M. Wiese-Szadkowska, S. Janciauskiene, T. Kosmalski, M. Kulik, A. Helmin-Basa. 2022. Latest Developments in Metal Complexes as Anticancer Agents. *Coordination Chemistry Reviews* 452: 214307–214327. https://doi.org/10.1016/j.ccr.2021.214307.
31. Ndagi, U., N. Mhlongo, M. E. Soliman. 2017. Metal Complexes in Cancer Therapy—An Update from Drug Design Perspective. *Drug Design, Development and Therapy* 11: 599–616. http://dx.doi.org/10.2147/DDDT.S119488.
32. Ghosh, P., S.K. Dey, M. H. Ara, K. Karim, A. B. M Islam. 2019. A Review on Synthesis and Versatile Applications of Some Selected Schiff Bases with Their Transition Metal Complexes. *Egyptian Journal of Chemistry* 62, 2: 523–547. https://doi.org/10.21608/ejchem.2019.13741.1852.
33. Qin, W., S. Long, M. Panunzio, and S. Biondi. 2013. Schiff Bases: A Short Survey on an Evergreen Chemistry Tool. *Molecules* 18, no. 10: 12264–12289. https://doi.org/10.3390/molecules181012264.
34. Kajal, A., S. Bala, S. Kamboj, N. Sharma, V. Saini.2013. Schiff Bases: A Versatile Pharmacophore. *Journal of Catalyst* 2013, Article ID 893512: 1–14 https://doi.org/10.1155/2013/893512.
35. Iacopetta, D., J. Ceramella, A. Catalano, C. Saturnino, M. G. Bonomo, C. Franchini, M. S. Sinicropi. 2021. Schiff Bases: Interesting Scaffolds with Promising Antitumoral Properties. *Appl. Sci.*, 11: 1877. https://doi.org/10.3390/app11041877.
36. Sridevi, G., S. A. Antony, R. Angayarkani. 2019. Schiff Base Metal Complexes as Anticancer Agents. *Asian Journal of Chemistry* 31, no. 3: 493–504. https://doi.org/10.14233/ajchem.2019.21697.
37. Li, L.-J, Q.-Q. Yan, G.-J. Liu, Z. Yuan, Z.-H. Lv, B. Fu, Y.-J. Han, J.-L. Du. 2017. Synthesis Characterization and Cytotoxicity Studies of Platinum(II) Complexes with Reduced Amino Pyridine Schiff Base and its Derivatives as Ligands. *Bioscience, Biotechnology, and Biochemistry* 81, no. 6: 1081–1089. https://doi.org/10.1080/09168451.2016.1259550.

38. Li, L. -J, C. Tian, C. Wang, G. Wang, L. Wang, J. Du. 2012. Platinum(II) Complexes with Tetradentate Schiff Bases As Ligands: Synthesis, Characterization and Detection of DNA Interaction by Differential Pulse Voltammetry. *E-Journal of Chemistry* 9, no. 3: 1422–1430.
39. Li, L.-J., B. Fu, Y. Qiao, C. Wang, Y.-Y. Huang, C.C. Liu, C. Tian, J. -L. Du. 2014. Synthesis, Characterization and Cytotoxicity Studies of Platinum(II) Complexes with Reduced Amino Acid Ester Schiff-bases as Ligands. *Inorganica Chimica Acta* 419: 135–140. http://dx.doi.org/10.1016/j.ica.2014.04.036.
40. Jean, S., K. Cormier, A. E. Patterson, C. M. Vogels, A. Decken, G. A. Robichaud, S. Turcotte, S. A. Westcott. Synthesis, Characterization, and Anticancer Properties of Organometallic Schiff Base Platinum Complexes. *Canadian Journal of Chemistry*. 93, no. 10: 1140–1146. https://doi.org/10.1139/cjc-2015-0157.
41. A. E. Patterson, J. J. Miller, B. A. Miles, E. L. Stewart, J-M E. J. Melanson, C. M. Vogels, A. M. Cockshutt, A. Decken, P. Morin Jr., S. A. Westcott. 2014. Synthesis, Characterization and Anticancer Properties of (Salicylaldiminato)platinum(II) Complexes. *Inorganica Chimica Acta* 415: 88–94. https://doi.org/10.1016/j.ica.2014.02.028.
42. Vojtek M, M. P. M. Marques, I.M.P.L.V.O. Ferreira, H. Mota-Filipe, C. Diniz, 2019. Anticancer Activity of Palladium-based Complexes against Triple-negative Breast Cancer. *Drug Discov Today* 24, no. 4: 1044–1058. doi: 10.1016/j.drudis.2019.02.012.
43. Caires A. C. 2007 Recent Advances Involving Palladium (II) Complexes for the Cancer Therapy. *Anticancer Agents Med Chem.* 7., no. 5 (Sep): 484–491. doi: 10.2174/187152007781668661. PMID: 17896909.
44. Fanelli, M., M. Formica, V. Fusi, L. Giorgi, M. Micheloni, P. Paoli. 2016. New Trends in Platinum and Palladium Complexes as Antineoplastic Agents. *Coordination Chemistry Reviews* 310: 41–79. https://doi.org/10.1016/j.ccr.2015.11.004.
45. Mbugua, S. N., N. R. S. Sibuyi, L. W. Njenga, R. A. Odhiambo, S. O. Wandiga, M. Meyer, R. A. Lalancette, M. O. Onani. 2020. New Palladium(II) and Platinum(II) Complexes Based on Pyrrole Schiff Bases: Synthesis, Characterization, X-ray Structure, and Anticancer Activity. *ACS Omega* 5: 14942–14954. https://dx.doi.org/10.1021/acsomega.0c00360.
46. Ali, M. A., A. H. Mirza, R. J. Butcher, M.T.H. Tarafder, T. B. Keat, A. M. Ali. 2002. B iological Activity of Palladium(II) and Platinum(II) Complexes of the Acetone Schiff Bases of *S*-methyl- and *S*-benzyldithiocarbazate and the X-ray Crystal Structure of the [Pd(asme)] (asme5anionic form of the 2 acetone Schiff base of *S*-methyldithiocarbazate) Complex. *Journal of Inorganic Biochemistry* 92: 141–148.
47. Kalaiarasi, G., S. Dharani, V. M. Lynch, R. Prabhakaran. 2019. Para Metallation of 3-acetyl-chromen-2-one Schiff Bases in Tetranuclear Palladacycles: Focus on Their Biomolecular Interaction and *In Vitro* Cytotoxicity. *Dalton Trans.* 48: 12496–12511. doi: 10.1039/C9DT02663K.
48. Cassini, A., R. Bonsignore, J. Oberkofler. 2018. Organometallic Gold-based Anticancer Therapeutics. *Reference Module in Chemistry, Molecular Sciences and Chemical Engineering*. 1–12. https://doi.org/10.1016/B978-0-12-409547-2.14350-1.
49. Tiekink, E. R.T. 2002. Gold Derivatives for the Treatment of Cancer, *Critical Reviews in Oncology/Hematology*, 42, no. 3: 225–248, https://doi.org/10.1016/S1040-8428(01)00216-5.
50. Ingo Ott. 2009. On the Medicinal Chemistry of Gold Complexes as Anticancer Drugs. *Coordination Chemistry Reviews* 253, no. 11–12: 1670–1681. https://doi.org/10.1016/j.ccr.2009.02.019.
51. Bian, M., X. Wang, Y. Sun, W. Liu. 2020. Synthesis and Biological Evaluation of Gold(III) Schiff Base Complexes for the Treatment of Hepatocellular Carcinoma through Attenuating TrxR Activity. *European Journal of Medicinal Chemistry* 193, May: 112234–57. https://doi.org/10.1016/j.ejmech.2020.112234.
52. Sankarganesh, M., J. D. Raja, N. Revathi, R. V. Solomon, R. S. Kumar. 2019. Gold(III) Complex from Pyrimidine and Morpholine Analogue Schiff Base Ligand: Synthesis, Characterization, DFT, TDDFT, Catalytic, Anticancer, Molecular Modeling with DNA and BSA and DNA Binding Studies. *Journal of Molecular Liquids* 294: 111655–111665.
53. Babgi, B.A.; J. Alsayari, H.M. Alenezi, M. H. Abdellatif, N.E. Eltayeb, A. H. M. Emwas, M. Jaremko, M.A. Hussien. 2021. Alteration of Anticancer and Protein-binding Properties of Gold(I) alkynyl by Phenolic Schiff Bases Moieties. *Pharmaceutics* 13: 461–473. https://doi.org/10.3390/pharmaceutics13040461.

54. Rademaker-Lakhai, J. M., D. van den Bongard, D. Pluim, J. H. Beijnen, J. H. M. Schellens. 2004. A Phase I and Pharmacological Study with Imidazolium-*trans*-DMSO-imidazole-tetrachlororuthenate, a novel Ruthenium Anticancer Agent. *Clin. Cancer Res.* 10, no. 11: 3717–27. https://doi.org/10.1158/1078-0432.CCR-03-0746.
55. Hartinger, C. G., M. A. Jakupec, S. Zorbas-Seifried, M. Groessl, A. Egger, W. Berger, H. Zorbas, P. J. Dyson, B. K. Keppler. KP1019, A New Redox-active Anticancer Agent—Preclinical Development and Results of a Clinical Phase I Study in Tumor Patients. *Chem. Biodiversity* **2008**, 5 no. 10: 2140–2155. https://doi.org/10.1002/cbdv.200890195.
56. Purkait, K., Ruturaj, A. Mukherjee, A. Gupta. 2019. ATP7B Binds Ruthenium(II) p-Cymene Half-sandwich Complexes: Role of Steric Hindrance and Ru–I Coordination in Rescuing the Sequestration. *Inorg. Chem.* 58:15659–70. doi: 10.1021/acs.inorgchem.9b02780.
57. Zhao, j., W. Li, S. Gou, S. Li, S. Lin, Q. Wei, G. Xu. 2018. Hypoxia-targeting Organometallic Ru(II)-Arene Complexes with Enhanced Anticancer Activity in Hypoxic Cancer Cells. *Inorganic Chemistry* 57, no. 14: 8396–8403. doi: 10.1021/acs.inorgchem.8b01070.
58. Parveen, S. 2020. Recent Advances in Anticancer Ruthenium Schiff Base Complexes. *Applied Organometallic Chem*istry 34, 8: 1–23, e5687. https://doi.org/10.1002/aoc.5687.
59. Alsalme, A., S. Laeeq, S. Dwivedi, M. S. Khan, K. A. Farhan, J. Musarrat, R. A. Khan, 2016. Synthesis, Characterization of α-amino Acid Schiff Base Derived Ru/Pt Complexes: Induces Cytotoxicity in HepG2 Cell Via Protein Binding and ROS Generation. *Spectrochimica Acta Part A: Molecular and Biomolecular Spectroscopy* 163: 1–7. http://dx.doi.org/10.1016/j.saa.2016.03.012.
60. Ejidike, I. P., P. A. Ajibade. 2016. Synthesis, Characterization, Anticancer, and Antioxidant Studies of Ru(III) Complexes of Monobasic Tridentate Schiff Bases. *Bioinorg. Chem. Appl.* 2016: 9672451. http://dx.doi.org/10.1155/2016/9672451.
61. Ejidike, I. P., P. A. Ajibade. 2016. Ruthenium(III) Complexes of Heterocyclic Tridentate (ONN) Schiff Base: Synthesis, Characterization and its Biological Properties as an Antiradical and Antiproliferative Agent. *Int. J. Mol. Sci.* 17: 60–70. doi:10.3390/ijms17010060.
62. Prakash, G., R. Manikandan, P. Viswanathamurthi, K. Velmurugan, R. Nandhakumar. 2014. Ruthenium(III) S-Methylisothiosemicarbazone Schiff Base Complexes Bearing PPh3/AsPh3 Coligand: Synthesis, Structure and Biological Investigations, Including Antioxidant, DNA and Protein Interaction, and In Vitro Anticancer Activities. *Journal of Photochemistry and Photobiology B: Biology* 138: 63–74. http://dx.doi.org/10.1016/j.jphotobiol.2014.04.019.
63. Zhang, Y., L. Lai, P. Cai, G.-Z. Cheng, X.-M. Xu, Y. Liu. 2015. Synthesis, Characterization and Anticancer Activity of Dinuclear Ruthenium (II) Complexes Linked by an Alkyl Chain. *New Journal of Chemistry* 39, no. 7: 5805–5812. http://dx.doi.org/10.1039/c5nj00582e.
64. Hu P., Y. Wang, Y. Zhang, H. Song, F. Gao, H. Lin, Z. Wang, L. Wei, F. Yang. 2016. Novel Mononuclear Ruthenium(II) Complexes as Potent and Low-toxicity Antitumour Agents: Synthesis, Characterization, Biological Evaluation and Mechanism of Action. *RSC Advances*, 6: 29963–29976. https://doi.org/10.1039/C6RA02571D.
65. Kahrovic, E., A. Zahirovic, S. K. Pavelic, E. Turkušic, A. Harej. 2017. In Vitro Anticancer Activity of Binuclear Ru(II) Complexes with Schiff Bases Derived from 5-Substituted Salicylaldehyde and 2-Aminopyridine with Notably Low IC50 Values. *J. Coord. Chem.* 70, 10: 1683–1697. https://doi.org/10.1080/00958972.2017.1308503.
66. Subbaiyana, S., I. Ponnusamy. 2019. Biological Investigations of Ruthenium(III) 3-(Benzothiazol-2-liminomethyl)-phenol Schiff Base Complexes Bearing PPh3 / AsPh3 Coligand. *Curr. Chem. Lett.* 8: 145–156. doi: 10.5267/j.ccl.2019.004.003.
67. Malik, M. A., M. K. Raza, O. A. Dar, M. Abid, M. Y. Wani, A. S. Al-Bogami, A. A. Hashmi. 2019. Probing the Antibacterial and Anticancer Potential of Tryptamine Based Mixed Ligand Schiff Base Ruthenium(III) Complexes. *Bioorganic Chemistry* 87: 773–782. https://doi.org/10.1016/j.bioorg.2019.03.080.
68. G. R. Jadhav, S. Sinha, M. Chhabra, P. Paira, 2016. Synthesis of Novel Anticancer Ruthenium–Arene Pyridinylmethylene Scaffolds via Three-component Reaction. *Bioorg. Med. Chem. Lett.* 26: 2695–2700. https://doi.org/10.1016/j.bmcl.2016.04.005.
69. Pitchaimani, J., M. R. C. Raja, S. Sujatha, S. K. Mahapatra, D. Moon, S. P. Anthony, V. Madhu. 2016. Arene Ruthenium (Ii) Complexes with chalcone, Aminoantipyrine and Aminopyrimidine Based Ligands: Synthesis, Structure and Preliminary Evaluation of Anti-leukemia Activity. *RSC advances* 6, no. 93: 90982–90992. https://doi.org/10.1039/C6RA18504E.

70. Pettinari, R., F. Marchetti, C. Di Nicola, C. Pettinari, A. Galindo, R. Petrelli, L. Cappellacci, M. Cuccioloni, L. Bonfili, A. M. Eleuteri, M. Fatima, C. Guedes da Silva, A. J. L. Pombeiro. 2018. Ligand Design for N, O-or N, N-Pyrazolone-based Hydrazones Ruthenium (II)-arene Complexes and Investigation of Their Anticancer Activity. *Inorganic Chemistry* 57, no. 22: 14123–14133. https://doi.org/10.1021/acs.inorgchem.8b01935.
71. Subarkhan, M. K. M., L. Ren, B. Xie, C. Chen, Y. Wang, H. Wang. 2019. Novel Tetranuclear Ruthenium(II) Arene Complexes Showing Potent Cytotoxic and Antimetastatic Activity as Well as Low Toxicity *In Vivo. Eur. J. Med. Chem.* 179: 246–256. https://doi.org/10.1016/j.ejmech.2019.06.061.
72. Subarkhan, M. K. M., S. Saranya, R. Rengan. 2018. Synthesis and Structure of New Binuclear Ruthenium(II) Arene Benzil Bis(benzoylhydrazone) Complexes: Investigation on Antiproliferative Activity and Apoptosis Induction. *Inorg. Chem. Front.* 5: 585–596. https://doi.org/10.1039/c7qi00761b.
73. Kumar, R. R., R. Ramesh, J. G. Małecki. 2018. Synthesis and Structure of Arene Ruthenium (II) Benzhydrazone Complexes: Antiproliferative activity, Apoptosis Induction and Cell Cycle Analysis. *Journal of Organometallic Chemistry* 862: 95–104. https://doi.org/10.1016/j.jorganchem.2018.03.013.
74. Gopalakrishnan D., S. Srinath, B. Baskar, N. S. P. Bhuvanesh, M. Ganeshpandian. 2019. Biological and Catalytic Evaluation of Ru(II)-pcymene Complexes of Schiff Base Ligands: Impact of Ligand Appended Moiety on Photo-induced DNA and Protein Cleavage, Cytotoxicity and C--H Activation. *Applied Organometallic Chemistry* 33: e4756. https://doi.org/10.1002/aoc.4756.
75. Yang, G. J., Wang, W., Mok, S.W.F., Wu, C., Law, B.Y.K., Miao, X.M., Wu, K.J., Zhong, H. J., Wong, C.Y., Wong, V.K.W., D.L. Ma, C.H. Leung. 2018. Selective Inhibition of Lysine-specific Demethylase 5A (KDM5A) Using a Rhodium (III) Complex for Triple-negative Breast Cancer Therapy. *Angew. Chem. Int. Ed.* 57, 40: 13091–13095. https://doi.org/10.1002/anie.201807305.
76. Threatt, S. D., T. W. Synold, J. Wu, J. K. Barton. 2020. In Vivo Anticancer Activity of a Rhodium Metalloinsertor in the HCT116 Xenograft Tumor Model. *Proceedings of the National Academy of Sciences of the United States of America*, 117, no. 30: 17535–17542. https://doi.org/10.1073/pnas.2006569117.
77. Liang, J., A. Levina, J. Jia, P. Kappen, C. Glover, B. Johannessen, P. A. Lay. 2019. Reactivity and Transformation of Antimetastatic and Cytotoxic Rhodium(III)-Dimethyl Sulfoxide Complexes in Biological Fluids: An XAS Speciation Study. *Inorg. Chem.* 58, no. 8: 4880–4893. https://doi.org/10.1021/acs.inorgchem.8b03477.
78. Fan, R., M. Bian, L. Hu, L. Wukun. 2019. A New Rhodium(I) NHC Complex Inhibits TrxR: In Vitro Cytotoxicity and In Vivo Hepatocellular Carcinoma Suppression. *Eur. J. Med. Chem.*183: 111721–111735. https://doi.org/10.1016/j.ejmech.2019.111721.
79. Song, X.-D., B.-B. Chen, S.-F. He, N.-L. Pan, J.-X. Liao, J.-X. Chen, G.-H. Wang, J. Sun. 2019. Guanidine-modified Cyclometalated Iridium(III) Complexes for Mitochondria-targeted Imaging and Photodynamic Therapy. *Eur. J. Med. Chem.* 179: 26–37. https://doi.org/10.1016/j.ejmech.2019.06.045.
80. Novohradsky, V., A. Rovira, C. Hally, A. Galindo, G. Vigueras, A. Gandioso, M. Svitelova, R. Bresolí-Obach, H. Kostrhunova, L. Markova, J. Kasparkova, S. Nonell, J. Ruiz, V. Brabec, V. Marchán. 2019. Towards Novel Photodynamic Anticancer Agents Generating Superoxide Anion Radicals: A Cyclometalated Ir-III Complex Conjugated to a Far-red Emitting Coumarin. *Angew. Chem. Int. Ed.* 58, no. 19: 6311–6315, https://doi.org/10.1002/anie.201901268.
81. Pérez-Arnaiz, C., M.I. Acuña, N. Busto, I. Echevarría, M. Martínez-Alonso, G. Espino, B. García, F. Domínguez. 2018. Thiabendazole-based Rh(III) and Ir(III) Biscyclometallated Complexes with Mitochondria-targeted Anticancer Activity and Metal-sensitive Photodynamic Activity. *Eur. J. Med. Chem.* 157: 279–293, https://doi.org/10.1016/j.ejmech.2018.07.065.
82. Aboura, W., L. K. Batchelor, A. Garcic, P. J. Dyson, B. Therrien. 2020. Reactivity and Biological Activity of N,N,S-Schiff-base Rhodium Pentamethylcyclopentadienyl Complexes. *Inorganica Chimica Acta* 501: 119265–119269. https://doi.org/10.1016/j.ica.2019.119265.
83. Yellol, G. S., A. Donaire, J. G. Yellol, V. Vasylyeva, C. Janiakb, J. Ruiz. On the Antitumor Properties of Novel Cyclometalated Benzimidazole Ru(II), Ir(III) and Rh(III) Complexes. *Chem. Commun.* 49: 11533–11535. https://doi.org/10.1039/c3cc46239k.
84. Mukhopadhyay, S., R. K. Gupta, R. P. Paitandi, N. K. Rana, G. Sharma, B. Koch, L. K. Rana, M. S. Hundal. 2015. Synthesis, Structure, DNA/Protein Binding, and Anticancer Activity of Some

Half-Sandwich Cyclometalated Rh(III) and Ir(III) Complexes. *Organometallics* 34, no. 18: 4491–4506. https://doi.org/10.1021/acs.organomet.5b00475.
85. Adhikari, S., D. Sutradhar, S. L. Shepherd, R. M. Phillips, A. K. Chandra, K. M. Rao. 2016. Synthesis, Structural, DFT Calculations and Biological Studies of Rhodium and Iridium Complexes Containing Azine Schiff-base Ligands. *Polyhedron* 117: 404–414. http://dx.doi.org/10.1016/j.poly.2016.06.001.
86. Radko, L., S. Stypula-Trebas, A. Posyniak, D. Zyro, J. Ochocki. 2019. Silver(I) Complexes of the Pharmaceutical Agents Metronidazole and 4-Hydroxymethylpyridine: Comparison of Cytotoxic Profile for Potential Clinical Application. *Molecules* 24, no. 10: 1949–1960. https://doi.org/10.3390/molecules24101949.
87. Ali, K. A., M. M. Abd-Elzaher, K. Mahmoud. 2013. Synthesis and Anticancer Properties of Silver(I) Complexes Containing 2,6 Bis(substituted) Pyridine Derivatives. *International Journal of Medicinal Chemistry* 2013, Article ID 256836: 1–7. http://dx.doi.org/10.1155/2013/256836.
88. Adeleke, A.A., S. J. Zamisa, M. S. Islam, K. Olofinsan, V. F. Salau, C. Mocktar, B. Omondi. 2021. Quinoline Functionalized Schiff Base Silver (I) Complexes: Interactions with Biomolecules and In Vitro Cytotoxicity, Antioxidant and Antimicrobial Activities. *Molecules* 26, 5: 1205–39. https://doi.org/10.3390/molecules26051205.
89. Rezaeivala, M., M. Ahmadi, B. Captain, M. Bayat, M. Saeidirad, S. S-Bolukba, B. Yıldız, R. W. Gable. 2020. Some New Morpholine-based Schiff-base Complexes; Synthesis, Characterization, Anticancer Activities and Theoretical Studies. *Inorganica Chimica Acta* 513: 119935–119946. https://doi.org/10.1016/j.ica.2020.119935.
90. Mazlan, N. A., T. B. S. A. Ravoof, E. R. T. Tiekink, M. I. M. Tahir, A. Veerakumarasivam, K. A. Crouse. 2014. Mixed-ligand Metal Complexes Containing an ONS Schiff Base and Imidazole/Benzimidazole Ligands: Synthesis, Characterization, Crystallography and Biological Activity. *Transition Metal Chemistry* 39, no. 6: 633–639. https://doi.org/10.1007/s11243-014-9842-9.
91. Bravo, C., M. P. Robalo, F. Marques, A.R. Fernandes, D.A. Sequeira, M.F.M. Piedade, M.H. Garcia, M. J. V. de Brito, T.S. Morais. 2019. First Heterobimetallic Cu(I)-dppf Complexes Designed for Anticancer Applications: Synthesis, Structural Characterization and Cytotoxicity. *New J. Chem.* 43 no. 31: 12308–12317. https://doi.org/10.1039/c9nj02068c.
92. Heffern, M. C., N. Yamamoto, R. J. Holbrook, A. L. Eckermann, T. J. Meade. 2013. Cobalt Derivatives as Promising Therapeutic Agents. *Current Opinion in Chemical Biology* 17, 2: 189–196. https://doi.org/10.1016/j.cbpa.2012.11.019.
93. Kumar, S., A.V. Trivedi. 2016. A Review on Role of Nickel in the Biological System. *Int. J. Curr. Microbiol. Appl. Sci.* 5, 3: 719–727. https://doi.org/10.20546/ijcmas.2016.503.084.
94. Ustun, E., A. Ozgur, K.A. Cos_kun, S. Demir, I. Ozdemir, Y. Tutar. 2016. CO-releasing Properties and Anticancer Activities of Manganese Complexes with Imidazole/ Benzimidazole Ligands, *J. Coord. Chem.* 69, 22: 3384–3394, https:// doi.org/10.1080/00958972.2016.1231921.
95. Refaat, H. M., D. A. N. El-Din. 2018. Chemical and Biological Evaluation of Moxifloxacin-benzimidazole Mixed Ligands Complexes: Anti-cancer and Anti-oxidant Activities. *J. Mol. Struct.* 1163: 103–113. https://doi.org/10.1016/j.molstruc.2018.02.116.
96. Mal, S. K., T. Chattopadhyay, A. Fathima, C. S. Purohit, M. S. Kiran, B. U. Nair, R. Ghosh. 2017. Synthesis and Structural Characterization of a Vanadium(V)- Pyridylbenzimidazole Complex: DNA Binding and Anticancer Activity. *Polyhedron* 126: 23–27. https://doi.org/10.1016/j.poly.2017.01.008.
97. Frieden E. 1974. The Evolution of Metals as Essential Elements (with Special Reference to Iron and Copper). *Advances in Experimental Medicine and Biology* 48: 1–29. https://doi.org/10.1007/978-1-4684-0943-7_1.
98. Czarnek, K., S. Terpiłowska, A. K. Siwicki. 2015. Selected Aspects of the Action of Cobalt Ions in the Human Body. *Cent. Eur. J. Immunol.* 40, no. 2: 236–242. https://doi.org/10.5114/ceji.2015.52837.
99. Romero, D. H., S. R.-Luna, A. L.-Monteon, A. C-Piña, N. P-Hernández, J. M.-Flores, A. C.-Navarro, G. P.-Gómez, D. M.-Morales, R. C.-Peralta. 2021. First-row Transition Metal Compounds Containing Benzimidazole Ligands: An Overview of Their Anticancer and Antitumor Activity. *Coordination Chemistry Reviews* 439: 213930–82. https://doi.org/10.1016/j.ccr.2021.213930.
100. M. M. Abd-Elzaher, A. A. Labib, H. A. Mousa, S. A. Moustafa, M. M. Ali, A. A. El-Rashedy. 2016. *Beni-Seuf Univ. J. Basic Appl. Sci.* 5: 85–96. https://doi.org/10.1016/j.bjbas.2016.01.001.

101. Prashanthi, Y., K. Kiranmai, Ira, K. S. Kumar, V.K. Chityala, Shivarj. 2012. Spectroscopic Characterization and Biological Activity of Mixed Ligand Complexes of Ni(II) with 1,10-Phenanthroline and Heterocyclic Schiff Bases. *Bioinorg. Chem. Appl.* 2012: 1–8. https://doi.org/10.1155/2012/948534.
102. Alorini, T. A., A. N. Al-Hakimi, S. El-Sayed Saeed, E. H. L. Alhamzi, A. E. A. E. Albadri. 2022. Synthesis, Characterization, and Anticancer Activity of Some Metal Complexes with a New Schiff Base Ligand. *Arabian Journal of Chemistry* 15: 103559–103576. https://doi.org/10.1016/j.arabjc.2021.103559.
103. Shafeeulla, R. M., G. Krishnamurthy, H. S. Bhojynaik, H. P. Shivarudrappa and Y. Shiralgi. 2017. Spectral Thermal Cytotoxic and Molecular Docking Studies of N'-2-Hydroxybenzoyl; Pyridine-4-carbohydrazide its Complexes. *Beni-Seuf Univ. J. Basic Appl. Sci.* 6: 332–144. https://doi.org/10.1016/j.bjbas.2017.06.001.
104. Faidallah, H. M., A. A. Saqer, K.A. Alamry, K.A. Khan, M. A. M. Zayed, S. A. Khan. 2014. Design, Synthesis and Biological Evaluation of some Novel Hexahydroquinoline-3-carbonitriles as Anticancer and Antimicrobial Agents. *Asian J. Chem.* 26, 23: 8139–8144. https://doi.org/10.14233/ajchem.2014.17616.
105. Tyagi, P., S. Chandra, B. S. Saraswat, D. Yadav. 2015. Design, Spectral Characterization, Thermal, DFT Studies and Anticancer Cell Line Activities of Co(II), Ni(II) and Cu(II) Complexes of Schiff Bases Derived from 4-amino-5-(pyridin-4-yl)-4H-1,2,4-triazole-3-thiol. *Spectrochim. Acta A Mol. Biomol. Spectrosc.* 145: 155–164. http://dx.doi.org/10.1016/j.saa.2015.03.034.
106. Matela, G. 2020. Schiff Bases and Complexes: A Review on Anti-Cancer Activity. *Anticancer Agents Med Chem.* 20, no. 16: 1908–1917. http://dx.doi.org/10.2174/1871520620666200507091207.
107. Wierełło, W. Z., D. Styburski, A. Maruszewska, K. Piorun, M. Skórka-Majewicz, M. Czerwin´ska, D. Maciejewska, I. Baranowska-Bosiacka, A. Krajewski, I. Gutowska. 2020. Bioelements in the Treatment of Burn Injuries: The Complex Review of Metabolism and Supplementation (Copper, Selenium, Zinc, Iron, Manganese, Chromium and Magnesium). *J. Trace Elem. Med. Bio.* 62: 126616. https://doi.org/10.1016/j.jtemb.2020.126616.
108. Bhattacharya, P. T., S. R. Misra, M. Hussain. 2016. Nutritional Aspects of Essential Trace Elements in Oral Health and Disease: An Extensive Review. *Scientifica* 2016: 1–12, https://doi.org/10.1155/2016/5464373.
109. Niu, M. J., Z. Li, G. L. Chang, X. J. Kong, M. Hong, Q. F. Zhang. 2015. Crystal Structure, Cytotoxicity and Interaction with DNA of Zinc (II) Complexes with o-Vanillin Schiff Base Ligands. *PloS one*, 10 no. 6: e0130922, 1–14. https://doi.org/10.1371/journal.pone.0130922.
110. Kuamr, K. S., C.P. Varma, V. N. Reena, K. K. Aravindakshan. 2017. Synthesis, Characterization, Cytotoxic, Anticancer and Antimicrobial Studies of Novel Schiff Base Ligand Derived from Vanillin and its Transition Metal Complexes. *J. Pharm. Sci. & Res.* 9, no. 8:1317–1323. https://doi.org/10.7897/2230-8407.0810201.
111. Alturiqi, A. S., Abdel-Nasser M. A. Alaghaz, R. A. Ammar, M. E. Zayed. 2018. Synthesis, Spectral Characterization, and Thermal and Cytotoxicity Studies of Cr(III), Ru(III), Mn(II), Co(II), Ni(II), Cu(II), and Zn(II) Complexes of Schiff Base Derived from 5-Hydroxymethylfuran-2-carbaldehyde. *Journal of Chemistry* 2018: 1–17. https://doi.org/10.1155/2018/5816906.
112. Kumar, K. S., V. K. Chityala, N. J. P. Subhashini, Y. Prashanthi, and Shivaraj. 2013. Synthesis, Characterization, and Biological and Cytotoxic Studies of Copper(II), Nickel(II), and Zinc(II) Binary Complexes of 3-Amino-5-methyl Isoxazole Schiff Base. *Inorganic Chemistry* 2013: 1–7. https://doi.org/10.1155/2013/562082.
113. Rajarajeswari, C., M. Ganeshpandian, M. Palaniandavar, A. Riyasdeen, M. A. Akbarsha. 2014. Mixed Ligand Copper(II) Complexes of 1,10-Phenanthroline with Tridentate Phenolate/Pyridyl/(Benz)Imidazolyl Schiff Base Ligands: Covalent vs Non-covalent DNA Binding, DNA Cleavage and Cytotoxicity. *Journal of Inorganic Biochemistry* 140: 2014, 255–268. https://doi.org/10.1016/j.jinorgbio.2014.07.016.
114. Jaividhya, P. and M. Ganeshpandian, R. Dhivya, M. A. Akbarsha, M. Palaniandavar. 2015. Fluorescent Mixed Ligand Copper(ii) Complexes of Anthracene-appended Schiff Bases: Studies on DNA Binding, Nuclease Activity and Cytotoxicity. *Dalton Trans.* 44, no. 26: 11997–12010. http://dx.doi.org/10.1039/C5DT00899A.

115. Jiao, J., M. Jiang, Y. T. Li, Z. Y. Wu, C. W. Yan. 2014. In Vitro Cytotoxic Activities, DNA-, and BSA-binding Studies of a New Dinuclear Copper(II) Complex with N-[3-(Dimethylamino)Propyl]-N'-(2-Carboxylatophenyl)-oxamide as Ligand. *Journal of Biochemical and Molecular Toxicology* 28, no. 2: 47–59. https://doi.org/10.1002/jbt.21535.
116. Song, W.-J., J.-P. Cheng, D.-H. Jiang, L. Guo, M. -F. Cai, H.-B. Yang, Q. -Y. Lin. 2014. Synthesis, Interaction with DNA and Antiproliferative Activities of Two Novel Cu(II) Complexes with Schiff Base of Benzimidazole. *Spectrochimica Acta Part A: Molecular and Biomolecular Spectroscopy* 121: 70–76. https://doi.org/10.1016/j.saa.2013.09.142.
117. Shokohi-pour, Z., H. Chiniforoshan, A. A. Momtazi-borojeni, Behrouz Notash. 2016. A Novel Schiff Base Derived from the Gabapentin Drug and Copper (II) Complex: Synthesis, Characterization, Interaction with DNA/protein and Cytotoxic Activity. *Journal of Photochemistry and Photobiology B: Biology* 162: 34–44. https://doi.org/10.1016/j.jphotobiol.2016.06.022.
118. P. Prasad, I. Pant, I. Khan, P. Kondaiah, A. R. Chakravarty. 2014. Mitochondria Targeted Photoinduced Anticancer Activity of Oxidovanadium(IV) Complexes of Curcumin in Visible Light. *Eur. J. Inorg. Chem.* 14: 2420–2431, https://doi.org/10.1002/ejic.201402001.
119. Geethalakshmi, V., N. Nalini, C. Theivarasu. 2020. Anticancer Activity of Morpholine Schiff Base Complexes. *AIP Conference Proceedings* 2270, no. 1: 100016–19. https://doi.org/10.1063/5.0024590.
120. Kaczmarek, M. T., M. Zabiszak, M. Nowak, R. Jastrzab. 2018. Lanthanides: Schiff Base Complexes, Applications in Cancer Diagnosis, Therapy, and Antibacterial Activity. *Coordination Chemistry Reviews* 370: 42–54. https://doi.org/10.1016/j.ccr.2018.05.012.
121. Kostova, I. 2005. Lanthanides as Anticancer Agents. *Current Medicinal Chemistry-Anti-Cancer Agents* 5, no. 6: 591–602. http://dx.doi.org/10.2174/156801105774574694.
122. Sathiyanarayanan, V., P.vV. Prasath, P.vC. Sekhar, K. Ravichandran, D. Easwaramoorthy, F. Mohammad, H. A. Al-Lohedan, W.vC. Oh, S. Sagadevan. 2020. Docking and In Vitro Molecular Biology Studies of p-Anisidine-appended 1-Hydroxy-2-acetonapthanone Schiff Base Lanthanum(III) Complexes. *RSC Advances* 10: 16457–16472. https://doi.org/10.1039/D0RA01936D.
123. Andiappan, K., A. Sanmugam, E. Deivanayagam, K. Karuppasamy, H. S. Kim, D. Vikraman 2018. *In vitro* Cytotoxicity Activity of Novel Schiff Base Ligand–lanthanide Complexes. *Scientific Reports* 8: 3054–65. https://doi.org/10.1038/s41598-018-21366-1.
124. Neelima, K. Poonia, S. Siddiqui, Md Arshad, D. Kumar, 2016. In vitro Anticancer Activities of Schiff Base and its Lanthanum Complex. *Spectrochimica Acta Part A: Molecular and Biomolecular Spectroscopy* 155: 146–154, https://doi.org/10.1016/j.saa.2015.10.015.
125. Nath, M., and P. K. Saini. 2011. Chemistry and Applications of Organotin(IV) Complexes of Schiff Bases. *Dalton Trans.* 40: 7077–7121. https://doi.org/10.1039/C0DT01426E.
126. Haezam, F. N., N. Awang, N. F. Kamaludin, M. M. Jotani and E. R. T. Tiekink. 2019. (*N,N*-Diisopropyldithiocarbamato)triphenyltin(IV): Crystal Structure, Hirshfeld Surface Analysis and Computational Study. *Acta Cryst.* E75: 1479–1485. https://doi.org/10.1107/S2056989019012490.
127. Tian, L., Y. Sun, B. Qian, G. Yang, Y. Yu, Z. Shang, X. Zheng. 2005. Synthesis, Characterization and Biological Activity of a Novel Binuclear Organotin Complex, $Ph_3Sn(HL){\cdot}Ph_2SnL$ [L = 3,5-Br_2-2-OC_6H_2CH=NCH(i-Pr)COO] *Appl. Organometallic Chemistry* 19, no. 10: 1127–1131. https://doi.org/10.1002/aoc.968.
128. Hong, M., H. Geng, M. Niu, F. Wang, D. Li, J. Liu, H. Yin. 2014. Organotin(IV) Complexes Derived from Schiff Base N′-[(1E)-(2-hydroxy-3-Methoxyphenyl)methylidene]pyridine-4-carbohydrazone: Synthesis, In Vitro Cytotoxicities and DNA/BSA Interaction. *European Journal of Medicinal Chemistry* 86: 550–561. https://doi.org/10.1016/j.ejmech.2014.08.070.
129. Sahoo, C. R., J. Sahoo, M. Mahapatra, D. Lenka, P. K. Sahu, B. Dehury, R. N. Padhy, S. K. Paidesetty. 2021. Coumarin Derivatives as Promising Antibacterial Agent(s). *Arabian Journal of Chemistry* 14, no. 2: 102922–102979. https://doi.org/10.1016/j.arabjc.2020.102922.
130. Xu, Z. Q. Chen, Y. Zhang, C. Liang, 2021. Coumarin-based Derivatives with Potential Anti-HIV Activity. *Fitoterapia* 150: 104863. https://doi.org/10.1016/j.fitote.2021.104863.
131. Prusty, J. S., A. Kumar. 2020. Coumarins: Antifungal Effectiveness and Future Therapeutic Scope. *Molecular Diversity* 24: 1367–1383. https://doi.org/10.1007/s11030-019-09992x.
132. Kontogiorgis C., A. Detsi, D. Hadjipavlou-Litina. 2012. Coumarin-based Drugs: A Patent Review (2008–Present). *Expert Opin Ther Pat.* 22: 437–454. https://doi.org/10.1517/13543776.2012.678835.

133. Thakur, A., R. Singla, V. Jaitak. 2015. Coumarins as Anticancer Agents: A Review on Synthetic Strategies, Mechanism of Action and SAR Studies. *European Journal of Medicinal Chemistry* 101: 476–495. https://doi.org/10.1016/j.ejmech.2015.07.010.
134. Creaven, B., M. Devereux, D. Karcz, A. Kellett, M. McCann, A. Noble, and M. Walsh. 2009. Copper(II) Complexes of Coumarin-derived Schiff Bases and Their Anti-Candida Activity. *Journal of Inorganic Biochemistry* 103: 1196–1203. http://doi.org/10.1016/j.jinorgbio.2009.05.017.
135. Creaven, B., E. Czeglédi, M. Devereux, E. A. Enyedy, A. Foltyn-Arfa Kia, D. Karcz., A. Kellett, S. McClean, Siobhán, N. V. Nagy, A. Noble, A. Rockenbauer, T. Szabó-Plánka, M. Walsh. 2010. Biological Activity and Coordination Modes of Copper(ii) Complexes of Schiff Base-derived Coumarin Ligands. *Dalton Trans*. 39, no. 45: 10854–10865. http://dx.doi.org/10.1039/C0DT00068J.
136. Şahin, O., U. O. Ozdemir, N. Seferoglu, Z. K. Genc, K. Kaya, B. Aydıner, S. Tekin, Z. Seferoglu. 2018. New Platinum (II) and Palladium (II) Complexes of Coumarin-thiazole Schiff Base with a Fluorescent Chemosensor Properties: Synthesis, Spectroscopic Characterization, X-ray Structure Determination, In Vitro Anticancer Activity on Various Human Carcinoma Cell Lines and Computational Studies. *Journal of Photochemistry and Photobiology B: Biology* 178: 428–439. https://doi.org/10.1016/j.jphotobiol.2017.11.030.
137. Mestizo. P. D., D. M. Narvaez. J. A. Pinzon-Ulloa. D. T. Di Bello. S. Franco-Ulloa. M. A. Macias, H. Groot. G. P. Miscione. L. Suescun. John J. Hurtado. 2021. Novel Complexes with ONNO Tetradentate Coumarin Schiff-base donor Ligands: X-Ray Structures, DFT Calculations, Molecular Dynamics and Potential Anticarcinogenic Activity. *Biometals* 34: 119–140. https://doi.org/10.1007/s10534-020-00268-8.
138. Ferraz de Paiva R. E., E. G. Vieira, D. Rodrigues da Silva, C. A. Wegermann, A. M. Costa Ferreira. 2021. Anticancer Compounds Based on Isatin-derivatives: Strategies to Ameliorate Selectivity and Efficiency. *Frontiers in Molecular Biosciences* 7: 1–24. https://doi.org/10.3389/fmolb.2020.627272.
139. Prakash, C. R., P. Theivendren, and S. Raja. 2012. Indolin-2-ones in Clinical Trials as Potential Kinase Inhibitors: A Review. *Pharmacology & Pharmacy*, 3: 62–71. https://doi.org/10.4236/pp.2012.31010.
140. Izzedine, H., I. Buhaescu, O. Rixe, and G. Deray. 2007. Sunitinib Malate. *Cancer Chemother. Pharmacol* 60: 357–364. https://doi.org/10.1007/s00280-006-0376-5.
141. Gomathi, R., A. Ramu, A. Murugan. 2014. Evaluation of DNA Binding, Cleavage, and Cytotoxic Activity of Cu(II), Co(II), and Ni(II) Schiff Base Complexes of 1-Phenylindoline-2,3-dione with Isonicotinohydrazide. *Bioinorganic Chemistry and Applications* 2014, Article ID 215392: 1–12. https://doi.org/10.1155/2014/215392.
142. Bulatov, E., R. Sayarova, R. Mingaleeva, R. Miftakhova, M. Gomzikova, Y. Ignatyev, A. Petukhov, P. Davidovich, A. Rizvanov and N. A. Barlev. 2018. Isatin-Schiff Base-copper (II) Complex Induces Cell Death in p53-Positive Tumors. *Cell Death Discovery* 4, 103–111. https://doi.org/10.1038/s41420-018-0120-z.

8 Advancement of Schiff Base Metal Complexes Interacting with DNA

*Mansi,[1] Pankaj Khanna,[2] and Leena Khanna[1]**
[1] University School of Basic and Applied Sciences, Guru Gobind Singh Indraprastha University, New Delhi, India
[2] Department of Chemistry, Acharya Narendra Dev College, University of Delhi, Kalkaji, New Delhi, India
*corresponding author
E-mail: mansipanghal72@gmail.com; pankajkhanna@andc.du.ac.in; leenakhanna@ipu.ac.in

CONTENTS

8.1 INTRODUCTION

Schiff bases have an imine group (-CH=N-), and are usually produced by combining amines with carbonyl compounds [1]. They can chelate with transition metal ions and form complexes with them. These compounds and their complexes are extremely important in coordination chemistry and also have biological uses. Various antibacterial, antifungal, and therapeutic actions are attributed to the presence of azomethine linkage [2,3,4,5]. The interaction of transition metal complexes with nucleic acids has got a lot of attention since it can aid the creation of novel therapies [6,7,8,9,10]. This has increased the curiosity in developing new transition metal-bound Schiff bases that can attach to duplex DNA with a high degree of selectivity. They can change the activities of DNA by attaching to it and interfering with the process of replication and protein synthesis.

An adduct is formed between metal complex and DNA, which can be maintained by a number of interactions such as stacking interactions, hydrogen bonding, van der Waals contacts, and electrostatic interactions. There are two types of binding with a nucleic acid: covalent and non-covalent. Many compounds bind to DNA covalently by generating adducts by inter- or intra-strand crosslinking through alkylation. This sort of binding is irreversible, resulting in total inhibition in DNA function and cell death. However, as illustrated in Figure 8.1, non-covalent binding of Schiff

DOI: 10.1201/9781003291459-8

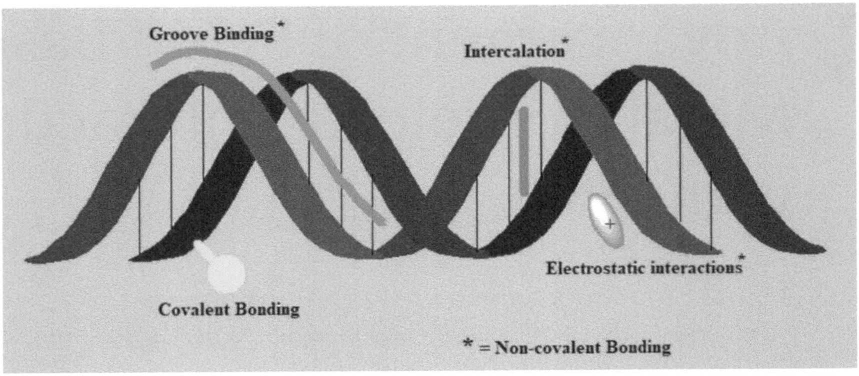

FIGURE 8.1 Types of interactions between molecules inside the DNA strand.

base complexes to DNA mostly involves intercalation between base pairs or binding to major and minor grooves via electrostatic interactions with the sugar-phosphate backbone. These interactions are often less harmful and even reversible, making them preferable to the covalent adducts. The non-covalent interactions result in structural distortion, interfering with DNA's usual processes of replication and protein synthesis. Intercalators often have planar heterocyclic moieties that stack vertically to the DNA backbone, strengthening, stiffening, unwinding, or even breaking the DNA double helix. Minor groove binders feature a concave aromatic structure and necessary topographies that aid in fitting into the convex DNA minor groove. They may contain electron-donating and electron-accepting groups capable of forming hydrogen bonds [11, 12, 13, 14, 15].

The investigation of these binding interactions has been known to employ a number of instrumental approaches and computational studies. UV-visible spectroscopy (metal complex titration with CT-DNA (calf thymus DNA)), fluorescence spectroscopy, viscosity measurements, DNA thermal denaturation investigations, and electrostatic potential measurements are some of the techniques used [16,17]. Each one of them is briefly described below.

8.2 UV-VISIBLE ABSORPTION SPECTROSCOPY STUDIES

The study is employed to investigate how DNA interacts with metal complexes. This method is based on observing changes in the UV-Visible spectrum of the either metal complex or DNA as a result of binding. Due to Π-Π^* transitions in the UV range, metal complexes show distinct absorption bands. In the visible area, they can exhibit ligand-to-metal charge transfer or d-d transitions. As a result, changes in the maxima of any of these bands can indicate a metal complex-DNA interaction, which can be further confirmed by altering nucleic acid concentrations.

Intercalation of metal complexes into DNA generally results in hypochromism with either bathochromism or hypsochromism, as shown in Figure 8.2. Groove (minor or major) binders, on the other hand, generate hyperchromism along the exterior of the DNA helix via H-bonding and electrostatic interactions. The intensity of the interaction is proportional to the magnitude of the changes in the metal complex spectrum [18,19].

Benesi-Hildebrand equation is used to calculate K_b, the intrinsic binding constant or association constant of the metal complex:

$$[DNA]/(\varepsilon A - \varepsilon F) = [DNA]/(\varepsilon B - \varepsilon F) + 1/K_b(\varepsilon B - \varepsilon F) \tag{8.1}$$

εA = extinction coefficient for observed/[complex]
εB = extinction coefficient for bound metal complex-DNA

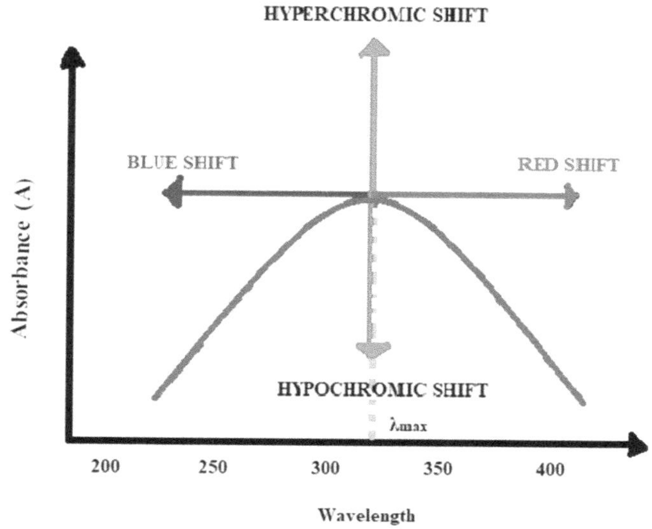

FIGURE 8.2 Type of shifts (Blue and Red shift).

εF = extinction coefficient for free metal complex
K_b = binding constant

A plot of [DNA]/(εA – εF) with [DNA] gives K_b as ratio of slope to the intercept. The values of order 10^5 M^{-1} for K_b shows good binding between DNA and metal complexes.

8.3 UV FLUORESCENCE STUDY

Fluorescence quenching may be used to investigate CT-DNA binding to metal complexes. Quenching is defined as a decrease in a sample's fluorescence intensity caused by a variety of causes. Based on their temperature dependency, it can be characterized as dynamic or static quenching. The fluorophore and quencher come into contact in dynamic quenching due to the temporary existence of the excited state, and at higher temperatures, the diffusion rate and molecule collisions are fast which results in an increase in quenching.

In static quenching fluorophore-quencher complex is formed and an increase in temperature destabilizes the complex. The fluorescence quenching spectra can be measured at different temperatures and concentrations of DNA in Tris–HCl buffer to study the mechanism [20, 21].

The fluorescence intensity of DNA spectra can be decreased by adding the quencher into the solution. Dynamic fluorescence quenching can be described by the Stern-Volmer equation:

$$F^0/F = 1 + K_q \tau^0 [Q] = 1 + K_{sv}[Q] \tag{8.2}$$

where F^0 = fluorescence intensity in the absence of quencher
F = fluorescence intensity in the presence of quencher
K_q = fluorophore quenching rate constant
K_{sv} = quenching constant, τ^0 = lifetime of the fluorophore in the absence of a quencher ($\tau^0 = 10^{-8}$s) [Q] = concentration of quencher

In static quenching, K_{sv} is decreased with increasing temperature.

EtBr can also be employed in competitive fluorescent quenching investigation, since it has a higher fluorescence when bound to DNA than free EtBr in buffer solution. The inclusion of a metal complex with a DNA affinity may result in a reduction in the EtBr-DNA adducts emission intensity. Both EtBr and ligands compete for binding sites on DNA, resulting in a conformational shift that may be evaluated by competitive fluorescence experiments. The emission spectra of the DNA-ligand complex are studied in the wavelength range of 530-680 nm and excitation at a wavelength of 500 nm is done. Subsequently, the quenching of the EtBr-DNA adduct fluorescence is analyzed, adding different metal complex concentrations to EtBr-DNA adduct fluorescence. At 590 nm fluorescence quenching can be defined by the Stern-Volmer equation:

$$I_0/I = 1 + K_{SV}[Q] = 1 + k_q\tau_0[Q] \tag{8.3}$$

where I_0 = fluorescence intensity of DNA-EtBr adduct in absence of quencher
I = fluorescence intensity of DNA-EtBr adduct in presence of quencher

The apparent binding constant K_{app} can be studied using an equation and it determines the strength of metal complex-DNA interaction:

$$K_{EtBr}[EtBr] = K_{app}[Q] \tag{8.4}$$

where $K_{EtBr} = 1 \times 10^7\ M^{-1}$, [EtBr] = EtBr concentration taken in the experiment
[Q] = quencher concentration causing a 50% decrease in initial fluorescence of EtBr-DNA

K_{app} values with a magnitude order of 10^5-10^6 M^{-1} show quite a good binding of complexes with DNA.

8.4 DNA VISCOSITY MEASUREMENT STUDY

The most straightforward method for determining the strength of metal complex-DNA interactions is to use hydrodynamic measurements [22, 23]. The interaction between the DNA strand and the ligand causes the viscosity of DNA to fluctuate with its length. An intercalative metal complex can cause base pair separation to fit into the DNA structure, lengthening the nucleic acid helix and increasing its viscosity. When binding occurs in the grooves, however, the metal complex may bend the DNA helix, resulting in a less noticeable change in DNA viscosity or no change at all. Under regulated temperature settings, specially constructed viscosimeters are employed. Data is interpreted as $(\eta/\eta_0)^{1/3}$ *vs* the ratio of the metal complex to DNA concentration.

η_0 = viscosity of free DNA
η = viscosity of DNA bound with metal complex

Viscosity values can be calculated using the observed flow time of a DNA solution (t) and corrected for the flow time of buffer alone (t_0): $\eta = t - t_0$. $(\eta/\eta_0)^{1/3}$, values can be plotted against the binding ratio r:

$$(r = [\text{Metal Complex}]/[\text{DNA}]) \tag{8.5}$$

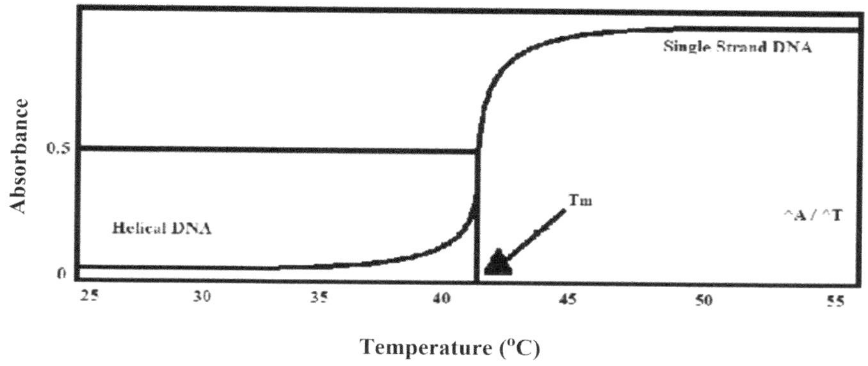

FIGURE 8.3 Thermal denaturation curve of DNA by studying UV spectroscopy.

8.5 DNA THERMAL DENATURATION STUDY

Melting temperature (Tm) is defined as the temperature at which half (50%) of DNA is denatured: that is, 50% of DNA becomes single-stranded. Tm determines the stability of double-helix DNA, which can be affected by binding with metal complexes that can disrupt its structure. These interactions may cause structural changes in DNA, increasing Tm. The degree of variance is determined by the nature and strength of the interactions [24, 25].

Tm is determined via a thermal denaturation method, and UV absorbance of the DNA strand is measured. In its UV vis absorption spectrum, DNA has a wide band at 200—350 nm with a maximum at 250–280 nm, and the absorbance is restricted due to H-bonding between base pairs. When double-helix DNA is uncoiled into single strands, the UV absorption rises (by around 40%) due to less base pair interaction. As a result, a hypochromic shift in the UV spectrum is detected, as illustrated in Figure 8.2.

Hyperchromic shift is happened during the process of DNA thermal denaturation, as shown in Figure 8.3. As the temperature rises, the uncoiling process of DNA accelerates, resulting in more single-stranded DNA and a higher maximum absorbance in the UV spectrum. The melting temperature Tm refers to the middle of the melting process, which is also the inflection point of the melting curve.

The DNA melting experiment involves recording the UV absorption spectra of CT-DNA between 200 and 1000 nm at temperatures ranging from 25 to 90°C, in the presence and absence of metal complexes, and adjusting the DNA to complex ratios. Tm can be determined as the first derivative by using the Savitsky-Golay algorithm. Also, ΔTm = Tm (DNA +complex) – TmDNA. ΔTm is the difference between the Tm value of free DNA (TmDNA) with the Tm of metal complex bound DNA (Tm (DNA +complex)).

Normally, a small ΔTm means the interaction is a groove binding type. However, a value of ΔTm over 10°C is evidence of intercalating binding mode due to stabilization of double-helical DNA by the metal complexes.

8.6 ELECTROCHEMICAL METHODS

Electrochemical methods are more sensitive and have high sensitivity and selectivity for electroactive molecules. The binding of metal complexes with DNA can be observed by the differential redox process that occurs in the presence and absence of DNA helix. The changes in the formal potential of redox couple as well as a reduction in peak current occurred due to a sudden drop in diffusion coefficient on binding with DNA [26]. Voltametric methods which involved the electrolysis of diffusion layer and measurements of current (I[A]) with the applied electrode potential (E[V]) are generally used. All DNA bases get oxidized at the glass carbon electrode and the electrochemical behavior of complexes of the solution to which DNA is added can be studied.

The valuable information provided by this method is: (a) diffusion coefficients of complex, DNA and adduct, (b) the binding affinity as binding constant (K), (c) mode of interactions, and (d) the size of the binding site at which metal complex-DNA interactions occur. The value of the binding constant (K) is determined using the following equation:

$$\log (1/[DNA]) = \log K + \log(I/I_0 - I) \quad (8.6)$$

I and I_0 are peak currents of the redox process in the presence and absence of DNA, respectively. The plot of log (1/[DNA]) versus log ($I/(I_0$-I)) gives the binding constant as the intercept. The high value of K shows intercalation, whereas a low value infers groove binding or electrostatic interactions.

8.7 MOLECULAR DOCKING STUDY ON DNA BINDING

Molecular docking has now been a pivotal computational tool for drug discovery. It helps to portray the behavior of ligand molecules at the binding site of target DNA whether it is groove binding or intercalation [3]. The ligand structure can be prepared by using ChemDraw in 2D conformation and converted to 3D structure and energy minimization can be carried out. Autodock and Schrodinger software can be used to form various binding conformations of nucleic DNA-ligand complex. These confirmations can have ionic and non-polar interactions, hydrogen bonding, and so on, which can be described in the form of binding energy (Kcal/mol). The lower the binding energy, the more stable will be the DNA-ligand complex. The binding conformation having the lowest binding energies can be visualized in the Discovery studio or PyMOL to get the DNA bases involved in binding with ligand in the complex.

Some of the recently developed Schiff base metal complexes with DNA binding potency have been summarized in this chapter, as seen in Figure 8.4 (1–20).

1. HNQ was complexed with Pd(II), Cr(III), VO(II) and Zn(II) ions (Figure 8.4(1)). Its chromium complex bound best with DNA. UV vis spectroscopy study showed K_b of Cr(III) complex was 5.40×10^5 M^{-1} with a hypochromic shift. A molecular docking study of this complex with 1BNA showed a binding energy value of -1.8-0.6 kcal/mol. All complexes due to planarity complexes intercalated well with DNA base pairs [27].
2. OVAP complexes with metals Cr(III), Fe(III), Co(II) and Cu(II) ions. Cu(II) complexes, Figure 8.4(2). All complexes bound well with DNA, Cu(II) complex has the highest K_b value of 5.46×10^5 M^{-1} and K_{sv} was 6.47×10^4 M^{-1} with a hypochromic shift in wavelength was observed in UV vis spectrum. Molecular docking with human DNA topo1 showed an internal binding energy value of -118.02 kcal/mol. Nuclease studies revealed the potency of complexes to cleave puc-19 A using oxidant H_2O_2 [28].
3. 2-methoxy-6-{(E)-[(thiophen-2-ylmethyl)imino]methyl}-phenol (HL) complexes, Figure 8.4(3) with Cu(II), Zn(II) and OV(IV) ions. Cu(II) complex bound at the minor groove of DNA with a hyperchromic shift in UV spectrum showing K_b value of 2.65×10^5 M^{-1} [29].
4. Four Schiff bases were prepared as HL1: 2-ethoxy-6-[(4-fluorophenyl)iminomethyl]phenol, HL2: 2-ethoxy-6-[(4-chlorophenylimino)methyl]phenol, HL3:2-ethoxy-6-[(4-bromophenyl) iminomethyl]Phenol and HL4:2-ethoxy-6-[(4-iodophenylimino)methyl]phenol. They have formed complexes with Cu (II) and Zn(II) ions Figure 8.4(4). A distorted tetrahedral geometry was observed for zinc(II) complexes [30]. Both Zn(II) and Cu(II) complexes showed good groove binding ability with DNA. The value of K_b for Cu(II) complex was 7.183 x 10^3

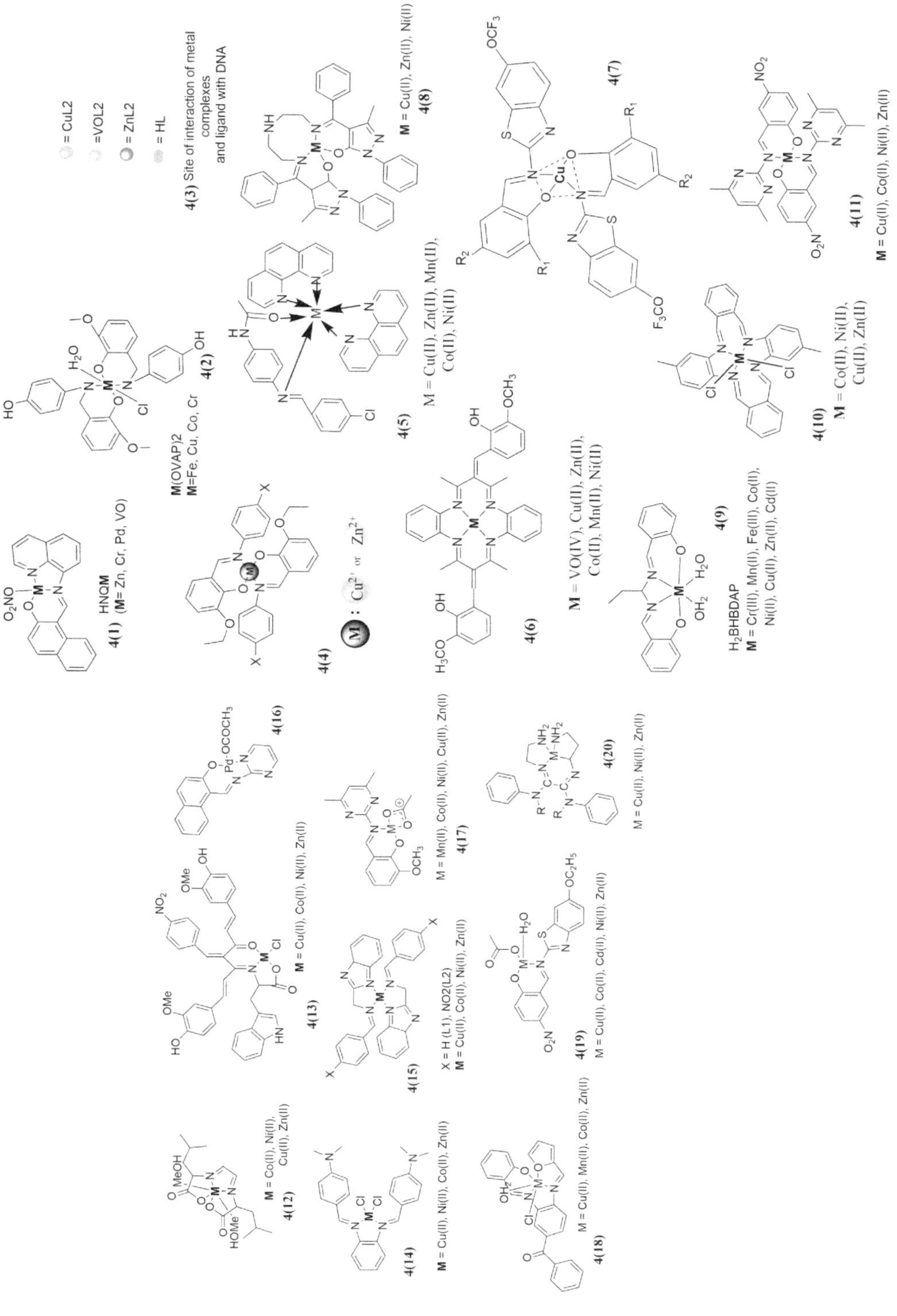

FIGURE 8.4 (1–20) Structure of Schiff-base metal complexes.

M^{-1} while of Zn(II) complex was 6.439 x 10^3 M^{-1}. A molecular docking study was performed with these complexes with the best binding energies obtained for Cu(HL4) as -7.03 kcal/mol and Zn(HL4) as -6.73 kcal/mol. The order of binding efficiency of complexes was as follows: CuHL4 > CuHL3 > ZnHL4 = ZnHL2 > CuHL1 > CuHL2 = ZnHL1 > ZnHL3.

5. N-(4-aminophenyl) acetamide derived Schiff base was complexed with metals Mn(II), Co(II), Ni(II), Cu(II), and Zn(II) ions Figure 8.4(5). All these complexes showed significant nuclease activity in the presence of oxidant H_2O_2. However, copper complexes showed the highest K_b value 1.4 x 10^6 M^{-1} for DNA binding with hypochromic shift in UV vis spectrum and acting as intercalators [31].
6. 3-(2-hydroxy-3-methoxybenzylidene)pentane-2,4-dione formed complex with Cu(II), Zn(II), VO(IV), Co(II), Mn(II) and Ni(II) ions Figure 8.4(6). Copper complexes were the best with the highest intercalation and K_b value of 1.2 x 10^4 M^{-1}. An intra-ligand π –π * transitions were observed in the electronic absorption spectrum of $[CuI]^{2+}$, recorded at λmax 375 in the absence and presence of DNA [32].
7. Copper complexes of 2-((E)-(6-(trifluoromethoxy)benzo[d]thiazol-2-ylimino)methyl)-4-methoxyphenol and 2-((E)- (6-(trifluoromethoxy)benzo[d]thiazol-2-ylimino)methyl)-4,6-dibromophenol, Figure 8.4(7) showed binding with DNA. UV spectroscopy analysis of DNA complex showed hypochromic and slight bathochromic shift with K_b and K_{sv} value of $(CuL_1)_2$ complex as 5.93 ±0.01 ×10^5 M^{-1} and 6.85 ±0.01 ×10^5 M^{-1} respectively [33].
8. 1,7-di(1-phenyl-3-methyl-5-pyrazolone-4 – benzylidene)diethenetriamine formed complex with Ni(II), Cu(II) and Mn(II) ions Figure 8.4(8). Ni(II) complexes showed a slight hypochromic shift in UV spectrum after binding with DNA and K_b value appeared as 7.893 x 10^5 M^{-1} [34].
9. N,N_0-bis(2-hydroxybenzylidene)-1,1- diaminopropane complexes with number of metal ions like Cr(III), Mn(II), Fe(III), Co(II), Ni(II), Cu(II), Zn(II) and Cd(II) ions Figure 8.4(9). On complexation, there was an increase in HOMO energy and a decrease in LUMO energy occurred leading to their stabilization. A molecular docking study was performed on the Zn(II) complex with PDB ID 2DND which showed the lowest binding energy value of –7.489 kcal/mol proving its strong binding interactions with DNA [35].
10. Schiff base dichloro[3,4,7,8,11,12,15,16-tetrabenzo-1,6,9,14-tetra-azacyclohexadecane-1,5,9,13-tetraene] complexes with Co(II), Ni(II), Cu(II) and Zn(II) ions as in Figure 8.4(10). The complexes showed minor groove binding with DNA and the binding constant follows the order as Cu(II)>Ni(II)>Co(II)> Zn(II). The complexes have octahedral geometry and steric hindrance by methyl groups prevented intercalation on binding with DNA base [36].
11. 2-(4,6-dimethylpyrimidin-2-ylimino)methyl)-4-Nitrophenol ligand was complexed with Cu(II), Co(II), Ni(II) and Zn(II) ions, as shown in Figure 8.4(11). Cu(II) complex has a good binding ability with DNA and showed groove binding via hydrophobic interaction of methyl groups and bulky ligands. The binding constants of Cu(II) complex have values: K_b as 5.61 × 10^5 M^{-1} and K_{sv} as 1.69 × 10^4 M^{-1}. Molecular docking study showed ΔG° value of -28.5 kJ/mol [37].
12. Mononuclear Schiff base formed by condensation of l-leucine and glyoxal were complexed with Co(II), Ni(II), Cu(II) and Zn(II) ions, as shown in Figure 8.4(12). Cu(II) complex have a distorted octahedral structure and bind with DNA as groove binder showing hyperchromic shift in UV spectroscopy [38].
13. Novel curcumin based Knoevenagel Schiff base complexes with Cu(II), Ni(II), Co(II) and Zn(II) ions, as in Figure 8.4(13). Hypochromic with bathochromic effect and intercalation was observed in UV spectrum of adduct formed between DNA and Cu(II) complex. The binding constant K_b was 3.4 ± 0.05 x 10^4 M^{-1}. The cleaving tendency of these complexes was quite good in the presence of H_2O_2 for PBR322 DNA [39].

14. Schiff base 4-formyl-N,N-dimethylaniline was complexed with Cu(II), Ni(II), Co(II) and Zn(II) ions, as in Figure 8.4(14). Cu(II) complex get intercalated between DNA base pairs with K_b value of 7.5×10^5 M^{-1} and showed hypochromism with red shift in UV spectrum [40].
15. (Z)-1-(1H-benzo[d]imidazol-2-yl)-N-benzylidene-methanamine (L1) and 1-(1H-benzo[d] imidazol-2-yl)-N-(4-nitrobenzylidene)methanamine(L2) were complexed with Cu(II), Ni(II), Co(II) and Zn(II) ions, as in Figure 8.4(15). Copper complex $Cu(L_2)_2$ bound well with DNA and showed intercalation with K_b value of 2.8×10^5 M^{-1}. UV spectrum of DNA complex revealed hypochromic effect in the range 8.5–26.9% and a slight red shift [41].
16. 1-(Pyrimidin-2-yliminomethyl)-naphthalen-2-ol] (HNP) was complexed with VO(II), Zn(II), Pd(II) and Cr(III) ions, as in Figure 8.4(16). Palladium complex HNPPd bound well with DNA, having highest binding constant. UV spectrum of metal complexes with HNP showed hyperchromic shift. The binding energy of -3.5 kJ/mol has been revealed by molecular modeling of HNPPd complex with PDB ID-3HB5 [42].
17. 2-[(4,6-dimethylpyrimidin-2-ylimino)methyl]-6-methoxyphenol and -OAc (Acetate) was complexed with Mn(II), Co(II), Ni(II), Cu(II) and Zn(II) ions, as seen in Figure 8.4(17). [Zn(L)(OAc)] showed highest binding energy with Ct-DNA having Kb value of 5.87×10^4 M^{-1} via groove binding mode. There was a hypochromic effect with slight red shift (2-5 nm) observed in UV spectrum. Molecular docking study displayed highest binding energy of -30.12 kJ/mol for Zn(II) complex, with PDB ID 355D. All complexes possessed square planar geometry [43].
18. 4-[(Furan-2-ylmethylene) amino]-3- [(2–hydroxy –benzylidene) amino]-phenyl}-phenyl methanone) get complexed with Mn(II), Co(II), Cu(II), Zn(II), as in Figure 8.4(18). The UV spectrum revealed hyperchromic effect and slight red shift (~2–3 nm) (15.83% for Mn(II) complex), having Kb value of $3.14 \pm 0.02 \times 10^4$ M^{-1} via intercalation binding mode. Molecular docking showed highest binding energy for Mn(II) complex of −6195.781 kcal/mol [44].
19. 2-((E)-(6-ethoxybenzo[d]thiazol-2-ylimino)methyl)-4-nitrophenol, has been complexed with Cu(II), Co(II), Ni(II), Zn(II), Cd(II) ions, as seen in Figure 8.4(19). DNA binding studies revealed a hypochromic shift for Cu(II) complex with K_b and K_{sv} value of 9.05×10^5 M^{-1} and 5.08×10^4 M^{-1} respectively. The binding with DNA involved strong $\pi - \pi*$stacking interaction [45].
20. A ligand L_2 [$C_{28}H_{20}N_6O_8$] was synthesized from condensation reaction of diethyloxalate and 3-hydroxy-4-nitrobenzaldehydenephenylhydrazine. It was then complexed with bis(ethylenediamine) Cu(II), Ni(II), and Zn(II) complexes to form $[(L_2)Cu(en)_2]Cl_2$, $[(L_2)Ni(en)_2]Cl_2$ and $[(L_2)Zn(en)_2]Cl_2$ complexes, as in Figure 8.4(20). DNA binding studies showed that Cu(II) complex having the highest binding constant with hypochromic effect and red shift via intercalation mode of binding. It was able to cleave DNA with both oxidative as well as hydrolytic processes [46].

8.8 CONCLUDING REMARKS

This chapter has briefly reviewed the application of Schiff base metal complexes as DNA binders. The major instrumental and computational techniques used to reveal the type and strength of these binding interactions has been discussed. The type of quantitative and qualitative information provided by these methods proved that Schiff base metal complexes can bind efficiently with DNA.

CONFLICT OF INTEREST

None to declare

ACKNOWLEDGMENT

The authors are thankful to Guru Gobind Singh Indraprastha University, New Delhi for providing funding under Faculty Research Grant Scheme (FRGS).

REFERENCES

(1) Mounika, K., Pragathi, A., Gyanakumari, C. Synthesis¸ Characterization and Biological Activity of a Schiff Base Derived from 3-Ethoxy Salicylaldehyde and 2-Amino Benzoic Acid and Its Transition Metal Complexes. *Journal of Scientific Research* **2010**, *2* (3), 513. https://doi.org/10.3329/jsr.v2i3.4899.

(2) Raman, N., Sakthivel, A., Rajasekaran, K. Synthesis and Spectral Characterization of Antifungal Sensitive Schiff Base Transition Metal Complexes. *Mycobiology* **2007**, *35* (3), 150. https://doi.org/10.4489/myco.2007.35.3.150.

(3) Chen, W., Li, Y., Cui, Y., Zhang, X., Zhu, H. L., Zeng, Q. Synthesis, Molecular Docking and Biological Evaluation of Schiff Base Transition Metal Complexes as Potential Urease Inhibitors. *European Journal of Medicinal Chemistry* **2010**, *45* (10), 4473–4478. https://doi.org/10.1016/j.ejmech.2010.07.007.

(4) Khan, M. I., Khan, A., Hussain, I., Khan, M. A., Gul, S., Iqbal, M., Inayat-Ur-Rahman, Khuda, F. Spectral, XRD, SEM and Biological Properties of New Mononuclear Schiff Base Transition Metal Complexes. *Inorganic Chemistry Communications* **2013**, *35*, 104–109. https://doi.org/10.1016/j.inoche.2013.06.014.

(5) Gwaram, N. S., Ali, H. M., Khaledi, H., Abdulla, M. A., Hadi, H. A., Lin, T. K., Ching, C. L., Ooi, C. L. Antibacterial Evaluation of Some Schiff Bases Derived from 2-Acetylpyridine and Their Metal Complexes. *Molecules* **2012**, *17* (5), 5952–5971. https://doi.org/10.3390/molecules17055952.

(6) Doğan, F., Ulusoy, M., Öztürk, Ö. F., Kaya, I., Salih, B. Synthesis, Characterization and Thermal Study of Some Tetradentate Schiff Base Transition Metal Complexes. *Journal of Thermal Analysis and Calorimetry* **2009**, *98* (3), 785–792. https://doi.org/10.1007/s10973-009-0205-2.

(7) Dhanaraj, C. J., Johnson, J., Joseph, J., Joseyphus, R. S. Quinoxaline-Based Schiff Base Transition Metal Complexes: Review. *Journal of Coordination Chemistry* **2013**, *66* (8), 1416–1450. https://doi.org/10.1080/00958972.2013.782008.

(8) Yamada, S. Advancement in Stereochemical Aspects of Schiff Base Metal Complexes. *Coordination Chemistry Reviews* **1999**, *190–192*, 537–555. https://doi.org/10.1016/S0010-8545(99)00099-5.

(9) Gupta, K. C., Sutar, A. K. Catalytic Activities of Schiff Base Transition Metal Complexes. *Coordination Chemistry Reviews* **2008**, *252* (12–14), 1420–1450. https://doi.org/10.1016/j.ccr.2007.09.005.

(10) Ramesh, G., Daravath, S., Ganji, N., Rambabu, A., Venkateswarlu, K., Shivaraj. Facile Synthesis, Structural Characterization, DNA Binding, Incision Evaluation, Antioxidant and Antimicrobial Activity Studies of Cobalt(II), Nickle(II) and Copper(II) Complexes of 3-Amino-5-(4-Fluorophenyl)Isoxazole Derivatives. *Journal of Molecular Structure* **2020**, *1202* (Ii), 127338. https://doi.org/10.1016/j.molstruc.2019.127338.

(11) Yarkandi, N. H., El-Ghamry, H. A., Gaber, M. Synthesis, Spectroscopic and DNA Binding Ability of CoII, NiII, CuII and ZnII Complexes of Schiff Base Ligand (E)-1-(((1H-Benzo[d]Imidazol-2-Yl)Methylimino)Methyl)Naphthalen-2-Ol. X-Ray Crystal Structure Determination of Cobalt (II) Complex. *Materials Science and Engineering C* **2017**, *75* (Ii), 1059–1067. https://doi.org/10.1016/j.msec.2017.02.171.

(12) Abdel-Rahman, L. H., Abu-Dief, A. M., Ismael, M., Mohamed, M. A. A., Hashem, N. A. Synthesis, Structure Elucidation, Biological Screening, Molecular Modeling and DNA Binding of Some Cu(II) Chelates Incorporating Imines Derived from Amino Acids. *Journal of Molecular Structure* **2016**, *1103*, 232–244. https://doi.org/10.1016/j.molstruc.2015.09.039.

(13) Mishra, M., Tiwari, K., Shukla, S., Mishra, R., Singh, V. P. Synthesis, Structural Investigation, DNA and Protein Binding Study of Some 3d-Metal Complexes with N′-(Phenyl-Pyridin-2-Yl-Methylene)-Thiophene-2- Carboxylic Acid Hydrazide. *Spectrochimica Acta – Part A: Molecular and Biomolecular Spectroscopy* **2014**, *132*, 452–464. https://doi.org/10.1016/j.saa.2014.05.007.

(14) Kavitha, B., Sravanthi, M., Saritha Reddy, P. DNA Interaction, Docking, Molecular Modelling and Biological Studies of o-Vanillin Derived Schiff Base Metal Complexes. *Journal of Molecular Structure* **2019**, *1185*, 153–167. https://doi.org/10.1016/j.molstruc.2019.02.093.

(15) Hussien, M. A., Nawar, N., Radwan, F. M., Hosny, N. M. Spectral Characterization, Optical Band Gap Calculations and DNA Binding of Some Binuclear Schiff-Base Metal Complexes Derived from 2-Amino-Ethanoic Acid and Acetylacetone. *Journal of Molecular Structure* **2015**, *1080*, 162–168. https://doi.org/10.1016/j.molstruc.2014.09.071.

(16) Ramotowska, S., Ciesielska, A., Makowski, M. What Can Electrochemical Methods Offer in Determining Dna–Drug Interactions? *Molecules* **2021**, *26* (11), 24. https://doi.org/10.3390/molecules26113478.

(17) Topală, T., Bodoki, A., Oprean, L., Oprean, R. Experimental Techniques Employed in the Study of Metal Complexes-DNA -Interactions. *Farmacia* **2014**, *62* (6), 1049–1061.

(18) Balakrishnan, C., Theetharappan, M., Kowsalya, P., Natarajan, S., Neelakantan, M. A., Mariappan, S. S. Biocatalysis, DNA–Protein Interactions, Cytotoxicity and Molecular Docking of Cu(II), Ni(II), Zn(II) and V(IV) Schiff Base Complexes. *Applied Organometallic Chemistry* **2017**, *31* (11), 1–16. https://doi.org/10.1002/aoc.3776.

(19) Syed Ali Fathima, S., Paulpandiyan, R., Nagarajan, E. R. Expatiating Biological Excellence of Aminoantipyrine Derived Novel Metal Complexes: Combined DNA Interaction, Antimicrobial, Free Radical Scavenging Studies and Molecular Docking Simulations. *Journal of Molecular Structure* **2019**, *1178*, 179–191. https://doi.org/10.1016/j.molstruc.2018.10.021.

(20) Ylldlz, M., Tan, E., Demir, N., Ylldlrlm, N., Ünver, H., Kiraz, A., Mestav, B. Synthesis and Spectral, Antimicrobial, Anion Sensing, and DNA Binding Properties of Schiff Base Podands and Their Metal Complexes. *Russian Journal of General Chemistry* **2015**, *85* (9), 2149–2162. https://doi.org/10.1134/S1070363215090200.

(21) Sumalatha, V., Daravath, S., Rambabu, A., Ramesh, G. Shivaraj. Antioxidant, Antimicrobial, DNA Binding and Cleavage Studies of Novel Co(II), Ni(II) and Cu(II) Complexes of N, O Donor Schiff Bases: Synthesis and Spectral Characterization. *Journal of Molecular Structure* **2021**, *1229* (Ii), 129606. https://doi.org/10.1016/j.molstruc.2020.129606.

(22) Jadoo, B., Booysen, I. N., Akerman, M. P., Rhyman, L., Ramasami, P. Novel Coumarin Rhenium(I) and -(V) Complexes: Formation, DFT and DNA Binding Studies. *Polyhedron* **2018**, *144*, 107–118. https://doi.org/10.1016/j.poly.2018.01.017.

(23) Dehkhodaei, M., Sahihi, M., Amiri Rudbari, H., Momenbeik, F. DNA and HSA Interaction of Vanadium (IV), Copper (II), and Zinc (II) Complexes Derived from an Asymmetric Bidentate Schiff-Base Ligand: Multi Spectroscopic, Viscosity Measurements, Molecular Docking, and ONIOM Studies. *Journal of Biological Inorganic Chemistry* **2018**, *23* (2), 181–192. https://doi.org/10.1007/s00775-017-1505-9.

(24) Selvarani, V., Annaraj, B., Neelakantan, M. A., Sundaramoorthy, S., Velmurugan, D. Synthesis, Characterization and Crystal Structures of Copper(II) and Nickel(II) Complexes of Propargyl Arm Containing N2O2 Ligands: Antimicrobial Activity and DNA Binding. *Polyhedron* **2013**, *54*, 74–83. https://doi.org/10.1016/j.poly.2013.02.030.

(25) Reddy, P. R., Shilpa, A., Raju, N., Raghavaiah, P. Synthesis, Structure, DNA Binding and Cleavage Properties of Ternary Amino Acid Schiff Base-Phen/Bipy Cu(II) Complexes. *Journal of Inorganic Biochemistry* **2011**, *105* (12), 1603–1612. https://doi.org/10.1016/j.jinorgbio.2011.08.022.

(26) Krishnamoorthy, P., Sathyadevi, P., Cowley, A. H., Butorac, R. R., Dharmaraj, N. Evaluation of DNA Binding, DNA Cleavage, Protein Binding and in Vitro Cytotoxic Activities of Bivalent Transition Metal Hydrazone Complexes. *European Journal of Medicinal Chemistry* **2011**, *46* (8), 3376–3387. https://doi.org/10.1016/j.ejmech.2011.05.001.

(27) Aljohani, F. S., Abu-Dief, A. M., El-Khatib, R. M., Al-Abdulkarim, H. A., Alharbi, A., Mahran, A., Khalifa, M. E., El-Metwaly, N. M. Structural Inspection for Novel Pd(II), VO(II), Zn(II) and Cr(III)-Azomethine Metal Chelates: DNA Interaction, Biological Screening and Theoretical Treatments. *Journal of Molecular Structure* **2021**, *1246*, 131139. https://doi.org/10.1016/j.molstruc.2021.131139.

(28) Bengi, K., Maddikayala, S., Pulimamidi, S. R. DNA Binding, Cleavage, Docking, Biological and Kinetic Studies of Cr(III), Fe(III), Co(II) and Cu(II) Complexes with Ortho-Vanillin Schiff Base Derivative. *Applied Organometallic Chemistry* **2022**, *36* (1), 1–18. https://doi.org/10.1002/aoc.6451.

(29) Rodríguez, M. R.; Lavecchia, M. J., Parajón-Costa, B. S., González-Baró, A. C., González-Baró, M. R., Cattáneo, E. R. DNA Cleavage Mechanism by Metal Complexes of Cu(II), Zn(II) and VO(IV) with a Schiff-Base Ligand. *Biochimie* **2021**, *186*, 43–50. https://doi.org/10.1016/j.biochi.2021.04.002.

(30) Kargar, H., Behjatmanesh-Ardakani, R., Torabi, V., Kashani, M., Chavoshpour-Natanzi, Z., Kazemi, Z., Mirkhani, V., Sahraei, A., Tahir, M. N., Ashfaq, M., Munawar, K. S. Synthesis, Characterization, Crystal Structures, DFT, TD-DFT, Molecular Docking and DNA Binding Studies of Novel Copper(II) and Zinc(II) Complexes Bearing Halogenated Bidentate N,O-Donor Schiff Base Ligands. *Polyhedron* **2021**, *195*, 114988. https://doi.org/10.1016/j.poly.2020.114988.

(31) Mahalakshmi, R., Raman, N. Enthused Research on DNA-Binding and DNA-Cleavage Aptitude of Mixed Ligand Metal Complexes. *Spectrochimica Acta – Part A: Molecular and Biomolecular Spectroscopy* **2013**, *112*, 198–205. https://doi.org/10.1016/j.saa.2013.04.054.

(32) Sakthivel, A., Thangagiri, B., Raman, N., Joseph, J., Guda, R., Kasula, M., Mitu, L. Spectroscopic, SOD, Anticancer, Antimicrobial, Molecular Docking and DNA Binding Properties of Bioactive VO(IV), Cu(II), Zn(II), Co(II), Mn(II) and Ni(II) Complexes Obtained from 3-(2-Hydroxy-3-Methoxybenzylidene)Pentane-2,4-Dione. *Journal of Biomolecular Structure and Dynamics* **2021**, *39* (17), 6500–6514. https://doi.org/10.1080/07391102.2020.1801508.

(33) Rambabu, A., Daravath, S., Shankar, D. S., Shivaraj. DNA-Binding, -Cleavage and Antimicrobial Investigation on Mononuclear Cu(II) Schiff Base Complexes Originated from Riluzole. *Journal of Molecular Structure* **2021**, *1244*, 131002. https://doi.org/10.1016/j.molstruc.2021.131002.

(34) Wang, Y., Yang, Z. Y. Synthesis, Characterization and DNA-Binding Properties of Three 3d Transition Metal Complexes of the Schiff Base Derived from Diethenetriamine with PMBP. *Transition Metal Chemistry* **2005**, *30* (7), 902–906. https://doi.org/10.1007/s11243-005-6298-y.

(35) Alaghaz, A. N. M. A., El-Sayed, B. A., El-Henawy, A. A., Ammar, R. A. A. Synthesis, Spectroscopic Characterization, Potentiometric Studies, Cytotoxic Studies and Molecular Docking Studies of DNA Binding of Transition Metal Complexes with 1,1-Diaminopropane-Schiff Base. *Journal of Molecular Structure* **2013**, *1035*, 83–93. https://doi.org/10.1016/j.molstruc.2012.09.032.

(36) Shakir, M.' Khanam, S., Azam, M., Aatif, M., Firdaus, F. Template Synthesis and Spectroscopic Characterization of 16-Membered [N4] Schiff-Base Macrocyclic Complexes of Co(II), Ni(II), Cu(II), and Zn(II): In Vitro DNA-Binding Studies. *Journal of Coordination Chemistry* **2011**, *64* (18), 3158–3168. https://doi.org/10.1080/00958972.2011.615394.

(37) Revathi, N., Sankarganesh, M., Rajesh, J., Raja, J. D. Biologically Active Cu(II), Co(II), Ni(II) and Zn(II) Complexes of Pyrimidine Derivative Schiff Base: DNA Binding, Antioxidant, Antibacterial and In Vitro Anticancer Studies. *Journal of Fluorescence* **2017**, *27* (5), 1801–1814. https://doi.org/10.1007/s10895-017-2118-y.

(38) Shakir, M., Shahid, N., Sami, N., Azam, M., Khan, A. U. Synthesis, Spectroscopic Characterization and Comparative DNA Binding Studies of Schiff Base Complexes Derived from l-Leucine and Glyoxal. *Spectrochimica Acta – Part A: Molecular and Biomolecular Spectroscopy* **2011**, *82* (1), 31–36. https://doi.org/10.1016/j.saa.2011.06.035.

(39) Chandrasekar, T., Raman, N. Exploration of Cellular DNA Lesion, DNA-Binding and Biocidal Ordeal of Novel Curcumin Based Knoevenagel Schiff Base Complexes Incorporating Tryptophan: Synthesis and Structural Validation. *Journal of Molecular Structure* **2016**, *1116*, 146–154. https://doi.org/10.1016/j.molstruc.2016.02.102.

(40) Packianathan, S., Raman, N. *Stimulated DNA Binding by Metalloinsertors Having the 4-Formyl-N,N-Dimethylaniline Schiff Base: Synthesis and Characterization*; Elsevier B.V., 2014; Vol. 45. https://doi.org/10.1016/j.inoche.2014.04.004.

(41) Kumaravel, G., Raman, N. A Treatise on Benzimidazole Based Schiff Base Metal(II) Complexes Accentuating Their Biological Efficacy: Spectroscopic Evaluation of DNA Interactions, DNA Cleavage and Antimicrobial Screening. *Materials Science and Engineering C* 2017, *70*, 184–194. https://doi.org/10.1016/j.msec.2016.08.069.

(42) Abu-Dief, A. M., El-khatib, R. M., Aljohani, F. S., Alzahrani, S. O., Mahran, A., Khalifa, M. E., El-Metwaly, N. M. Synthesis and Intensive Characterization for Novel Zn(II), Pd(II), Cr(III) and VO(II)-Schiff Base Complexes; DNA-Interaction, DFT, Drug-Likeness and Molecular Docking Studies. *Journal of Molecular Structure* **2021**, *1242*, 130693. https://doi.org/10.1016/j.molstruc.2021.130693.

(43) Senthilkumar, G. S., Sankarganesh, M., Dhaveethu Raja, J., Sakthivel, A., Vijay Solomon, R., Mitu, L. Novel Metal(II) Complexes with Pyrimidine Derivative Ligand: Synthesis, Multi-Spectroscopic, DNA Binding/Cleavage, Molecular Docking with DNA/BSA, and Antimicrobial Studies. *Monatshefte fur Chemie* **2021**, *152* (2), 251–261. https://doi.org/10.1007/s00706-021-02737-3.

(44) Aldulmani, S. A. A. Spectral, Modeling, Dna Binding/Cleavage and Biological Activity Studies on the Newly Synthesized 4-[(Furan-2-Ylmethylene)Amino]-3-[(2-hydroxy-benzylidene)Amino]-Phenyl}-Phenyl-Methanone and Some Bivalent Metal(II) Chelates. *Journal of Molecular Structure* **2021**, *1226*, 129356. https://doi.org/10.1016/j.molstruc.2020.129356.

(45) Rao, N. N., kishan, E., Gopichand, K., Nagaraju, R., Ganai, A. M., Rao, P. V. Design, Synthesis, Spectral Characterization, DNA Binding, Photo Cleavage and Antibacterial Studies of Transition Metal Complexes of Benzothiazole Schiff Base. *Chemical Data Collections* **2020**, *27* . https://doi.org/10.1016/j.cdc.2020.100368.

(46) Raman, N., Selvan, A., Sudharsan, S. Metallation of Ethylenediamine Based Schiff Base with Biologically Active Cu(II), Ni(II) and Zn(II) Ions: Synthesis, Spectroscopic Characterization, Electrochemical Behaviour, DNA Binding, Photonuclease Activity and in Vitro Antimicrobial Efficacy. *Spectrochimica Acta – Part A: Molecular and Biomolecular Spectroscopy* **2011**, *79* (5), 873–883. https://doi.org/10.1016/j.saa.2011.03.017.

9 Schiff Base Metal Complexes in Nano-Medicines

Anuradha[*1] *and Jagvir Singh*[2]
[1]Department of Zoology, Raghuveer Singh Government P.G. College-Lalitpur, Uttar Pradesh, India
[2]Department of Chemistry, Nehru Mahavidyalaya-Lalitpur, Uttar Pradesh, India
*E-mail: singhjagvir0143@gmail.com

CONTENTS

9.1 INTRODUCTION

The astral world is indeed completely different and fascinating, but if we explore beyond the astral scale, there is a deep and relatively unknown world beyond which the human eye cannot see [1–5]. Beyond the microscopic scale, that is, at a nanoscopic level, a billion times smaller than the average scale we are working on today. This level is the manipulation of atoms and molecules, which means the actual international science and technology at the nanoscale. Nanotechnology, that is, the technological know-how of producing miniature, is very small [6, 7]. Measuring its size in nanotechnology is difficult to imagine given how small it is. This has main benefits as well as capability dangers [8–12].

Nanotechnology can create many new substances and gadgets with a great variety of programs; however, it raises a few of the same troubles as any new technology [13, 14]. The smallest form of computation is considered to be atomic, whereas many physical systems are created by studying substances at the molecular and atomic level, which is called nanotechnology. In brand-new technology, nanotechnology has become an essential part of our lifestyles. even though this generation has been found in our midst since the early instances. In the last few decades, studies have started on this in its simplest form [15–19]. It would no longer be a stretch to refer to the 21st century

DOI: 10.1201/9781003291459-9

as the nano century since nanotechnology has evolved along with technological advancement. In today's scenario, large-scale research is being done all over the world regarding nanotechnology and its medicinal use [20–22]. Nanotechnology has immense potential in many fields such as electronics, medicine, auto, bioscience, petroleum, forensics, and defense, due to its microscopic size, unmatched strength, and durability [23, 24].

Nanotechnology is a type of applied science (Figure 9.1) that we can consider to be the set of all the techniques and related sciences used and studied on a scale from 1 to 100 nano (i.e. 10^{-9} m or a billionth of a meter) [25–30]. It can additionally be called the engineering of molecules and atoms. This generation combines physical, chemistry, biotechnology, and information technology. Nanotechnology is the technology whereby any molecule can be modified at the atomic, molecular and supramolecular stages. It includes debris starting from one to one hundred nanometers. Pharmaceutical nanotechnology has furnished extra pleasant-tuned analysis and targeted remedy of ailment at a molecular degree [31, 32]. Pharmaceutical nanotechnology gives numerous opportunities to combat many diseases, inclusive of most cancers, diabetes mellitus, neurodegenerative illnesses [33–35].

These compounds play very important roles in diverse biological structures and are useful in applications in industries such as polymers, dyes, medicine, agriculture and industry [36, 37]. They're also used as analytical or separation reagents. The most important uses and applications of such compounds are due to their early accessibility and structural diversity. These compounds have been considered models of biological systems. Their biological activity is one of the fundamental bases for inorganic biochemistry [38, 39]. These ligands are derived from one-of-a-kind resources and their metal complexes show antibacterial, antifungal, antitumor, and antiviral activities and they work also as therapeutic retailers toward biological disorders like cancer, infection, and hypersensitive reaction. Schiff bases are pronounced to make contributions to dyes and polymer industries [40, 42]. Schiff bases with their complexes have become the most effective antimalarial agent in a few of the different synthesized bases [43, 44]. Bacteria is the foundation reason for many infectious illnesses and is accountable for the growth in mortality rate.

Walaa H. Mahmoud, Ahmed M. Refaat, and Gehad G. Mohamed (2021) synthesized a series of naphthalene-1,8-diylbis azanylylidene-bis-methanylylidene dibenzaldehyde containing complexes

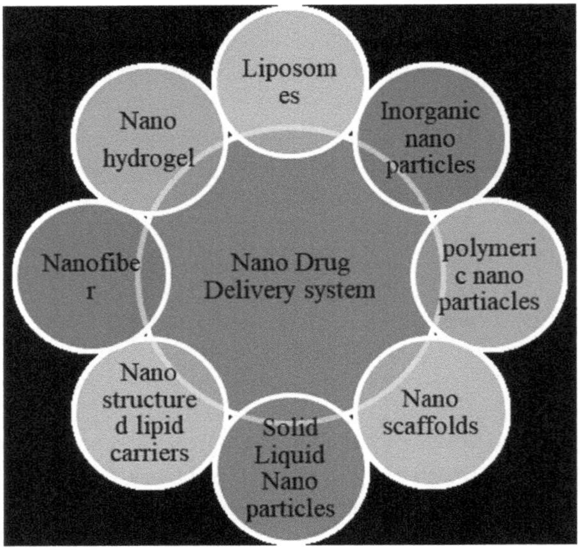

FIGURE 9.1 Different fields of application of nano drug delivery system.

[45]. They have taken 1,8- diaminonaphthalene (1): phthalaldehydic acid (2) molar ratio for preparation of such nano Schiff base. Using the different physicochemical and techniques they characterized the nano Schiff base with transition metal complexes. The newly synthesized compounds were biologically active against S. aureus (PDB ID: 3Q8U), Streptococcus pneumoniae (4YW4), and Escherichia coli (PDB ID: 3T88). They evaluated anticancer activity mutant oxidoreductase (PDB ID: 3HB5) and reported that IC_{50} of the cadmium complex was 17.6 mg/mL which was the highest activities to overall nano metal complexes [46].

Rashad et al. (2020) used the reflux and sonochemical methods for the preparation of nano Schiff base by 2-((pyridin-3-ylmethylene) amino) phenol and their nickel-metal complexes [47]. This nickel oxide was in the form of nanopowder with cubic-like microstructure obtained from nickel-nano Schiff complexes series at 650°C for 2 h after calcination. These complexes have potent activities against breast carcinoma (MCF-7) and these values were observed for IC50 of Ni(II) metal complexes 5.5, 12.5, and 9.6. The observed magnetic behavior was paramagnetic in nature at room temperature and magnetization value 0.47 emu/g and coercive force data were 42.68 Oe [48].

Parsaee et al. (2016) proposed a novel nano-sized binuclear hexadentate imine Schiff base ligand and its nickel (II) complexes, by using an ultrasonic method 2-amino-3,5-dibromobenzaldehyde, and 5-bromo-2-hydroxyaniline to synthesize nano Schiff base ligand [49]. The investigation showed that these newel compounds had good protection activities against antimicrobial organisms like gram-positive and gram-negative bacteria.

Firuzabadi et al. (2018) synthesized and characterized novel nano uranyl Schiff base ligand and its complexes. Diamines and 2-hydroxy-1-naphthaldehyde were used for synthesis, utilizing the reflection way of nano Schiff base. All the reactions of metal complexes showed quasi reversible redox reactions as their electrochemistry [50]. These compounds have also investigated their anticancer activities against cancer cell lines and metal complexes were very active in that case. Kannaiyan et al. (2021) used greenway for prepared the nano Schiff base and its nanosilver particles. It was an efficient method to develop phenothiazinium Schiff base ligand and its nanosilver with the help of the white part of the egg (lysozyme) as a catalyst. UV, IR, XRD, and NMR spectral and biological tested against gram-positive and gram-negative bacteria were used for characterization of newly synthesized nano Schiff base, nanosilver, and its metal complexes. The resultant investigations showed that the nanosilver complexes exhibit very good enhanced antimicrobial activity [51].

Prafullya Kumar Mudi and et al. (2021) reported novel hydrazine which was synthesized through condensation reaction. The head-tail interlocking aromatic rings of hydrazine ligand were used for the development of nano-aggregates with blue emission [52]. The monomeric and aggregated forms of nano Schiff base ligand photophysical studied have been searched out. The aggregate form of Schiff base exhibited blue shift with about twenty-fold high intensity of fluorescence, as compared to the monomeric form, 0.99 ns J-type aggregation developed [53].

Swami et al. (2021) has studied Schiff base and nanosilver nanoparticles and used them as qualitative for the detection of fluoride. These compounds sensing for fluoride were tested through UV-visible, FSEM, and FSEM-EDAX spectral strategies [54]. Tohidiyan et al. (2017) synthesized a tridentate 5-bromo-2-hydroxybezenylideneamino) methyl)-four-bromophenol novel Schiff base with Cu(II) complexes characterized through FT-IR, UV-Vis, FESEM, X-ray diffraction, DFT and molar conductivity. Schiff base complexes are of special interest in the field of medicine and are discussed in detail with their applications [55].

9.2 SCHIFF'S BASE LIGAND

Schiff bases are compounds found through Hugo Schiff; depending on their structure, they can be positioned within the imine group, being either ketamine or aldimine. The imine is additionally called azomethine refers principally to aldimine (Scheme 1). Schiff bases are compounds derived from amino and carbonyl teams that are getting used as ligands to make coordination complexes with

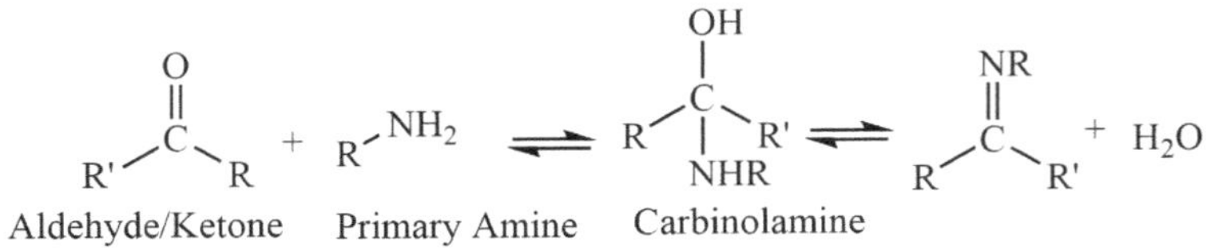

SCHEME 9.1 Synthesis Route of Schiff's Base Ligand with the help of hydrolysis reaction between aldehydic or ketonic and amine-containing compounds.

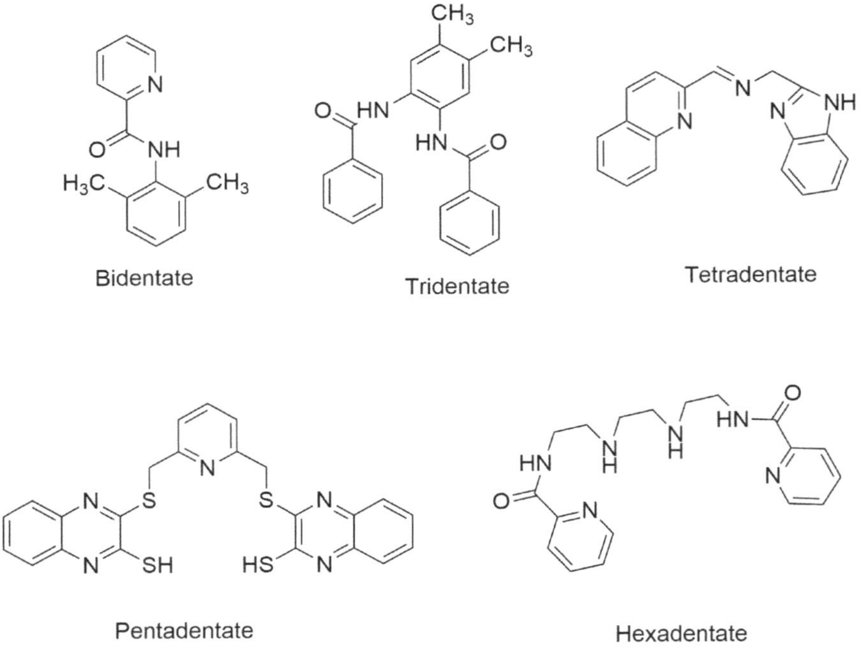

FIGURE 9.2 Different types of Schiff's base ligands.

metal ions [56, 57]. The azomethine group of Schiff base coordinates with metal ions. The Schiff base reaction recreates an important role in the appearance of >C=N- bonds in organic synthesis. Schiff base ligands have the property of chelation. These ligands contain oxygen, nitrogen, and sulphur donor atoms to form metal complexes. The presence of the imine cluster in the Schiff base is liable for several biological properties. They're getting used as chemical compound stabilizers, corrosion inhibitors, dyes, pigments, catalysts, antioxidants, antitumor activity, and antimicrobials [58].

Due to their biological chemistry, clinical and pharmacologic properties, they're helpful within the diagnosis of varied varieties of pathogens and tumors.

9.2.1 Types of Schiff Base Ligands

The main basis of the classification can be explained on the basis of its bonding groups or its dental types (Figure 9.2). Typically, these are classified as bidentate, tridentate, tetradentate, or polydentate ligands as shown in Figure 9.1. Their second basis of classification can be that according to the type of molecules that these ligands are made of, they take their classification, such as Pyridial, Chitosan type [59].

The molecular or ionic species that are permanently attached to the central metal atom or ion in a hybrid compound is called a ligand. For example, the CN^- ion in $K_4[Fe(CN)_6]$ is the ligand because it is directly attached to the central Fe^{2+} ion in the hybrid. $[Cu(NH_3)_4]^{2+}$ is another hybrid ion in which the Cu^{2+} ion is attached to four NH_3 ligands whilst a polyvalent ligand attaches itself to an important steel ion via two or extra donor atoms in one of these manners that a five or six membered cycle is fashioned with the critical ion, and this impact is called the chelate impact [60]. The chelates provide stability to the hybrid compound that complexes increase their affinity with DNA and increase its biomedical uses.

9.2.1.1 Nanotube Based Schiff Base

A carbon nanotube is a cylindrical molecule made of ring strips of one-layered carbon molecules (graphene). Their diameter is smaller than 1 nm. These compounds are 1/6 lighter than the metal, however, these are 400 times ductile in nature [61]. They have been first created separately in 1993 by using Ijima, Ichihashi, Bethune, and their team. Carbon nanotubes can be used as a frame by companies to store drugs and antigens as fillers. The main role of carbon nanotubes in medicine and therapy is to transport capsules, biomolecules, genes to cells or organs, aid in tissue regeneration, and test and analyze as biochemical sensors. Carbon nanotubes are used for the unique transportation of chemical substances and capsules inside the remedy of most cancer tumors [62, 63]. In the case of diabetes, carbon nanotubes are tremendously touchy and accurate in measuring glucose ranges because of the availability of high electrochemical surface place, excessive electrical conductivity, and useful structural residences, and they can be used as blood capillaries. Carbon nanotubes are often biodegradable and these may be decomposed into graphene and its different derivatives utilizing bacteria, fungi, and different types of microorganisms.

Daraie et al. (2021) reported halloysite-based Schiff base consisted of immobilization of copper iodide nanoparticles. Click reaction is the type of organic reaction where Copper (I) nanoparticles are used as catalysts. Amiri et al. (2011) have reported nano Schiff base and cobalt-salophen complexes and investigated them as drugs. The results of cyclic voltammetry and polarization were irreversibly oxidized into a one-electron system. These novel Schiff bases and their complexes have been found applicable and used as mediators. These compounds have an advantage due to the long-time stability of their renewability and reproducibilities as their special characters [64–66].

9.2.1.2 Pyridyl Schiff Bases

Pyridine is an organic compound that belongs to any of a category of fragrant heterocyclic compounds containing five carbon atoms and one nitrogen atom (Scheme 2). Especially the simplest one, C_5H_{5n} while the pyridyl is any of three isomeric dissimilar radicals, C_5H_{4n}-, derived from pyridine; corresponding to phenyl. Pyridine is used to synthesize coordination compounds and use them as precursors to pharmaceuticals [67–69]. Biologically it's also a vital solvent and reagent which performs a vital position in curing diseases by means of being appropriate for the formation of various ligands.

These ligands include pyridylimines and their derivatives such as 2,6-diiminopyridines, hydrazines, thiosemicarbazones, and other associated ligands have been synthesized in the past at nano scale [70–73]. The material is synthesized with non-transition and transition metals such as 3d, 4d, and 5d transition series [74]. Pyridyl 2-imine with trimethylaluminum gives pyridyl amide series compounds as **1**. In sharp comparison, 2-(1-(2,6-R2-phenylamino)propyl)quinoline-8-ols form mononuclear as **2** and ε-caprolactone alongside dinuclear as **3**. Stiba-alkene is type **4** and trimethylaluminum yields as **5**. Triethyl indium is **6**, 2-Pyridinecarboxaldehyde and pyridine carboxylic acid is as **7** (Figure 9.3).

Li et al. (2013) have synthesized Mn (II) metal complexes containg *o*-hydroxyl-aromatic amines and 2,6-pyridinedicarbaldehyde. The catalitic activities were tested in cyclohexene autoxidation and were found to be good catalysts for epoxidation of cyclohexene.

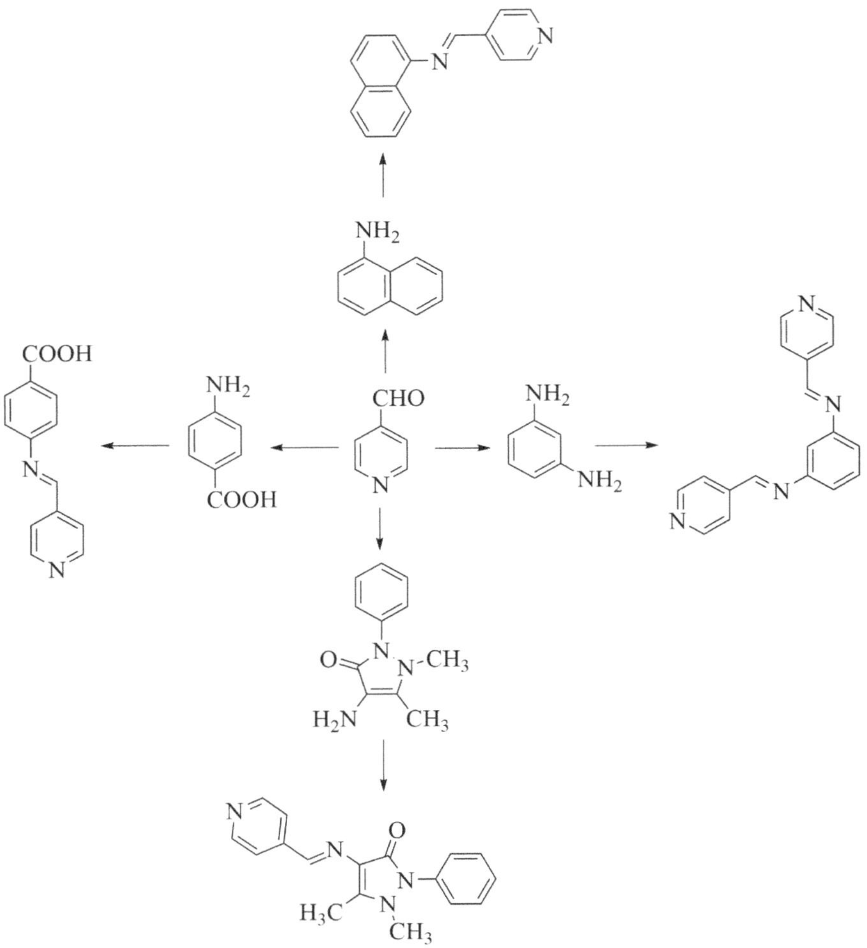

SCHEME 9.2 Synthesis route of Schiff bases of pyridine-4-carbaldehyde.

9.2.1.3 Chitosan Schiff Bases

Natural cellulose or fiber has very comparable properties with human fibrin, which is a part of blood coagulability. "Chitosan" is capable of suppressing most cancer cells; it regulates the pH in the body, which prevents the spread of metastases [75–77].

"Chitosan" is a drug that may lower blood strain, improve micro function in tissues, control sugar levels in urine, adsorb, and put off heavy metals from body salts [78–80]. It promotes rapid healing of burns and wound surfaces, leaving no scarring, and has anesthetic and hemostatic effects (Figure 9.4).

9.3 METAL COMPLEXES

These ligands combine with metal ions to form a ring-type structure called a chelate compound. These compounds are used in a variety of medical procedures known as chelation therapy. Ring-type-based ligands or Schiff bases are used in medical processes to remove polluted and harmful metal ions from the human body [81, 82]. These chelating agents prevent metal ions from reacting in the body by forming stable coordination complexes with metal ions and also prevent these metal

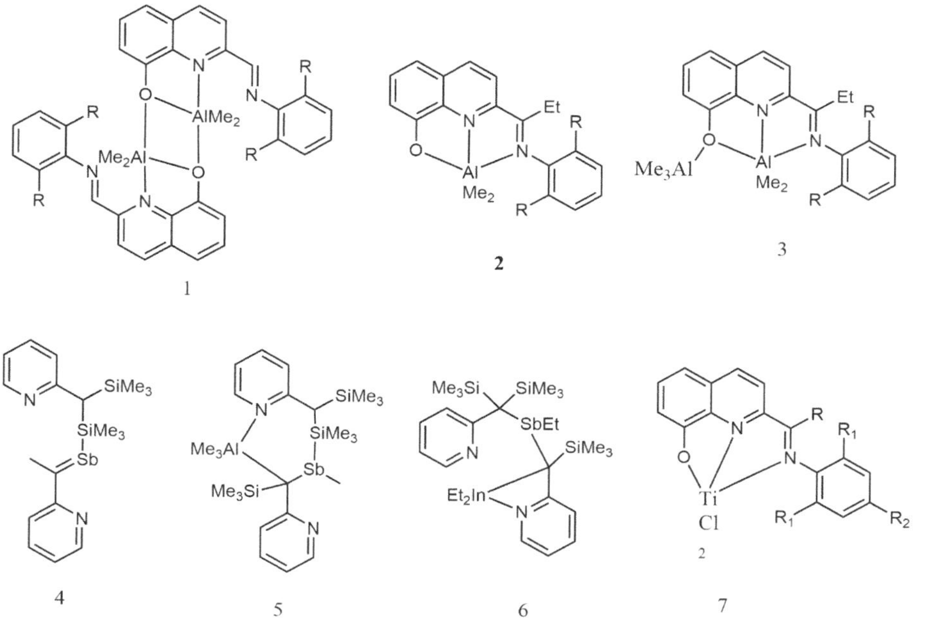

FIGURE 9.3 Different types series of pyridyl Schiff bases and their metal complexes where R = H, halogens Me, NO_2, n-Bu, Ph.

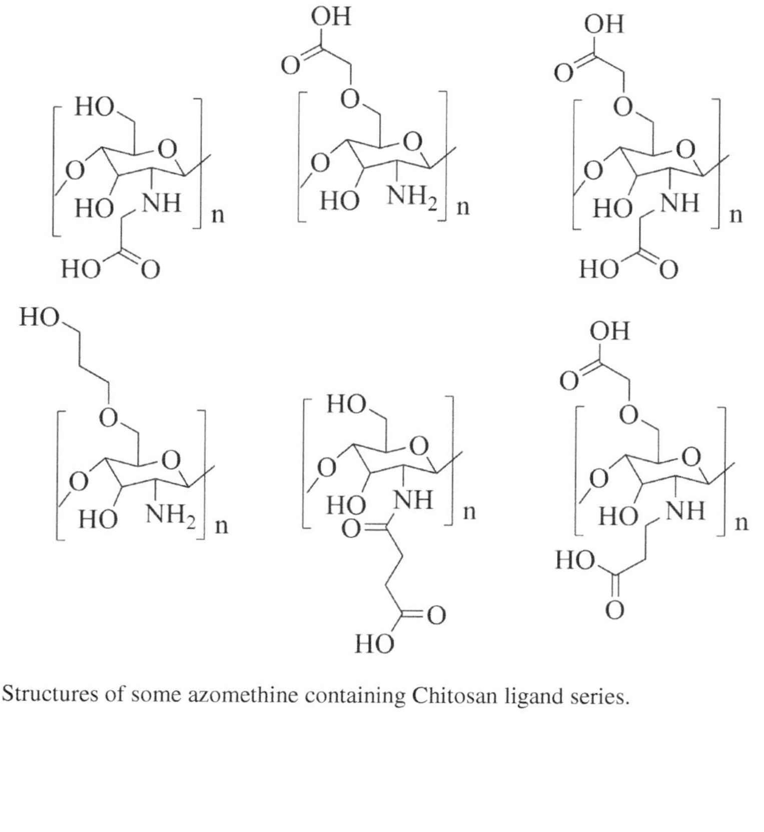

FIGURE 9.4 Structures of some azomethine containing Chitosan ligand series.

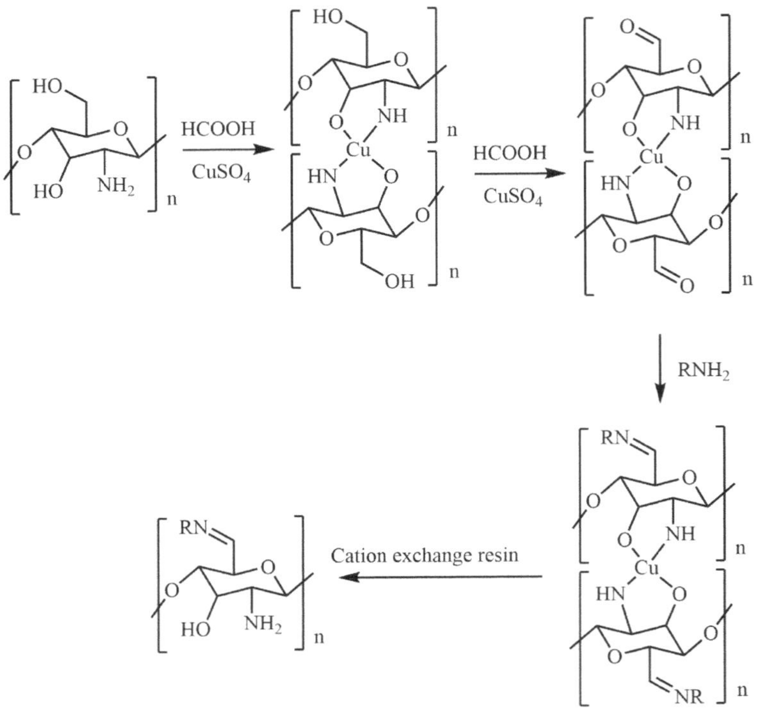

SCHEME 9.3 Modified chitosans used for the synthesis of water-soluble chitosan-based Schiff base.

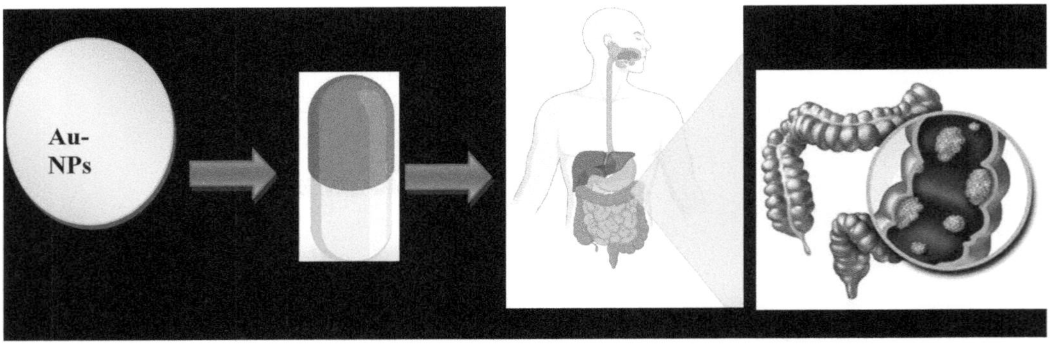

FIGURE 9.5 Application of Schiff base containing Gold nanoparticles in the treatment of colorectal tumor development cell in human.

ions from dissolving in human blood. It is known by the type of poisoning: Arsenic Poisoning, Beryllium Poisoning, Cadmium Poisoning, Copper Poisoning, Iron Poisoning, Lead Poisoning, Mercury Poisoning. They are used as a diagnosis for cancer and similar life-threatening diseases (Figure 9.5).

9.4 MEDICINAL APPLICATION OF NANO SCHIFF BASE

The nano Schiff base complexes hold the greatest interest and are attractive in the field of drugs resistance. Some are as follows.

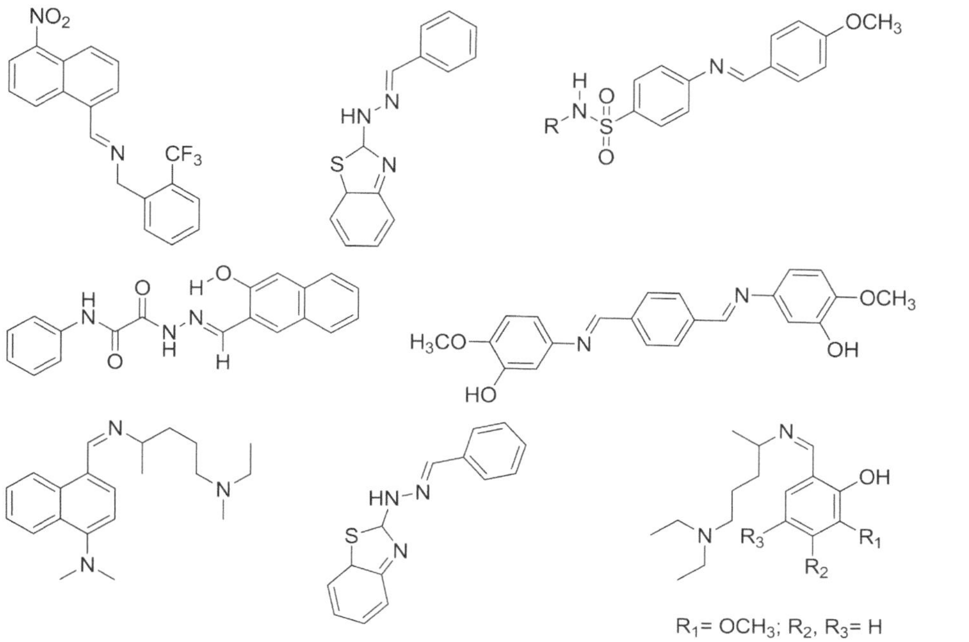

FIGURE 9.6 Structures of Schiff base and their metal complexes have been shown potent Antimalarial activity.

9.4.1 Anti-Malaria Activities

Malaria is a challenge for low-income countries to the health of their people. It is a tropical disease that is spread by mosquitoes and can be fatal if it is not diagnosed and treated early. Deblina Roy and et al. have evaluated the newly synthesized quinoline-imidazole containing nano Schiff base and its derivatives and found that these were potent against ant-malarial activities. The quinoline ring of these derivatives shows minimal cytotoxicity, high selectivity, and index and improves the antimalarial activity [83]. Karnatak and et al. (2022) synthesized hydrazone derivatives of N-amino-11-azaartemisinin (Figure 9.6) and tested their multidrug-resistant power against antimalarial activities. Plasmodium yoelii was chosen in a Swiss mice model and it was found to provide hydrazine compounds via an oral route which 100% protected the infected mice at the dose of 24 mg/kg × four days while artemisinin was transferred via the intramuscular route [84].

Savir et al. (2021) synthesized Schiff base metal complexes of nickel. In all complexes, the ligand was joined with metal ions by N, O, and S atoms [85].

Bailey et al. (2021) reported triazolopyrimidine class of Schiff base ligands 2-(N-phenyl carboxamide) triazolopyrimidine and screened identified their potent antimalarial activities. P. falciparum and P. knowlesi were taken as tested and it was found that P. falciparum was more potent as compared to P. knowlesi. Scott et al. (2008) synthesized novel Schiff base-phenol and naphthalene-amine ligands and tested their antimalarial activity against Plasmodium falciparum [86].

9.4.2 Antibacterial and Antifungal Activities

Antibacterial substances are substances that kill microorganisms such as bacteria, mold, yeast, and fungi, inhibiting their growth and reproduction. There are more bacteria in the human body than there are cells. They can infect the throat, lungs, skin, intestine, or any part of the body. Joseyphus

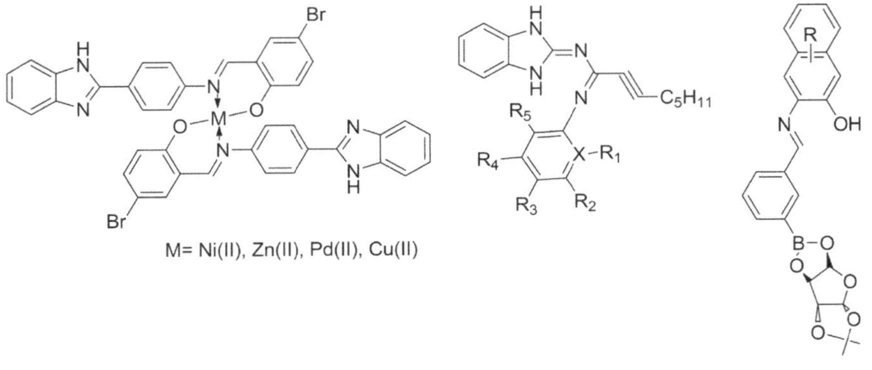

FIGURE 9.7 Structures of Schiff base and their metal complexes have been shown potent antimicrobial activities.

and Nair (2008) prepared nano Schiff base and their zinc (II) complexes [87]. After the antimicrobial activities result, they concluded that the azomethine group (>C=N-) participated in biological processes and resulted in interference with normal cells. The observations of their studies showed that the heteroatom has higher inhibition microbial growth which was due to uncoordinated atoms, but Zn(II)complexes have the highest potent effect.

Goszczynska, Kwiecien, corresponding author, and Fijalkowski (2015) synthesized a series of alkyl 2-(2-formylphenoxy)alkanoates Schiff base and reported (Figure 9.7) that these new ligands had the potential capacity as the antibacterial agents, gram-positive and gram-negative bacteria. The microbial growth has been measured by the minimal inhibitory concentration method using the Schiff base. The lowest concentrations of all compounds were more inhibited against microbial growth [88–90].

Thierry Youmbi Fonkui (2019) synthesized benzimidazole nano Schiff base derivatives amide and primary amines. Imidazole and benzimidazole are the class of nitrogen associated with the class of heterocyclic compounds that are very important and applicable as mean pharmacohores. The benzo derivatives of benzimidazole like imidazole, estimizole, mebendazole, carbendazime have attraction in the development of therapeutic drugs [91]. In their experiment, micro dilution was used to determine antifungal and antibacterial activities against *Aspergillus*, *Fusarium*, *Plasmodium falciparum*, *Trypanosoma brucei*, *Klebsiella pneumonia*, and *Escherichia coli*.

9.4.3 Anticancer Activities

Most cancers are a sickness of abnormal increase of a cellular. Usually, the cells of our body develop and divide in a managed way. When regular cells are damaged or cells turn out to be unique, they die and are replaced by healthy cells. In cancer, the indicators that manipulate cell growth do not now work properly. Cancer cells continue to grow and multiply until measures are taken to stop them. In other words, most cancer cells don't follow the same rules as healthy cells. Each cell can be controlled with the help of genes that dictate the way cells develop and divide [92–95].

Cancer develops from the frame's personal cells. It starts off evolving with an alternate inside the genes of a single cellular. Genetic adjustments or mutations (changes or manipulations in the cellular) can intervene with the everyday characteristic of a cell. Most changes are not intended to harm; however, occasionally they interfere with the regulation of cell growth. Maximum of the manipulations manifest spontaneously, that too whilst cells divide into distinctive parts. Unlike cancer in adults, most cancers in children aren't due to lifestyle or environment. As an alternative, the maximum cases of most cancers in children are the end result of a mutation in a gene that takes place by means of danger [96–98].

There are more than a hundred different varieties of cancer in children. How cells look underneath a microscope and the molecular and genetic characteristics of cells provide records to help decide the particular type of cancer.

Alorini et al. (2022) have reported very interesting and important research. In this study they synthesized some metal complexes with nano Schiff base ligand. These compounds showed anticancer activity against PC-3, HeLa, SKOV3, tumor cell lines. In this study, IC50 values of copper complexes were observed at 0.161, 0.063, and 0.087 μg/mL which were the highest values. For the above tumor cells newly prepared nickel complexes lowest potential and the observed IC50 values were 1.8287, 1.2502, and 3.8453 μg/mL. The cisplatin, estramustine, and etoposide were used as reference drugs that are IC50 values 2.4 μg/mL, 0.35–1.3 μg/mL, 17.4 μg/mL, respectively. Abd-Elzaher et al. (2016) reported thiazole containing Schiff base ligands and metal complexes.

The four types of cancer cells, namely, Liver cancer HepG2, Breast cancer MCF-7, Lung carcinoma A549, and Colorectal cancer HCT116 cells, have been used for the investigation against newly prepared nano compounds. Doxorubicin drug is used as the standard or reference drug. The newly synthesized compounds were tested and it was found that zinc (II) complexes have higher potent inhibitory results against HepG2 (80%), MCF7 (70%), A549 (60%), and HCT116 80 (64%) ratio.

Ikechukwu et al. (2016) synthesized, characterized, and evaluated anticancer activities of Ru(III) complexes containing Schiff base ligand. Renal cancer cells (TK-10), Melanoma cancer cells (UACC-62), and Breast cancer cells (MCF-7) were chosen for the evaluation of anticancer activities and in this study Sulforhodamine was used as a reference drug. 3.57, 6.44, and 9.06 μM IC50 values for Ru(III) complexes. The results were best used as a reference drug value.

9.4.4 Antitubercular

Desai et al. (2021) synthesized a quinoline-based dihydropyridines series of compounds that were evaluated against M. tuberculosis H37Ra. Mycobacterium tuberculosis is the main cause of tuberculosis which is a type of bacteria [99]. The lungs are most affected by this disease, which gives rise to pulmonary TB and extra-pulmonary TB in other organs. There are about 200 types of bacteria in the Mycobacteriaceae family [100, 103]. This type of bacteria is also the cause of leprosy.

9.5 CONCLUSION

The biological chemistry of nano Schiff's bases complexes have clinical and pharmacologic properties, they're helpful within the diagnosis of varied varieties of pathogens and tumors. Nanotechnology may be used as part of tissue engineering to help regenerate, repair, or reshape broken tissue by the use of appropriate nanomaterial-based total scaffolds and boom elements. Tissue engineering, if successful, could update traditional treatments consisting of organ transplantation or prosthetic implants. Even though work in this area is not complete, it is now very possible to destroy actual cancer cells using nanoparticles of Schiff base metal complexes. These compounds may be used for neighborhood diagnosis and remedy of cancer cells with minimum harm to healthy cells. In contrast to conventional radiation and high doses of chemotherapy, which can have numerous side effects and harm healthy cells, a new approach utilizing gold-lipid nanoparticles has been developed. These nanoparticles are activated by light and can be delivered to specific locations within the body to release the drug. Unlike traditional treatments, these nanoparticles are not harmful to the body. By converting light into heat, they can elevate the temperature of cancer cells, ultimately killing them. This innovative approach is capable of targeting and destroying a significant portion of cancer cells.

REFERENCES

1 Lone, I.H., Khan, K.Z., Fozdar, B.I., Hussain, F. 2013. Synthesis antimicrobial and antioxidant studies of new oximes of steroidal chalcones. Steroids 78(9):945–950.
2. The PyMOL Molecular Graphics System 2009. Version 1504, Schrodinger, LLC.
3. Bauer, A.W., Kirby, W.M.M., Sherris, J.C., Truck, M. 1966. Antibiotic susceptibility testing by a standardized single disk method. Am J Clin Pathol 45(4):493–496.
4. Mosmann, T. 1983. Rapid colorimetric assay for cellular growth and survival: Application to proliferation and cytotoxicity assays. J Immunol Methods 65(1–2):55–63.
5. Klein, R.A. 2006. Modified van der Waals atomic radii for hydrogen bonding based on electron density topology. Chem Phys Lett 425(1):128–133.
6. Ghogomu, J.N., Nkungli, N.K. 2016. A DFT study of some structural and spectral properties of 4-methoxyacetophenone thiosemicarbazone and its complexes with some transition metal chlorides: Potent antimicrobial agents. Adv Chem 2016:1–15.
7. Kulkarni, A.D., Truhlar, D.G. 2011. Performance of DFT and Moller-Plesset second-order perturbation theory for structural parameters in complexes of Ru. J Chem Theory Comput 7(7):2325–2332.
8. Karnan, M., Balachandran, V., Murugan, M., Murali, M.K., Nataraj, A. 2013. Vibrational FT-IR and FT-Raman spectra, NBO, HOMO-LUMO, molecular electrostatic potential surface and computational analysis of 4-(trifluoromethyl)benzylbromide. Spectrochim Acta A 116:84–95.
9. Fleming, I. 2010. Molecular orbitals and organic chemical reactions, 1st edn. Wiley, Chichester.
10. Mahmoud, W.H., Omar, M.M., Sayed, F.N., Mohamed, G.G. 2018. Synthesis, characterization, spectroscopic and theoretical studies of transition metal complexes of new nano Schiff base derived from L-histidine and 2-acetylferrocene and evaluation of biological and anticancer activities. Appl Organomet Chem 32(7):4386–4390.
11. Geary, J. 1971. The use of conductivity measurements in organic solvents for the characterization of coordination compounds. Coord Chem Rev 7(1):81–122.
12. Nawar, N., Hosny, N.M. 2000. Synthesis, spectral and antimicrobial activity studies of o-aminoacetophenone, o-hydroxybenzoylhydrazone complexes. Transition Met Chem 25(1):1–8.
13. El-Baradie, K.Y., El-Wakiel, N.A., El-Ghamry, H.A. 2014. Synthesis, characterization and corrosion inhibition in acid medium of L-histidine Schiff base complexes. Appl Organomet Chem 29(3):117–125.
14. Shakir, M., Abbasi, A., Khan, A.U., Khan, S.N. 2011. Synthesis and spectroscopic studies on the Schiff base ligand derived from condensation of 2-furaldehyde and diaminobenzidene, and its complexes with Co(II), Ni(II), Cu(II) and Zn(II): Comparative DNA binding studies of ligand and its Cu(II) and Zn(II) complexes. Spectrochim Acta A 78(1):29–35.
15. Refat, M.S., El- Sayed, M.Y., Adam, A.M.A. 2013. Cu(II), Co(II) and Ni(II) complexes of new Schiff base ligand: Synthesis, thermal and spectroscopic characterizations. J Mol Struct 1038:62–72.
16. Sallam, S.A. 2006. Binuclear Cu(II), Ni(II) and Co(II) complexes with N_2O_2 chromophores of glycylglycine Schiff bases of acetylacetone, benzoylacetone and thenoyltrifluoroacetone. Transition Met Chem 31(1):46.
17. Muller, J., Schubi, D., Mossmer, C.M., Strahle, J., Weser, U. 1999. Structure-function correlation of Cu(II) and Cu(I) di-Schiff-base complexes during the catalysis of superoxide dismutation. Inorg Chim Acta 75(1):63–69.
18. Raja, D., Bhuvanesh, N.S.P., Natarajan, K. 2012. A novel water soluble ligand bridged Co(II) coordination polymer of 2-oxo-1,2-dihydroquinoline-3-carbaldehyde (isonicotinic) hydrazone: Evaluation of the (DNA) binding, protein interaction, radical scavenging and anticancer activity. Dalton Trans 41(15):4365.
19. Manimohan, M., Pugalmani, S., Sithique, M.A. 2019. Biologically active novel NNO donor tridentate water soluble hydrazide based O-carboxymethyl chitosan Schiff base Cu (II) metal complexes: Synthesis and characterisation. Int J Biol Macromol 136:738–754.
20. Inan, A., Ikiz, M., Tayhan, S.E., Bilgin, S., Genc, N., Sayın, K., Ceyhan, G., Kose, M., Dag, A., Ispir E. 2018. Antiproliferative, antioxidant, computational and electrochemical studies of new azo-containing Schiff base ruthenium (II) complexes. New J Chem 42(4):2952–2963.
21. Pravin, N., Kumaravel, G., Senthilkumar, R., Raman, N. 2017 Mixed ligand Cu(II) complexes containing a Schiff base and diimine. Appl Organomet Chem 31(10):3714–3739.

22. Zhang, P., Huang, H. 2018. Future potential of osmium complexes as anticancer drug candidates, photosensitizers and organelle-targeted probes. Dalton Trans 47(42):14841–14854.
23. Sharma, V.K., Doong, R., Kim, H., Varma, R.S., Dionysiou, D.D. 2016. ACS symposium series, ferrites and ferrates: Chemistry and applications in sustainable energy and environmental remediation silica-coated magnetic nano-particles: Application in Catalysis. ACS symposium series 1238:1–38.
24. Diana S., 2018. Nanotechnology in Targeted Drug Delivery and Therapeutics. Nanoscience and Nanotechnology in Drug Delivery: 357-409.
25. Yousef, T.A. 2020. Structural, optical, morphology characterization and DFT studies of nano sized Cu(II) complexes containing Schiff base using green synthesis. Journal of Molecular Structure 1215:128180.
26. Sadimenko, A. P. 2012. Advances in heterocyclic chemistry advances in heterocyclic chemistry organometallic complexes of pyridyl Schiff bases. Advances in Heterocyclic Chemistry 107:133–218.
27. Ghanghas, P., Choudhary, A., Kumar, D., Poonia, K. 2021. Coordination metal complexes with Schiff bases: Useful pharmacophores with comprehensive biological applications. Inorganic Chemistry Communications 130:108710.
28. Roberts, K., Elizabeth S., Benvenuto, M.A. 2018. Environmental Chemistry: Undergraduate and Graduate Classroom, Laboratory, and Local Community Learning Experiences Synthesis of a Novel Series of Nitrogen-Containing Ligands for Use as Water Remediators, All Incorporating Long-Chain Aliphatic Moieties. Environmental Chemistry 1276:81–87.
29. Marchesan, S., Prato, M. 2013. Nanomaterials for (Nano) medicine. ACS Medicinal Chemistry Letters 4(2):147–149.
30. Morgan, Sarah E., Lochhead, Robert Y. 2010. Polymeric delivery of therapeuticsl polymers in nano pharmaceutical materials. ACS Symposium Series 1053:25–45.
31. Prabhakar, U.P.S. Helina, J., Sagayaraj, Y.V. 2021. Synthesis, characterization and biological activity of 2-[(2-hydroxy-phenylimino)-methyl]-6-methoxy-phenol (MSAP) and nano sized of Co (II), Ni (II), Cu (II) and Zn (II) metal complexes. Materials Today: Proceedings.
32. Wahab, S., Alshahrani, M.Y., Ahmad, M.F., Abbas, H. 2021. Current trends and future perspectives of nanomedicine for the management of colon cancer. European Journal of Pharmacology 910:174464.
33. Andreia C., 2018. Nanoparticles as delivery systems in cancer therapy: focus on gold nanoparticles and drugs. Nanoscience and Nanotechnology in Drug Delivery, 257–295.
34. Pervaiz, M., Sadiq, S., Sadiq, A., Younas, U., Ashraf, A., Saeed, Z., Adnan, A. 2021. Azo-Schiff base derivatives of transition metal complexes as antimicrobial agents. Coordination Chemistry Reviews 447:214128.
35. Gurunathan, R.J., Varjani, B.K.S., Ngo, S., Gnansounou, H.H. 2022. Advancements in heavy metals removal from effluents employing nano-adsorbents: Way towards cleaner production. Environmental Research 203:111815.
36. Adabi, M., Naghibzadeh, M., Adabi, M., Zarrinfard, M.A., Esnaashari, S., Seifalian, A.M. 2017. Biocompatibility and nanostructured materials: Applications in nanomedicine. Artif. Cells Nanomed. Biotechnol 45:833–842.
37. Agrahari, V., Hiremath, P. 2017. Challenges associated and approaches for successful translation of nanomedicines into commercial products. Nanomedicine 12:819–823.
38. Albanese, A., Tang, P.S., Chan, W.C. 2012. The effect of nanoparticle size, shape, and surface chemistry on biological systems. Annu. Rev. Biomed. Eng 14:1–16.
39. Bastogne, T. 2017. Quality-by-design of nano pharmaceuticals -a state of the art. Nanomedicine 13:2151–2157.
40. Boverhof, D.R., Bramante, C.M., Butala, J.H., Clancy, S.F., Lafranconi, M., West, J. 2015. Comparative assessment of nanomaterial definitions and safety evaluation considerations. Regul. Toxicol. Pharmacol. 73, 137–150.
41. Choi, H., Han, H. 2018. Nanomedicines: Current status and future perspectives in aspect of drug delivery and pharmacokinetics. J. Pharm. Investig 48:43.
42. Collins, A.R., Annangi, B., Rubio, L., Marcos, R., Dorn, M., Merker, C. et al. 2017. High throughput toxicity screening and intracellular detection of nanomaterials. Wiley Interdiscip. Rev. Nanomed. Nanobiotechnol 9:e1413.

43. De Jong, W.H., Hagens, W.I., Krystek, P., Burger, M.C., Sips, A.J., Geertsma, R.E. 2008. Particle size-dependent organ distribution of gold nanoparticles after intravenous administration. Biomaterials 29:1912–1919.
44. Desai, N. 2012. Challenges in development of nanoparticle-based therapeutics. AAPS J 14:282–295.
45. Fadeel, B. 2013. Nanosafety: Towards safer design of nanomedicines. J. Intern. Med 274:578–580.
46. Fadeel, B., Fornara, A., Toprak, M.S., Bhattacharya, K. 2015. Keeping it real: The importance of material characterization in nanotoxicology. Biochem. Biophys. Res. Commun 468:498–503.
47. Gaspar, R. 2010. Therapeutic products: Regulating drugs and medical devices, in International Handbook on Regulating Nanotechnologies, eds. G. A. Hodge, D. M. Browman, and A. D. Maynard (Cheltenham: Edward Elgar) 291–320.
48. Hussaarts, L., Muhlebach, S., Shah, V.P., McNeil, S., Borchard, G., Fluhmann, B. 2017. Equivalence of complex drug products: advances in and challenges for current regulatory frameworks. Ann. N. Y. Acad. Sci. 1407:39–49.
49. Hussain, S.M., Warheit, D.B., Ng, S.P., Comfort, K.K., Grabinski, C.M., Braydich-Stolle, L. K. 2015. At the crossroads of nanotoxicology in vitro: Past achievements and current challenges. Toxicol. Sci. 147:5–16.
50. Landsiedel, R., Ma- Hock, L., Wiench, K., Wohlleben, W., Sauer, U.G. 2017. Safety assessment of nanomaterials uses an advanced decision-making framework, the DF4 nano grouping. J. Nanopart. Res. 19:171.
51. McCall, M.J., Coleman, V.A., Herrmann, J., Kirby, J.K., Gardner, I.R., Brent, P.J. et al. 2013. A tiered approach. Nat. Nanotechnol 8:307–308.
52. Monopoli, M.P., Aberg, C., Salvati, A., Dawson, K. A. 2012. Biomolecular coronas provide the biological identity of nanosized materials. Nat. Nanotechnol 7:779–786.
53. Mühlebach, S. 2018. Regulatory challenges of nanomedicines and their follow-on versions: a generic or similar approach. Adv. Drug Deliv. Rev.
54. Muhlebach, S., Borchard, G., Yildiz, S. 2015. Regulatory challenges and approaches to characterize nanomedicines and their follow-on similars. Nanomedicine 10:659–674.
55. Müller, R.H., Gohla, S., Keck, C.M. 2011. State of the art of nanocrystals-special features, production, nanotoxicology aspects and intracellular delivery. Eur. J. Pharm. Biopharm 78:1–9.
56. Oksel, C., Ma, Y.C., Wang, Z.X. 2015. Structure-activity relationship models for hazard assessment and risk management of engineered nanomaterials. Proc. Eng. 102:1500–1510.
57. Pita, R., Ehmann, F., Papaluca, M. 2016. Nanomedicines in the EU-regulatory overview. AAPS J. 18:1576–1582.
58. Sainz, V., Conniot, J., Matos, A.I., Peres, C., Zupancic, E., Moura, L. et al. 2015. Regulatory aspects on nanomedicines. Biochem. Biophys. Res. Commun. 468:504–510.
59. Tinkle, S., McNeil, S.E., Mühlebach, S., Bawa, R., Borchard, G., Barenholz, Y.C. et al. 2014. Nanomedicines: Addressing the scientific and regulatory gap. Ann. N. Y. Acad. Sci. 1313:35–56.
60. Williams, D. 2003. Revisiting the definition of biocompatibility. Med. Device Technol. 14:10–13.
61. Zhao, Y., Chen, C. 2016. Nano on reflection. Nat. Nanotechnol. 11:828–834.
62. Cristina, G.M. 2020. Engineering Drug Delivery Systems Nano- and microparticles as drug carriers. National Library of Medicine 71–110.
63. Aizat, M.A. 2019. Nanotechnology in water and wastewater treatment chitosan nanocomposite. Application in wastewater treatments. Theory and application. Micro and Nano Technol. 243–265.
64. Solis Maldonado, C. 2018. Direct synthesis of metal complexes applications of heterometallic complexes in catalysis. Direct Synthesis of Metal Complexes 369–377.
65. Parsaee, Z. 2017. Sonochemical synthesis and DFT studies of nano novel Schiff base cadmium complexes: Green, efficient, recyclable catalysts and precursors of Cd NPs. Journal of Molecular Structure S002228601730830X.
66. Sun, Y., Lu, Y., Bian, M., Yang, Z., Ma, X., Liu, W. 2021. Pt(II) and Au(III) complexes containing Schiff-base ligands: A promising source for antitumor treatment. European Journal of Medicinal Chemistry 211:113098.
67. Sun, Y., Jiang, X., Liu, Y., Liu, D., Chen, C., Lu, C., Liu, J. 2021. Recent advances in Cu(II)/Cu(I)-MOFs based nano-platforms for developing new nano-medicines. Journal of Inorganic Biochemistry 225:111599.

68. Antony, R., Arun, T., Manickam, S.T.D. 2019. A review on applications of chitosan-based Schiff bases. International Journal of Biological Macromolecules, S0141813018359920.
69. Mitchell, D.T., Lee, S.B., Trofin, L., Li, N., Nevanen, T.K., Soderlund, H., Martin, C.R. 2002. Smart nanotubes for bioseparations and biocatalysis. Journal of the American Chemical Society 124(40):11864–11865.
70. Kustin, K., Pessoa, J.C., Crans, D.C. 2007. Vanadium: The versatile metal volume 974 vanadium Schiff base complexes: Chemistry, properties, and concerns about possible therapeutic applications 0974:340–351.
71. Jia, J.G., Zheng, L.M. 2020. Metal-organic nanotubes: Designs, structures and functions. Coordination Chemistry Reviews 403:213083.
72. Al Zoubi, W., Al Mohanna, N. 2014. Membrane sensors based on Schiff bases as chelating ionophores: A review. Spectrochimica Acta Part A: Molecular and Biomolecular Spectroscopy 132:854–870.
73. Yin, L. 2020. Nanotechnology improves delivery efficiency and bioavailability of tea polyphenols. J Food Biochem. 13380.
74. Alothman, A.A., Almarhoon, Z.M. 2020. Nano-sized some transition metal complexes of Schiff base ligand based on 1-aminoquinolin-2(1H)-one. Journal of Molecular Structure 1206:127704.
75. Barea, M.J., Jenkins, M.J., Gaber, M.H., Bridson, R.H. 2010. Evaluation of liposomes coated with a pH responsive polymer. Int. J. Pharm 402:89–94.
76. Benmohammed, M., Kenidra, B. 2020. An ultra-fast method for clustering of big genomic data. Int. J. Appl. Metaheuristic Comput. (IJAMC) 11:45–60.
77. Bertrand, N., Wu, J., Xu, X., Kamaly, N., Farokhzad, O.C. 2014. Cancer nanotechnology: The impact of passive and active targeting in the era of modern cancer biology. Adv. Drug Deliv. Rev., 2014 Feb. 66:2–25.
78. Chen, C.M., Peng, E.H. 2003. Nanopore sequencing of polynucleotides assisted by a rotating electric field. Appl. Phys. Letter 82:1308–1310.
79. Jia, S.Z., Sun, H.Y., Wang, Q.L. 2002. Nanopore technology and its applications. Prog. Biochem. Biophys 29:202–205.
80. Grayson, A.C.R., Choi, I.S., Tyler, B.M., Wang, P.P., Brem, H., Cima, M.J., Langer, R. 2003. Multi-pulse drug delivery from a resorbable polymeric microchip device. Nat. Mat 2:767–772.
81. Curiel, D.T., Douglas, J.T. 2002. Vector Targeting for Therapeutic Gene Delivery. Wiley-Liss Inc. Hoboken, New Jersey.
82. Findeis, M.A. 2001. Nonviral vectors for gene therapy. Method. Mol. Med. 65 Humana Press, Totowa, New Jersey.
83. Kabanov, A.V., Felgner, P.L., Seymour, L.W. 1998. Self-Assembling Complexes for Gene Delivery: From Laboratory to Clinical Trial. John Wiley & Sons Ltd. Chichester, West Sussex.
84. Bhaskaran, N.A., Kumar, L. 2021. Treating colon cancers with a non-conventional yet strategic approach: an overview of various nanoparticulate systems. J. Contr. Release. 10;336:16–39.
85. Bhattacharya, S. 2020. Fabrication and characterization of chitosan-based polymeric nanoparticles of Imatinib for colorectal cancer targeting application. Int. J. Biol. Macromol 151:104–115.
86. Chen, Y., Gao, D. 2015. In vivo delivery of miRNAs for cancer therapy: Challenges and strategies. Adv Drug Deliv Rev, 81:128–4187.
87. Chen, J., Li, X., Chen, L., Xie, F. 2018. Starch film-coated microparticles for oral colon-specific drug delivery. Carbohydr. Polym 191:242–254.
88. Chien, C.-W., Hou, P.C., Wu, H.C., Chang, Y.L., Lin, S.C., Lin, B.W., Lee, J.C., Chang, Y.J., Sun, H.S., Tsai, S.J. 2016. Targeting $TYRO_3$ inhibits epithelial-mesenchymal transition and increases drug sensitivity in colon cancer. Oncogene 35:5872–5881.
89. Cho, H., Lai, T.C., Tomoda, K., Kwon, G.S. 2014. Polymeric micelles for multi-drug delivery in cancer. AAPS Pharm Sci Tech 16:10–20.
90. Cho, T.J., Mac Cuspie, R.I., Gigault, J., Gorham, J.M., Elliott, J.T., Hackley, V.A. 2014. Highly stable positively charged dendron-encapsulated gold nanoparticles. Langmuir 30:3883–3893.
91. Doroudian, M., MacLoughlin, R., , Poynton, F., Prina-Mello, A., Prina-Mello, A., Prina-Mello, A., Donnelly, S.C. 2019. Nanotechnology based therapeutics for lung disease. Thorax 74:965–976.
92. Santos, A.M., Meneguin, A.B., Akhter, D.T., Fletcher, N., Houston, Z.H., Bell, C., Thurecht, K.J., Gremiao, M.P.D. 2021. Understanding the role of colon-specific microparticles based on retrograded

starch/pectin in the delivery of chitosan nanoparticles along the gastrointestinal tract. Eur. J. Pharm. Biopharm. 158:371–378.

93. Fletcher, R., Wang, Y., Schoen, R., Finn, O. J.Y. 2018. Undefined, colorectal cancer prevention: Immune modulation taking the stage. Biochim Biophys Acta. 2018 Apr; 1869(2): 138–148.
94. Fridman, W.H. Zitvogel, L. Sautes-Fridman, C. Kroemer, G. 2017. The immune contexture in cancer prognosis and treatment. Nat. Rev. Clin. Oncol. 14(12):717–734.
95. Handali, S., Moghimipour, E., Rezaei, M., Ramezani, Z., Kouchak, M., Amini, M., Angali, K.A., Saremy, S., Dorkoosh, F.A. 2018. A novel 5-Fluorouracil targeted delivery to colon cancer using folic acid conjugated liposomes. Biomed. Pharmacother 108:1259–1273.
96. Alibolandi, M., Hoseini, F., Mohammadi, M., Ramezani, P., Einafshar, E., Taghdisi, S.M., Ramezani, M., Abnous, K. 2018. Curcumin-entrapped MUC-1 aptamer targeted dendrimer-gold hybrid nanostructure as a theranostic system for colon adenocarcinoma. Int. J. Pharm 549:67–75.
97. Aneja, P., Rahman, M., Beg, S., Aneja, S., Dhingra, V., Chugh, R. 2015. Cancer targeted magic bullets for effective treatment of cancer. Recent pat. Antiinfect. Drug Discov 9:121–135.
98. Zhang, J., Zhang, H., Da, Yao, Y.F., Zhong, S.L., Zhao, J.H., Tang, J.H. 2015. β-elemene reverses chemoresistance of breast cancer cells by reducing resistance transmission via exosomes. Cell. Physiol. Biochem. 36:2274–2286.
99. Zhang, C., Huang, P., Zhang, Y., Chen, J., Shentu, W., Sun, Y., Yang, Z., Chen, S. 2014. Anti-tumor efficacy of ultrasonic cavitation is potentiated by concurrent delivery of anti-angiogenic drug in colon cancer. Canc. Lett. 347, 105–113.
100. Yang, G., Zhao, Y., Zhang, Y., Dang, B., Liu, Y., Feng, N. 2015. Enhanced oral bioavailability of silymarin using liposomes containing a bile salt: Preparation by supercritical fluid technology and evaluation in vitro and in vivo. Int. J. Nanomedicine 10:6633–6644.
101. Wu, X., He, C., Wu, Y., Chen, X. 2016. Synergistic therapeutic effects of Schiff's base cross-linked injectable hydrogels for local co-delivery of metformin and 5-fluorouracil in a mouse colon carcinoma model. Biomaterials 75:148–162.
102. Taylor, K.M.L., Rieter, W.J., Lin, W. 2008. Manganese-based nanoscale metal-organic frameworks for magnetic resonance imaging. J. Am. Chem. Soc 130:14358–14359.
103. Jeong, S., Kim, S.H., Joo, I., Ahn, S.J., Han, J.K. 2016. Usefulness of hydrogel-CT for detecting and staging of rectosigmoid colon cancer. Eur. J. Radiol 85:1020–1026.

10 Schiff Base Metal Complexes Based Sensors for Application

Sapna Raghav[*1] *and Jyoti Raghav*[2]
[1]Science Block, NBGSM College, Sohna, Gurugram University, Gurugram, India
*[1] Email: sapnaraghav04@gmail.com
[2] Schools of Engineering and Applied Science, Bennett University, Greater Noida, India
Email: jr7582@bennett.edu.in

CONTENTS

10.1 INTRODUCTION

The condensation product of carbonyl (aldehyde or keto) with the primary amines forms the Schiff bases and reacts with metal form SBM complexes (shown in Scheme 10.1). Hugo Schiff reported the first Schiff base in 1864. The general formula of Schiff base is RHC=NR' and known as azomethine group, where R group is aryl, alkyl, or heterocyclic group [1].

Schiff base complexes have a broad application range in dye, food, and medicine industries. These have very good biological, fungicidal, catalysis and agrochemical properties. But in recent years, more attention has been toward metal complexes of Schiff bases, that is, SBM complexes, due to its remarkable applications in pharmacological and biological application. Including biological properties, SBM complexes also have the ability to act as electrode materials, sensors, environmental sensors, solar cells, and energy storage devices [3–7].

Schiff bases have a strong tendency to form SBM complexes with metal with their different oxidation states. These have heterocyclic moieties, and hence are an important class of organic compounds. SBM complexes have been utilized for the removal of pollutants and heavy metal ion sensing. These are excellent flourimetric and spectrophotometric agents [8, 9]. Additionally, they have been utilized in electrochemical sensors and many chromatographic methods, so as to utilize them as selective and sensitive detection [10, 11]. Many complexes of Schiff bases had been synthesized and have provided an enormously rich world of chemistry.

DOI: 10.1201/9781003291459-10

In this chapter, we have focused on the sensor application of Schiff base metal complexes special emphasis based on electrochemical sensing [12], fluorescent, and colorimetric chemosensors [13], and optical chemical sensors [14].

10.2 SCHIFF BASE COMPLEXES AS CHEMOSENSORS

Colorimetric and spectrofluorometric chemosensors display very high selectivity and sensitivity toward natural species, anions and cations. SBCSs showed outstanding implementation for the sensing of metal cations as well as anions, due to facile capability to synchronize with all metal ions and cost-effective synthesis routes. A library of Schiff base chemosensors has been reported which were utilized as fluorescent chemosensors [18–32] (Table 10.1) and colorimetric chemosensors [33–36] (Table 10.2) for detection of metal cations by colorimetric detection. Many SBCSs reported both fluorescent and colorimetric chemosensors as tabulated in Table 10.3 [37–43]. Before using Schiff

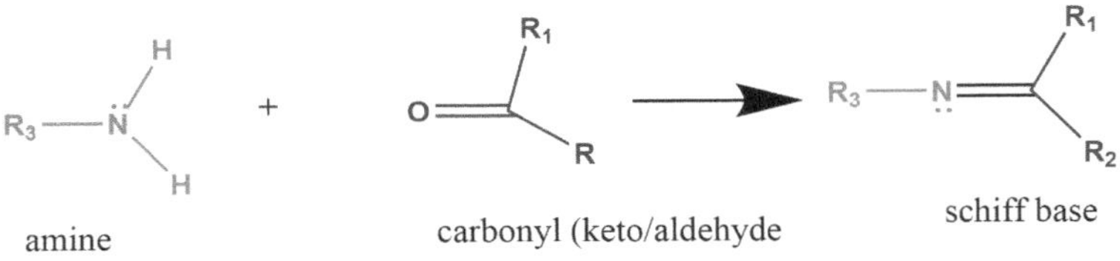

SCHEME 10.1 General route of Schiff base synthesis.

TABLE 10.1
List of fluorescence chemosensors

Chemosensors	Analyte/ sensor mechanism	Sensing mechanism	M:L ratio & LOD	Ref.
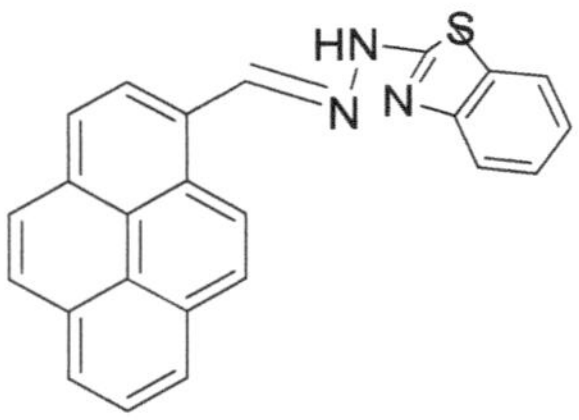	Ag^{+} off-on	Chelation and photo-induced electron transfer	1:1& 0.0048	[18]
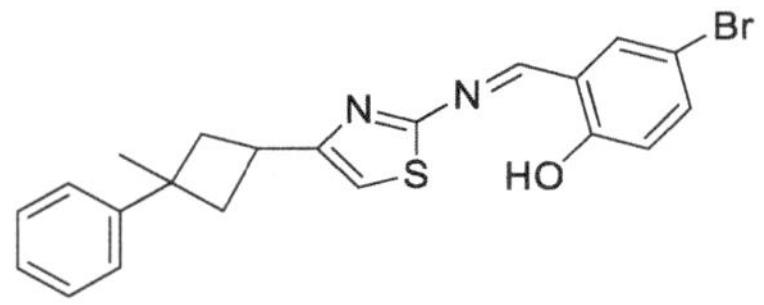	Cu^{2+} on-off	Chelation	1:2 & 0.008	[19]
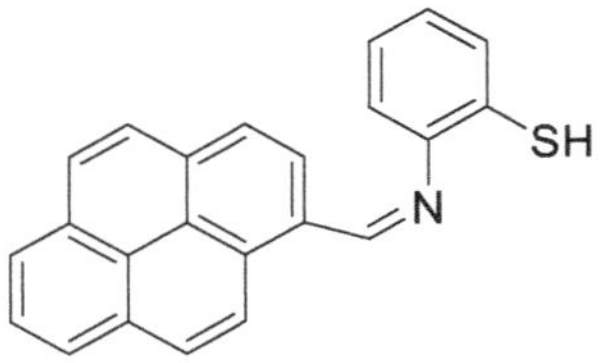	Hg^{2+} turn-on	Chelation and chelation-enhanced fluorescence	2:1 & 2.82	[20]

TABLE 10.1 (Continued)
List of fluorescence chemosensors

Chemosensors	Analyte/ sensor mechanism	Sensing mechanism	M:L ratio & LOD	Ref.
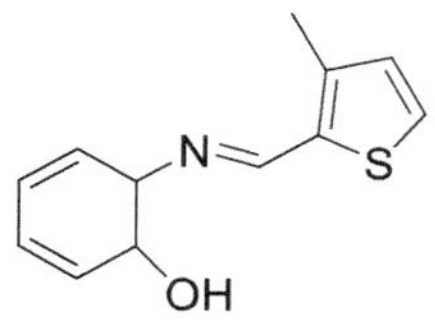	Hg^{2+} turn-on	chelation-enhanced fluorescence	1:2 & 20	[21]
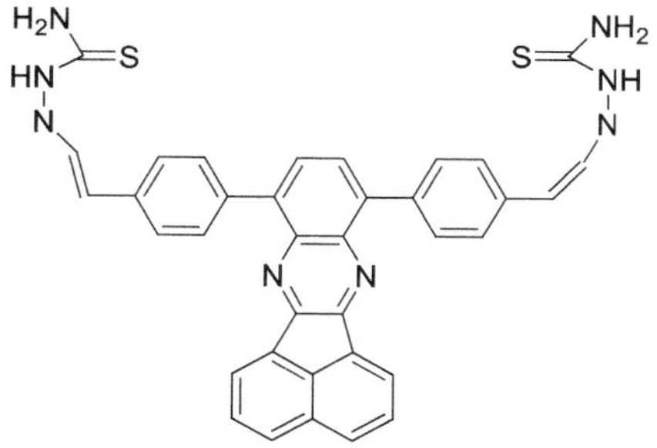	Hg^{2+} on-off	Chelation	2:1 & 0.907	[22]
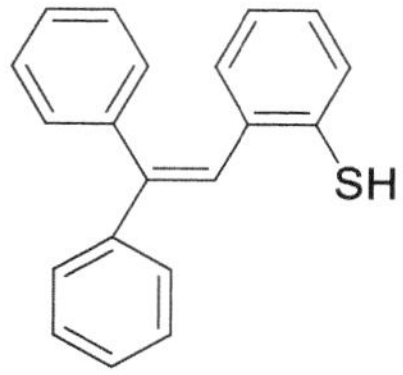	Hg^{2+} turn-on	Chelation	1:2 & 0.022	[23]
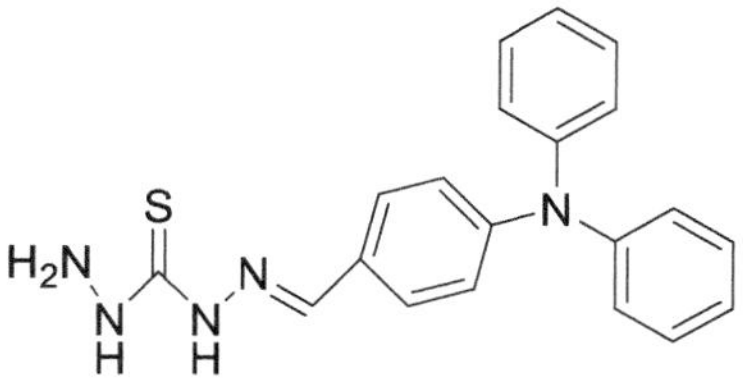	Hg^{2+}-	Chelation, chelation enhancement quenching effect & photo-induced electron transfer	1:2 & 0.031	[24]
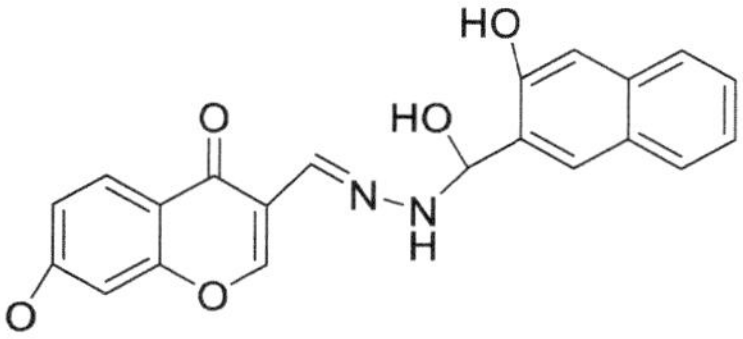	Zn^{2+} on-off	photo-induced electron transfer	1:1 & 0.1	[25]
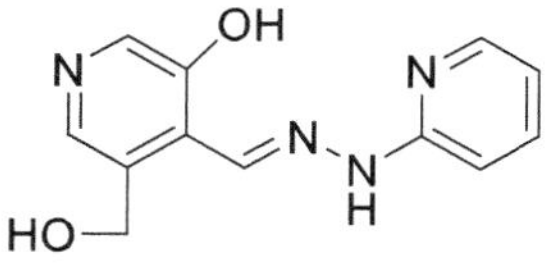	Zn^{+2} turn-on	-	1:1 & 0.07	[26]

(*continued*)

TABLE 10.1 (Continued)
List of fluorescence chemosensors

Chemosensors	Analyte/ sensor mechanism	Sensing mechanism	M:L ratio & LOD	Ref.
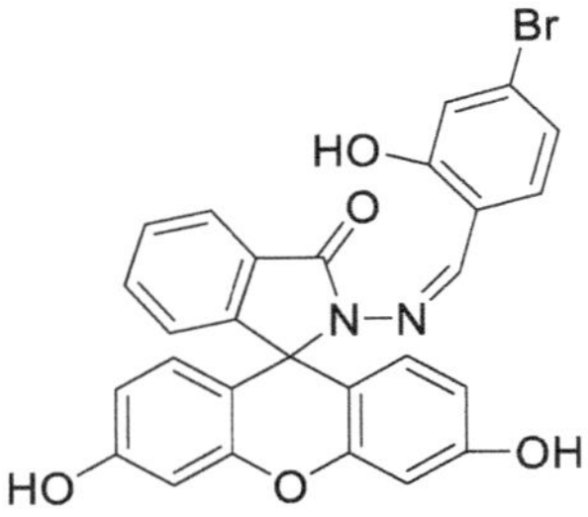	Zn^{+2} turn-on	chelation-enhanced fluorescence and photo-induced electron transfer	1:1 & 1.59	[27]
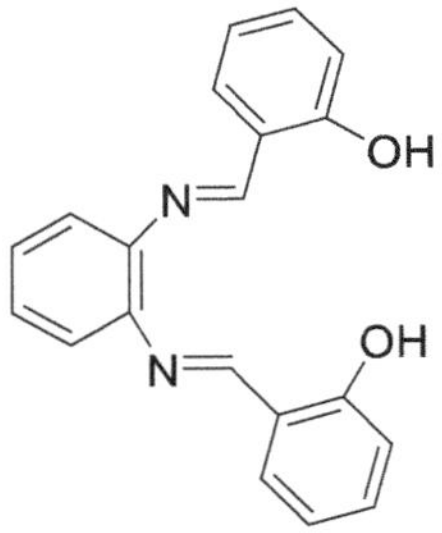	Ni^{+2}	chelation-enhanced fluorescence	1:1 & 0.1	[28]
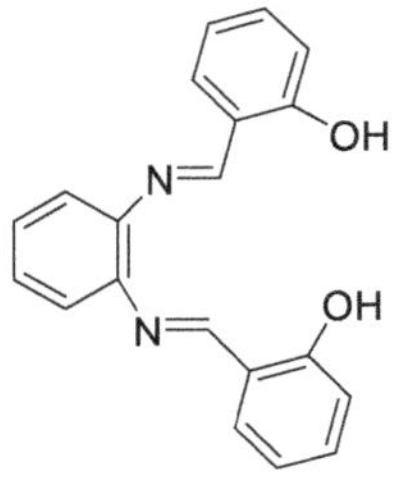	Pb^{2+} turn-on	Chelation	2:1 & .3	[29]
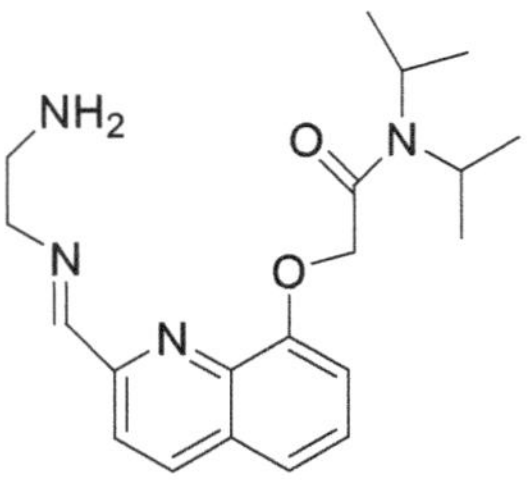	Cd^{2+} turn-on	Chelation	1:1 & 0.0024	[30]
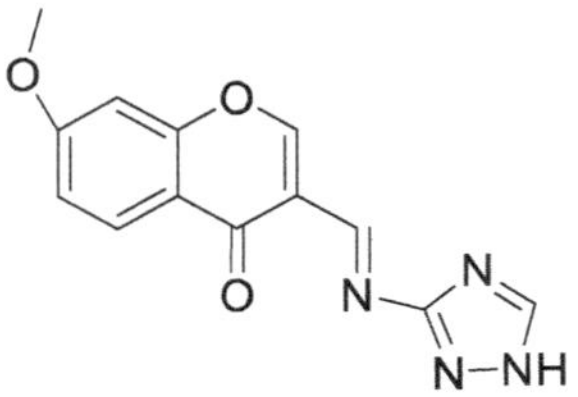	Cd^{2+} turn-on	Chelation	1:1 & 1.0	[31]
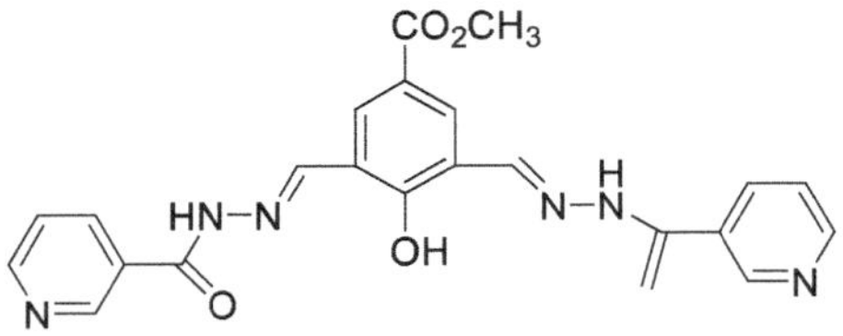	Al^{3+} turn-on	Chelation and intra/ inter-molecular charge transfer	1:1 & 0.1	[32]

TABLE 10.2
List of colorimetric chemosensors

Sensor	Analyte & Color change	Mechanism	M:L ratio & LOD
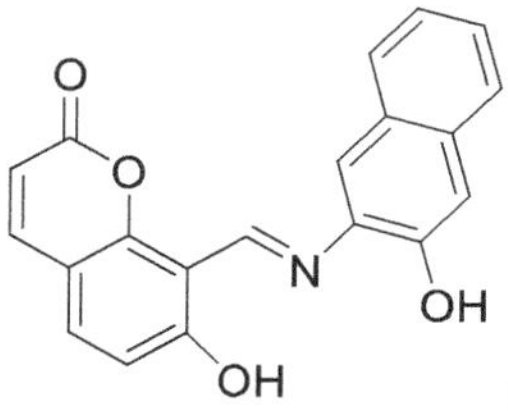 [33]	Ag^+ & Colorless to red	Chelation	1:2 & 0.00086
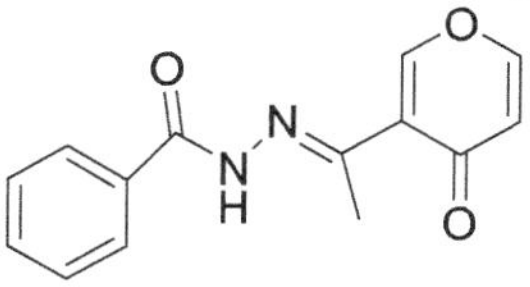 [34]	Cu^{2+} & Light yellow to bluish green	Chelation & intra/inter-molecular charge transfer	1:2 & 0.156
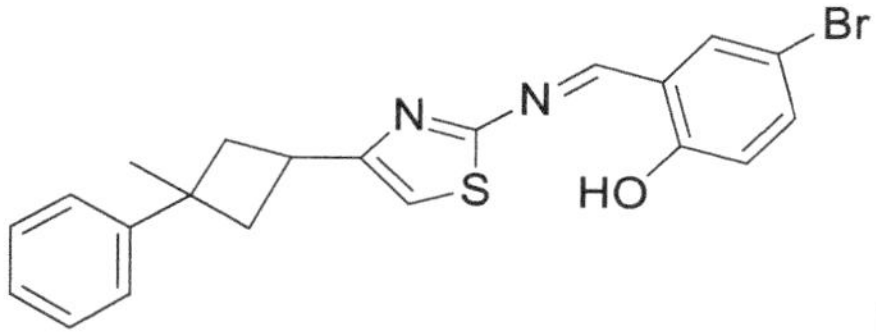 [35]	Ni^{2+} & Yellow to red	Chelation	1:1
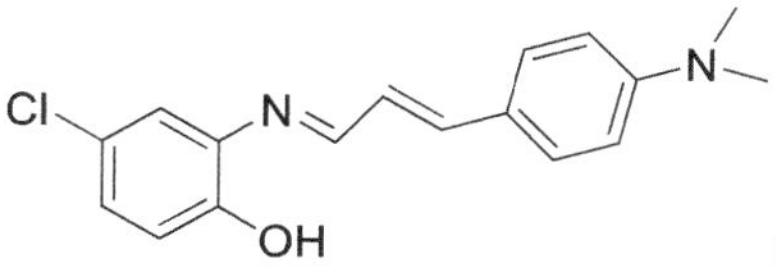 [36]	Ni^{2+} Faded yellow to deep orange	Chelation	1:1 & 0.1

TABLE 10.3
List of colorimetric and fluorescence chemosensors

Sensor	Analyte & Type	Color change	Mechanism	M:L ratio & LOD
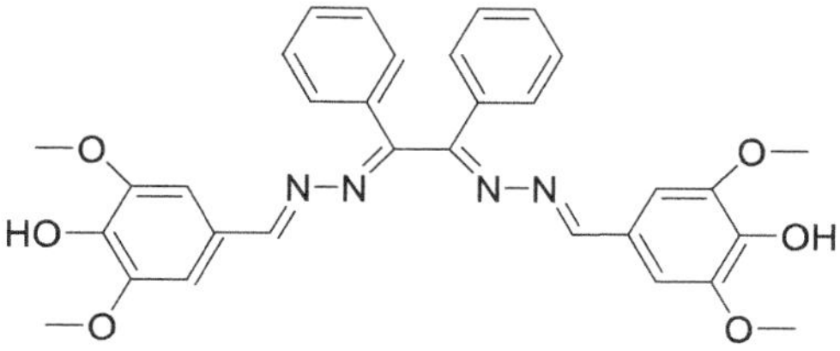 [37]	Pb^{2+} Turn-on	Colorless to yellow	Chelation-enhanced fluorescence & photo-induced electron transfer	1:2 & 0.008
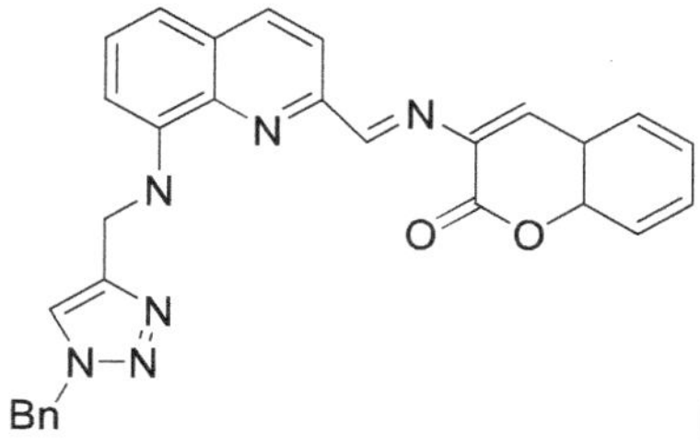 [38]	Pb^{2+} Off-on	Colorless to yellow	photo-induced electron transfer	1:1 & 0.036

(continued)

TABLE 10.3 (Continued)
List of colorimetric and fluorescence chemosensors

Sensor	Analyte & Type	Color change	Mechanism	M:L ratio & LOD
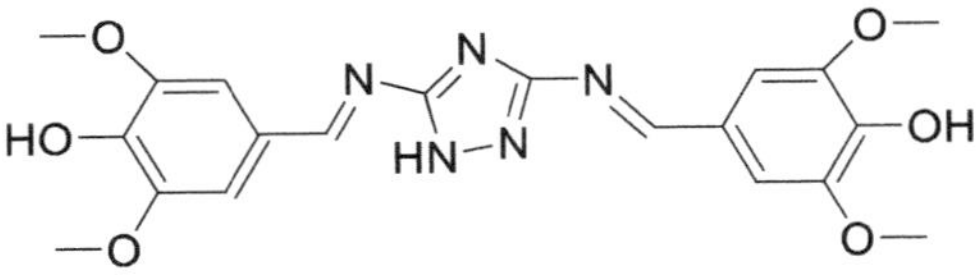 [39]	Pb^{2+}	Colorless to light yellow	& intra/inter-molecular charge transfer	1:1 & 0.9
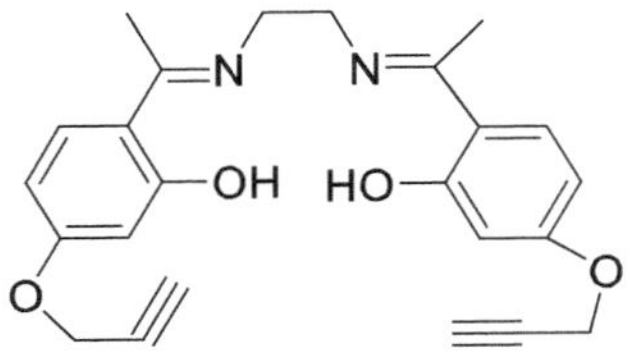 [40]	Cd^{2+} Turn-on	Pale yellow to green	photo-induced electron transfer & chelation-enhanced fluorescence	1:1 & 0.0129
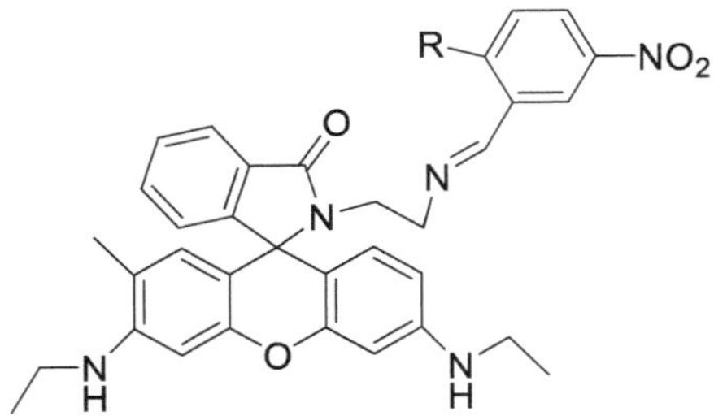 [41]	Pb^{2+} Turn-on	Colorless to Maroon	Chelation	1:1 & 0.55
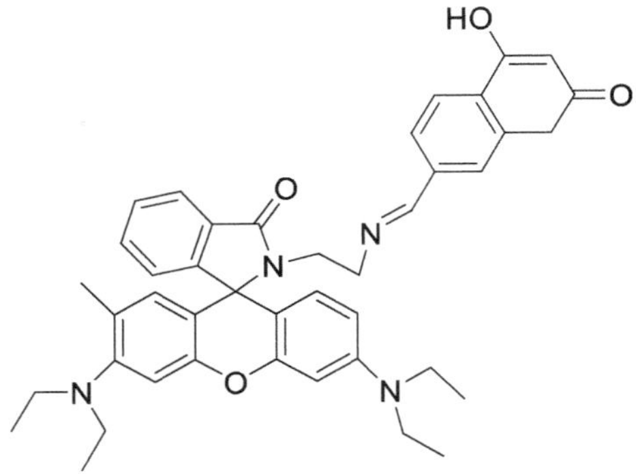 [42]	Pb^{2+} Turn-on	Straw to pink	Chelation	1:1 & 0.018
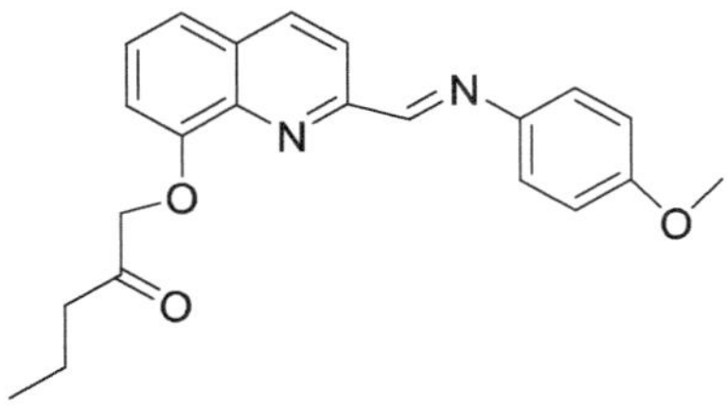 [43]	Fe^{3+} turn-on	Colorless to brown	Chelation & photo-induced electron transfer	1:1 & 0.048

base as chemosensors some concepts should be kept in mind, that is, in aqueous medium incorporation of bulky chain is favored to achieve heterogeneous system. These chemosensors systems are easy and cost-effective separation.

Moreover, in area of chemo-sensing, UV/Visible spectroscopic techniques and fluorescent spectroscopic study are utilized [15, 16].

10.3 SENSING MECHANISM OF CHEMOSENSORS

A number of factors are responsible for the higher sensitivity and selectivity of Schiff bases chemosensors:

- atomic radii of cation/anion
- charge on an ion
- electronic configuration
- electronic configuration of the metal ion, and
- binding capacity of chemosensors with the ion [17].

Accordingly, diverse mechanisms had reported for the interaction among chemosensors and metal ion and vice-versa, such as;

- chelation enhancement quenching effect (CHEQ) mechanisms
- chelation-enhanced fluorescence (CHEF) mechanisms
- intra/inter-molecular charge transfer (ICT) mechanisms
- photo-induced electron transfer (PET) mechanisms
- ring-opening mechanisms, and
- hydrolysis mechanisms.

10.4 COMPARATIVE STUDY OF CHEMOSENSORS

A library of colorimetric and spectrofluorometric chemosensors were tested for the detection of various metal cations from the different sources [44–53]. Table 10.4 tabulated the comparative properties of other reported organic chemosensors with SBCSs [54–58].

10.4.1 Electrochemical Sensors

The electrochemical sensors work on the methodology of electrochemical methods: in this method an electrochemical measurable signal, which is in the form of chemical information, changes into an analytical signal. In an electrochemical sensor there are two basic units:

Receptor: converts chemical information into energy form
Transducer: convert energy form which carrying chemical information into signal

A wide-ranging electrochemical technique is utilized for this, but most commonly, utilized devices are potentiometry, voltammetry, amperometry, amperometry, and conductometry. The signal is measured in the term of potential difference among reference and indicator electrodes. The indicator electrodes may be metal-metal oxide electrode or ion-selective electrode. The ion-selective electrodes are synthesized generally from polymeric material and Schiff base which are utilized for sensing purposes. In potentiometric sensors, ion-selective electrodes bids interesting advantages, such as very high selectivity, fast response time, cheap, simplicity, low detection limit and broad

TABLE 10.4
Some other organic chemosensors with SBCSs

Analyte	Method	Sensor	LOD (μM)	Organic Sensor	LOD
Ag+	Colorimetric & fluorescence spectroscopy	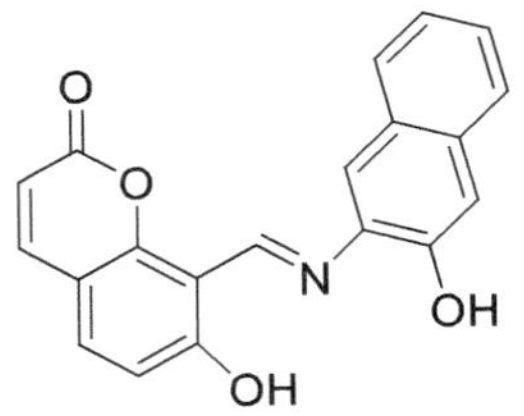	0.00086	cellulose-based colorimetric chemosensors containing thiourea moiety [54]	1.0
Cu2+	Fluorescence spectroscopy	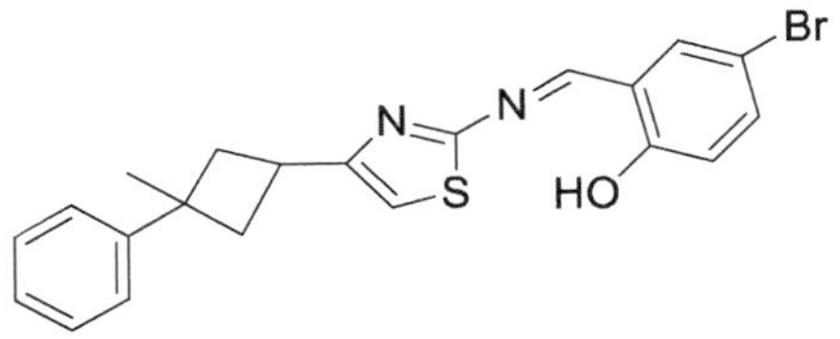	0.0088	pyrene appended bis-triazolylated 1,4-dihydropyridine [55]	0.53
Ni2+	Colorimetric	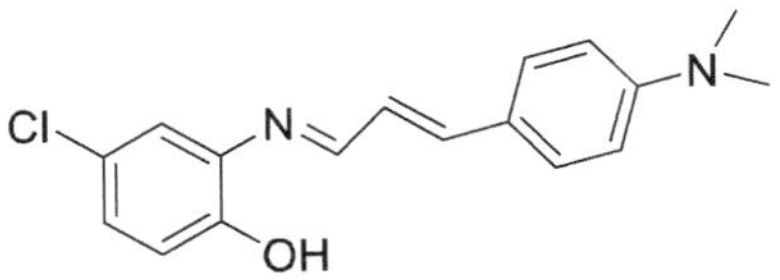	0.1	anthrapyridone-triazole-based probe [56]	0.5
Pb2+	Colorimetric & fluorescence spectroscopy	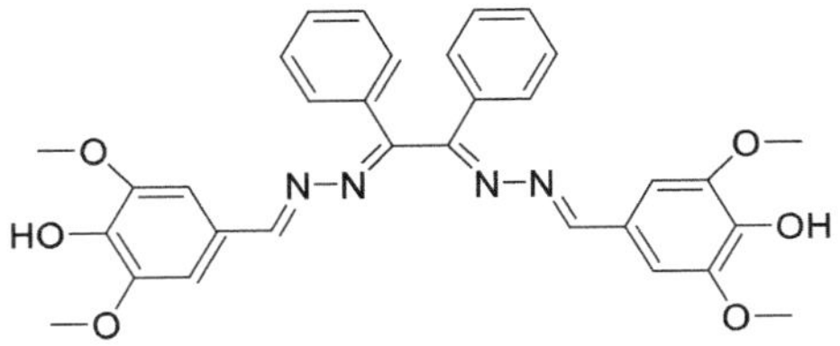	0.008	Rhodamine [57]	0.025
Pb2+	Colorimetric & fluorescence spectroscopy	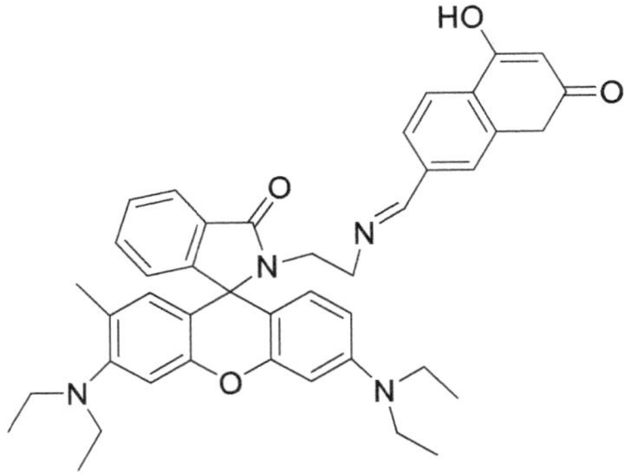	0.018	3-(1-isoquinolinyl) imidazo[5, 1-a] isoquinoline [58]	0.21

linear dynamic range. In this section, we have discussed some Schiff base metal complexes for the potentiometric sensors.

A PVC membrane based sensor has been reported by Gupta et al., having structure 1 shown in Figure 10.1 which is an ionophore. In this ionophore sensor oleic acid acts as anion excluder and ortho nitro phenyl octyl ether as a plasticizer. The reported sensor has excellent selectivity for the heavy metal ions and representative metals. The sensor is pH independent for the response. This was utilized as the ion-selective electrode in potentiometric titration of anions such as carbonates, fluoride, and oxalate [59].

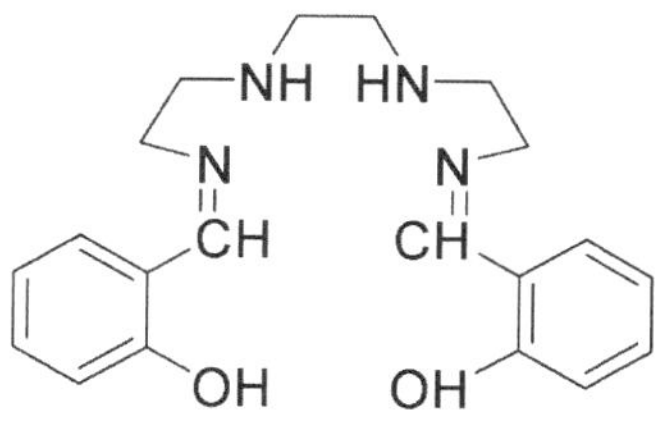

FIGURE 10.1 N,N-bis[2-(salicylidene amino) ethy1] ethane-1,2-diamine.

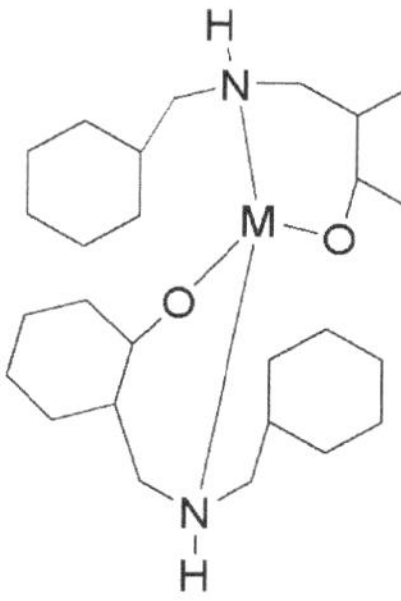

Structure 2

FIGURE 10.2 Structure of ionophore M(SBA)2 M=Cu(II), Zn(II), Cd(II).

10.4.2 [M(SBA)$_2$] Sensor

Wang et al. reported another PVC based membrane electrode sensor shown in structure 2 (Figure 10.2), as ion-selective electrode sensor for thiocyanate. The purported sensor had a very fast response time of 5–10 seconds and a very good efficiency up to three months. The proposed sensor was effectively utilized to detect thiocyanate in the saliva and urine sample of humans and wastewater too [60].

10.4.3 SNSB Sensor

Ganjali and his co-worker reported a series of PVC membrane based sensors for heavy metal ion, such as Tb(III) [61], Sm(III) [62], Yb(III) [63], Tb(III) [64], La(III) [65], and Eu(III) [66–68].

Ganjali further reported a new sensor for Eu(III) shown in structure 2 (Figure 10.3), in which 30% PVC membrane, 5% SNSB, 63% NPOE, and 5% potassium tetrakis(p-chlorophenyl) borate (KTpClPB) [68]. The synthesized sensor is pH independent, higher reproducibility, fast response time, and enhanced discriminating capacity for Eu(III) ions in assessment to other metal cations such as Mg, Ca, Na, Cu, Ni, Co, Zn, Ce, Gd, Sm, Yb, Pb, Tb, Nb, Fe, and Cr.

Askuner et al. reported the synthesis of new sensing membrane by trapping of PVC on a new ligand 4-(1-phenyl-1-methylcyclobutane-3-yl)-2-(2-hydroxy-5-bromobenzylidene)aminothiazole (showing in structure 2) (Figure 10.4). The membrane composition is 240 mg DOP, 120 mg PVC, 1.30 mg KTpClPB, 1.12 mg PCT dye, and 1.5 mL of THF. The accurateness of sensor was established by standard reference materials of peach leaves and water. The purported sensor effectively senses the copper presence in tea samples and tap water [69].

Mashhadizadeh et al. reported a sensor which was highly selective for Ni(II). The sensor is based on PVC electrode membrane having Schiff base NDBBD shown in structure 2 (Figure 10.5), work as carrier. Sensor has a response time of less than 10 seconds and is extendable with up to

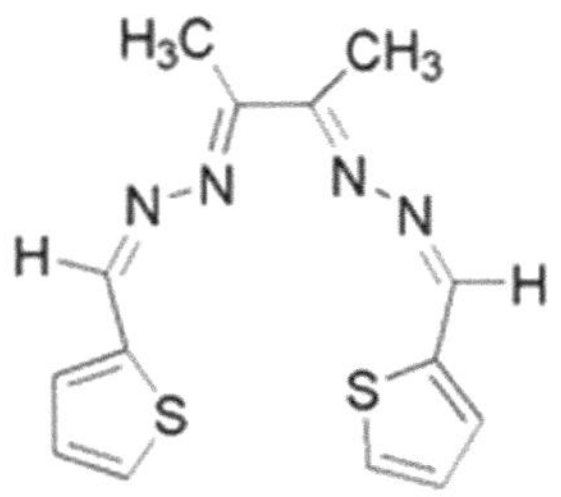

FIGURE 10.3 Bis(thiophenol) butane–2,3-dihydrazone.

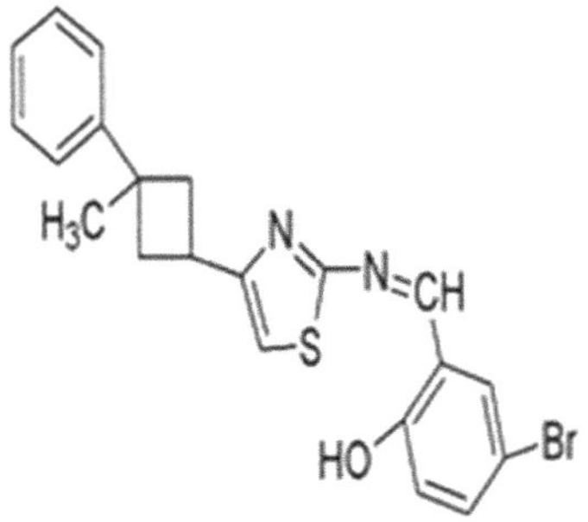

FIGURE 10.4 4-(1-phenyl-1-methylcyClobutane-3-y1)-2-(2-hydroxy-5-bromobenzylidene) aminothiazole.

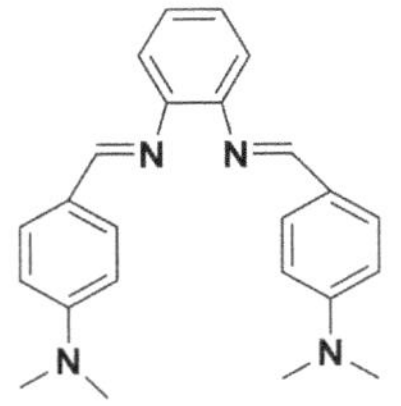

FIGURE 10.5 N,N-bis-(4-dimethylamino-benzene-1,2-diamine) (NDBBD).

two months' stability. The sensor is pH dependent, and works in the pH range of 4.5 to 9.0. The synthesized sensor shows excellent discriminating ability for Ni(II). The sensor electrode was successfully utilized in the potentiometric titration as an indicator electrode for Ni(II) ion [70].

Gupta and his co-authors reported the PVC electrode membrane shown in Figure 10.6 for the detection for Co(II). The sensor has anion excluder as NaTPB and solvent mediator DOP fabrication a new PVC membrane electrode for Co (II). The sensor electrode was effectively work in the Co(II) potentometric titration as indicator electrode [71].

Ardakani et al. reported copper Schiff base metal complex sensor structure 7 (Figure 10.7) for perchlorate detection in the real sample as an indicator electrode for sensing application. The sensor has a fast response time of less than 15 seconds and is used up to three months. The selectivity coefficient for the anions was examined by the fixed primary ion method and interference method [72].

Fu et al. synthesized Schiff base selective electrode for zinc, shown in structure 8 (Figure 10.8). The synthesized electrode has excellent selectivity for Zn in comparison to a broad variety of metal cations. The electrode has the advantage of a very fast response time, low resistance, and long life time. The electrode is pH independent; the test solution performs in a range of 3.4–5.8 pH [73].

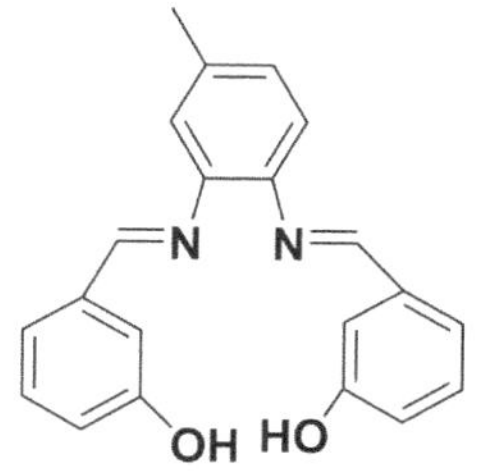

FIGURE 10.6 N, N-bis(salicylidene)-3,4-diaminotoluene.

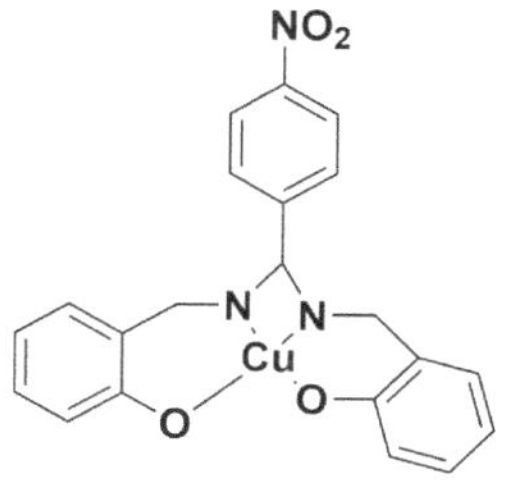

FIGURE 10.7 N,N-4-nitro-phenylmethanebis-(salicylaidiminato) copper (II).

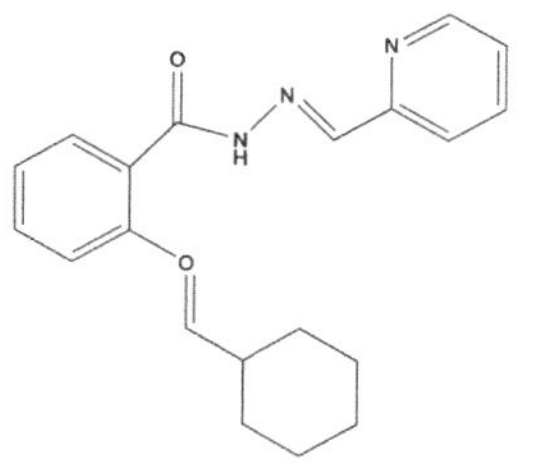

FIGURE 10.8 Schiff base.

Further, Sindhu et al. reported a Cu(II)-selective electrode made up by PVC membrane by using TADOBSCD shown in structure 9 (Figure 10.9) Schiff base complex as a neutral carrier. The synthesized electrode has very fast response time, that is, less than 10 seconds for the Cu concentration of 0.001 M and lees than 15 seconds for 10-6 M concentration of Cu(II). The reported electrode can be used up to eight months. The purported electrode has excellent selectivity and higher sensitivity for copper among other bivalent, trivalent, and tetravalent metal ions. The electrode has been utilized in the pH range of 3.5–8.0. The synthesized electrode was effectively utilized in potentiometric titration as an indicator electrode for the detection of many herbal drugs [74].

Chandra et al. reported SBM complexes based PVC membrane electrode for the Cu(II) ion. The Schiff base used in the electrode was zinc complex of acetophenone thio semi carbazone (ZATSC), which acts as a neutral carrier and the synthesis route is shown in Scheme 10.2. The electrode has a response time of fewer 10 seconds for 10^{-4} M concentration and less than 15 seconds for 10^{-6} M concentration. The synthesized electrode can be used up to nine months without any considerable divergence in potentials. The synthesized electrode work in a pH range of 1.5–12.3 and effectively examine the copper concentration in Yamuna river water at different area in Delhi [75].

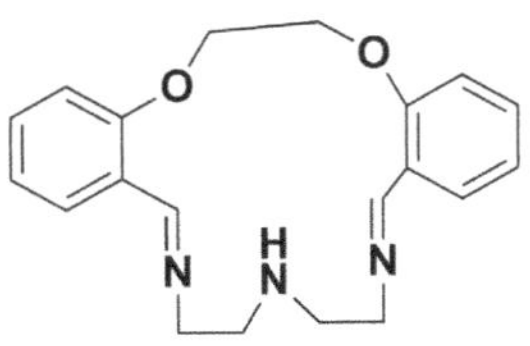

FIGURE 10.9 8.11.14-triaza-1.4-dioxo, 5(6), 16(17)-dibenzocycioseptadecane.

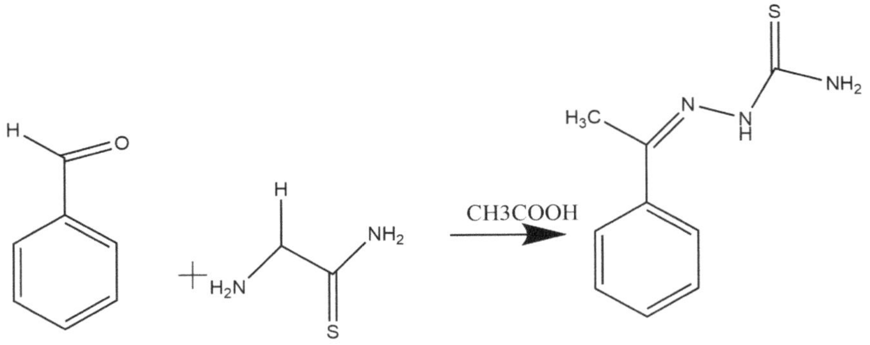

SCHEME 10.2 Synthesis of actophenonethiosemicarbazone.

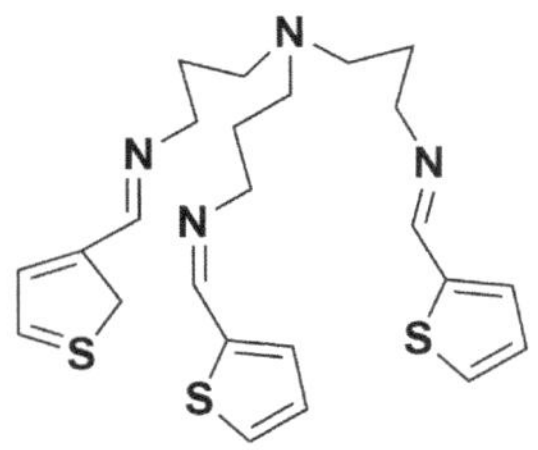

FIGURE 10.10 (tris (3-thiophenal) propyl)amine) (TTA).

Bandi et al. synthesized four novel Schiff base copper selective electrodes by polymeric membranes shown in structure 9a, 9b, 10a, and 10d (shown in below Table 10.5) [76, 77]. These electrodes were synthesized by using different anionic excluders and plasticizers. The polymeric membrane electrode shows good results in potentiometric data. The PME electrode further modified by coating graphite and examined as selective electrode for Cu(II). The coated electrode was effectively utilized in potentiometric determination of metal ion against EDTA. The quantity of Cu(II) was examined from the water, vegetable, soil, medicinal plants, and edible oil. The composition of electrodes and their structures are tabulated in Table 10.5.

Ayman reported a new PVC membrane based fluorescence sensor for Zn(II) detection from human hair, powdered milk, and in a few pharmaceuticals. The fluorescence sensor was prepared by using microwave irradiation via simple condensation between salicylaldehyde and 2,3-diaminonaphthalene. The sensor is pH independent and works in a range of 6–8, and has good selectivity for Zn(II) [78].

Zamani reported heptadentate Schiff's base (TTA, Figure 10.11) for the selective detection of Fe (III) ions. TTA is utilized as a polymeric membrane sensor, utilized in the pH range of 2.4–4.3 with a response time of less than 10 seconds. The proposed sensor was very selective for Fe(III) ion among transition and heavy metal ions. The electrode sensor life time was minimum ten weeks [79].

TABLE 10.5
List of Schiff's base based sensors

Sensor electrode	Structure
Structure 9a	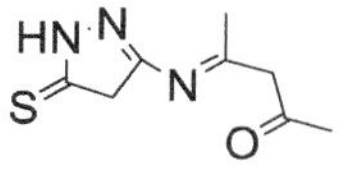 Structure 9a: 4-(mercapto-1.3.4-thudiazol-2-ylimino) pentan-2-one
Structure 9b	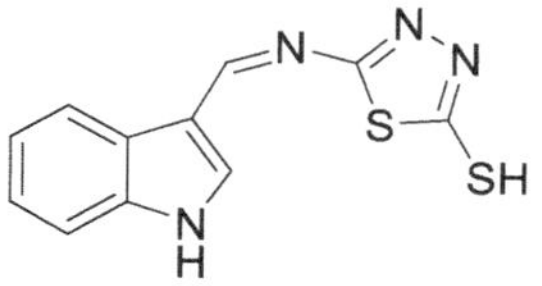 Structure 9b: (2-(indol-3-yl)vinyl)-1.3.4-thiadiazole-2-thiol Composition: NaTPB: TBP: PVC:: 4: 58: 36
Structure 10a	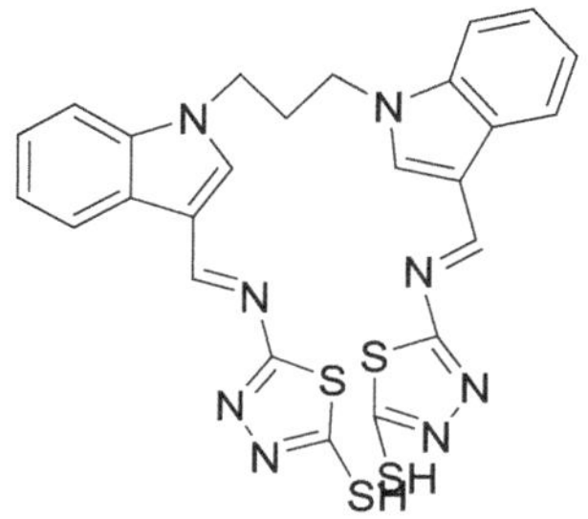 Structure 10a:5-((1-(3-(3-((5-mercapto-1.3.4-thiadiazole-2-ylimino)methyl)-1H-indole-1-yl)propyl)-1Hindol-3-yl)methyleneamino)1.3.4-thiadiazole-2-thiol(S1) Composition: KTpClPB: NPOE: PVC:: 4: 2: 58: 36
Structure 10b	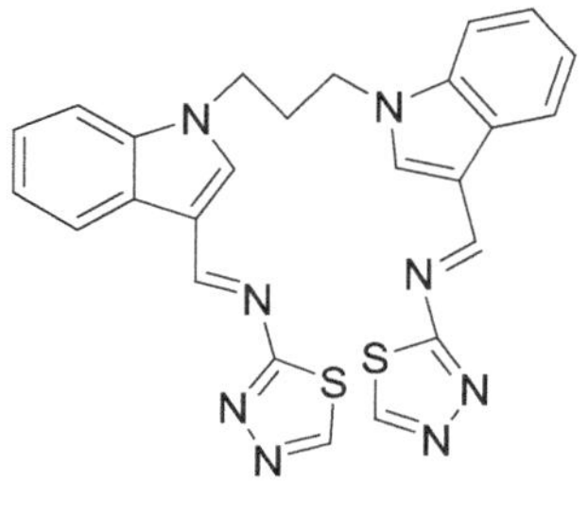 Structure 10b:N-((1-(3-(3-((thiazole-2-ylimino)methyl)-1H-indol-1-yl)propyl)-1H-indol-3-yl)methylene)thiazol-2-amine Composition: KTpClPB: NPOE: PVC:: 6: 2: 55: 37

Yuan reported an SBM complex of cobalt having general formula $[Co(L)_2](ClO_4).(C_3H_6O).(H_2O)$, where C_3H_6O is acetone and L is 2-((E)-(3-aminopyridin-4-ylimino)methyl)-phenol. Reported SBM complex has a broad pH range of 4–13 for the examination of Pb(II) ion over the broad variety of metal cations [80].

Ghanei-Motlagh and his co-authors reported several Schiff base complexes of polymeric membrane electrode (PME) for the selective detection of iodide ion and Cu(II) ion shown in structure 12 [81] and shown in structure 13 [82] (Figure 10.11 & 10.12), respectively. Structure 12 is an SBM complex of molybdenum which is a PME coated with a platinum disk and acts as a selective

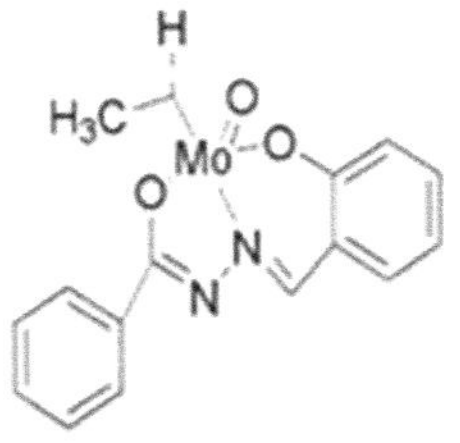

FIGURE 10.11 Methanol [E-N-(2-hydroxybenzlidene) benzohydrazido] dioxidomolybdeum (VI) [Mo (MHBD)].

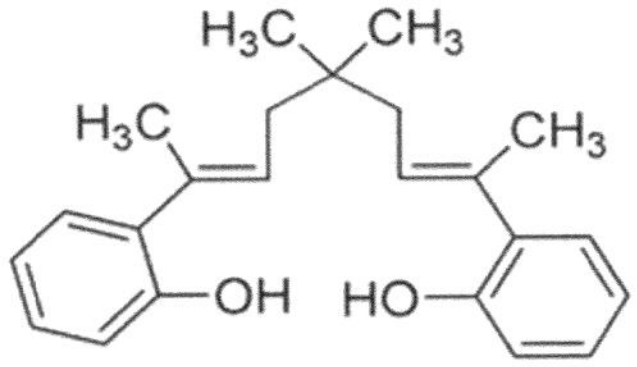

FIGURE 10.12 N.N1-(2.2-dimethylpropane-1 3-dily)-bis(dihyroxyacetophenone).

electrode for iodide ion. Ghanei-Motlagh et al. also studied the membrane composition, interfering ions, pH (independent of pH and work in broad range 2.0–8.5 pH), and response time (less than 10 seconds). The synthesized sensors were efficaciously functional to examination of iodide sample in the presence of interfering anions in the waste water and also work in precipitation titration as indicator electrode [81].

Figure 10.12 shows a PVC membrane sensor (NDHA) for ISE for Cu(II). The composition of PME is 30% PVC, 65% DBP, 1% KTpCIPB, and 4% NDHA. The synthesized electrode works as a potentiometric sensor in non-aqueous solution of acetone, ethanol, and methanol. The synthesized electrodes work in potentiometric titrations against EDTA for Cu(II) ion investigation. The detection ability of NDHA for metal ions was investigated by utilizing ab initio theoretical calculations for calcium, magnesium, aluminium, potassium, copper, cobalt, mercury, silver, zinc, cadmium, and lead. NDHA was complexed to M(NDHA)x for selective determination of metal ion, and binding capacity was calculated.

Figure 10.13 shows a novel platinum coated ISE for the detection of Cu (II) ion. The proposed coated electrode has a lesser response time of 9 seconds and stability up to 3.5 months. The proposed electrode has very high selectivity among a broad class of metal ion for the Cu(II). The sample for the determination of analyte was from environmental samples. The detection method was same potentiometric titration against EDTA [83].

Figure 10.14 shows a new membrane electrode reported by Hosseini which is highly selective for zinc ion based on N, N- Phenylene bis (salicylate aminato) Schiff base. The proposed ISE has a composition of PVC, NPOE, KTpCIB and a ligand L (shown in Figure 10.15) as a sensing substance. The proposed sensor electrode has less than a 10-second response time and works in the pH range of 3.0–7.0 with a stability period of two months. The ion-selective electrode worked even in the sample of wastewater which has multiple interfering polyvalent cations, signifying higher selectivity of the proposed electrode toward Zn(II) [84].

Zamani reported a novel ISE for the Tb(III) synthesized by using Schiff base N,N-bis-(5-nitro salicylidene)-2-amino benzylamine, ionophore, PVC, solvent mediator, NaTPB in the ration of 3: 30: 65: 2 (in percentage), respectively. The proposed electrode worked in the pH range of 2.6 to 9.4, with a response time of approx. 10 seconds. The electrode sensor has stability at least up to two

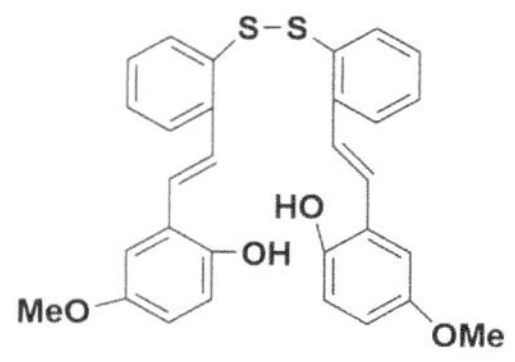

FIGURE 10.13 2-02-(2-(2-(2-hydroxy-5-methoxybenzylideneamino)phenyl)disufany1)-phenylimino) methyl)-4-methoxyphenol.

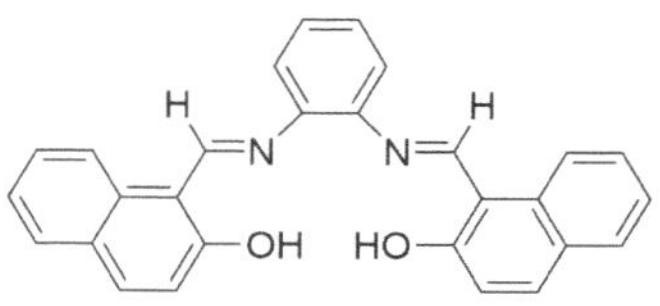

FIGURE 10.14 NN'-phenylene bis (salicylide arninato).

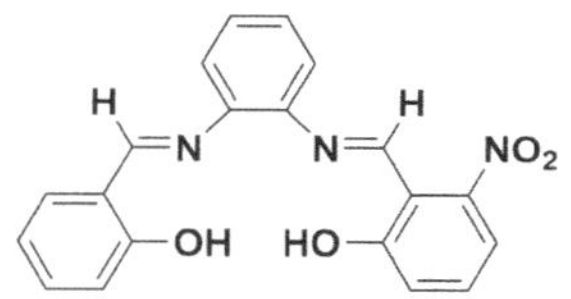

FIGURE 10.15 N.N'-bis (5-nitrosalicylidene)-2-aminobenzylamine.

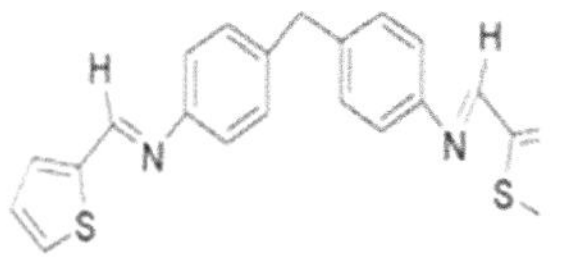

FIGURE 10.16 N-[(Z)-1-(2-thienyl) methylidene]-N-[4-4-{[(Z)-1-(2-thienyl) methylidene] amino} benzyl) phenyl] amine (TBPA).

months. The ISE is selective to Tb(III) ion among the representative metals, transition, and heavy metal cations. To examine the electrode analytical applicability for the sensor ability, the electrode was standardized by certified reference material [85].

Zamani et al. reported a new PVC based sensor of Schiff base TBPA (Figure 10.16) for the selective electrode for Lu(III). The synthesized sensor was very much selective for lutetium among the lanthanides specially and also from representative metals and transition metals. The proposed sensor works in the pH range of 2.7–10.6 and stability of two months at least with very fast response time of ten seconds. The synthesized sensor was successfully utilized as an indicator ISE electrode in Lu(III) potentiometric titration against EDTA. The assembled sensor accurateness was examined by observing Lu (III) ion in combinations of two and three different metal ions [86].

Kamal reported a novel electro active material for potentiometric and voltammetric sensor (VS) for the detection of Fe(II) ion. The electroactive material was synthesized from rhodamine based Schiff base, namely dimethyl-imino cinnamyl linked rhodamine (RC). VS worked in the concentration range of 2×10^{-6} to 3×10^{-4} M with a detection limit of 1.6×10^{-7} M. Reported sensors have

TABLE 10.6
List of electrochemical sensors which works as ion-selective electrode in potentiometrically

Sensor	Analyte	Source	Response time & stability
PVC based dibenzo(perhydrotriazino) aza-14-crown-4 ethers [89] Composition: Schiff base: PVC: NaTPB: TBP:: 5: 30: 3: 65 in w/w; mg	Be(II)		15 second & 4 months
GNPs-SPE and MSPE [90]	Ce(III)		4 s and 7s & 7 and 5 months
4-aminoantipyrine based fluorescent sensors [92]	Al(III)		5 second
oligoether linked Schiff bases [93]	Transition metal cations [Ca(II), Mg(II), Mn(II)]		
PVC based 2-(2aminothiazol-4-yl) phenol and PVC based [L2] 2-(4-phenyl-1,3-thiazol-2-yliminomethyl)phenol [90]	Gd(III)	Mineral sample	8 s & 3 months
M1: PVC: NPOE: CTAB 3.5: 30.0: 63.0: 3.5 [94]	fluoride	Tea sample	2.5 moths
PVC based 2-((*E*)-((1*R*,2*S*)-2-((*E*)-5-*tert*butyl-2-hydroxybenzylideneamino) cyclohexylimino)-methyl)-4-*tert*-butyl phenol [95]	Co(II) ion- selective electrode	water, medicinal plants soil, beer,	
2-((*E*)-((1*R*,2*S*)-2-((*E*)-3,5-di-*tert*-butylhydroxybenzylideneamino) cyclohexylimino)-methyl)-4,6-di-*tert*-butylphenol [95]	Co(II) ion- selective electrode	water, medicinal plants soil, beer,	

negligible interference from the other metal cations (from alkali and alkaline metals, and also from transition metals) [87].

Gupta et al. synthesized several Schiff base-based sensors for the detection of metal cations and anions from different samples [88–94], by utilizing different techniques as tabulated in Table 10.6. A novel amended screen printed electrode was synthesized by using gold nano particles which are self-assembled on 1,4-bis-(8-mercaptooctyloxy)-benzene (I) for the purpose of Ce (III) ion potentiometrically. The detection limit of AuNPs screen printed sensor was 3.25 × 10–10 at the pH range of 2.8–8.5 with a response time of 4 seconds, while the modified screen printed sensor possessed a detection limit of 9.5 × 10–8 mol L–1 at pH range of 3.5–7.5 with response time 7 seconds [90].

10.5 CONCLUSION

In this chapter, we have concluded the sensor application of Schiff base complexes, with a special emphasis on chemosensors and electrochemical sensors. On the literature study of sensor application of Schiff base complexes, it was concluded that these show more enhanced results in sensing with less LOD, fast response time, extra stability, and some chemosensors act as both fluorescent as well as colorimetric sensors. These complexes have broad way applications in many areas, and this area needs much more research and many fascinating outcomes are left.

ABBREVIATIONS

ISE	Ion-selective electrode
PVC	Poly vinyl chloride
SBM	Schiff base metal complexes
SBCS	Schiff base chemosensors
THF	Tetrahydrofuran
(PME)	Polymeric membrane electrode
EDTA	Ethylenediamine tetraacetic acid
LOD	Limit of detection
KTpCIB	Potassium tetrakis(p-chlorophenyl) borate
AuNps	Gold nanoparticles
NaTPB	Sodium tetraphenylborate
CHEQ	Chelation enhancement quenching effect
CHEF	Chelation-enhanced fluorescence
ICT	Intra/inter-molecular charge transfer
PET	Photo-induced electron transfer
o-NPOE	O-nitro pheny loctyl ether

ACKNOWLEDGMENTS

Dr Sapna Raghav is thankful to the Department of Chemistry, NBGSM College, Sohna, and Jyoti Raghav is grateful to Bennett University, Greater Noida, India.

REFERENCES

[1] Cohen, Allen B. 1975. The interaction of α-1-antitrypsin with chymotrypsin, trypsin and elastase. *Biochimica et Biophysica Acta (BBA)-Enzymology* 391:193–200.

[2] Abu-Dief, A. M., Mohamed, I. M. 2015. A review on versatile applications of transition metal complexes incorporating Schiff bases. *Beni-suef University Journal of Basic and Applied Sciences*, 4(2), 119–133.

[3] Rao, R., Patra, A. K., Chetana, P. R. 2007. DNA binding and oxidative cleavage activity of ternary (L-proline) copper(II) complexes of heterocyclic bases. *Polyhedron* 26:5331–5338.

[4] Fu, Y., Li, P., Bu, L., Wang, T., Xie, Q., Chen, J., Yao, S. 2011. Exploiting metal-organic coordination polymers as highly efficient immobilization matrixes of enzymes for sensitive electrochemical biosensing. *Analytical Chemistry* 83(17):6511–6517.

[5] Kimmel, D. W., LeBlanc, G., Meschievitz, M. E., Cliffel, D. E. 2012. Electrochemical sensors and biosensors. *Analytical Chemistry* 84:685–707.

[6] Tian, H., Yu, Z., Hagfeldt, A., Kloo, L., Su, L. 2011. Organic redox couples and organic counter electrode for efficient organic dye-sensitized solar cells. *Journal of American Chemical Society* 133:9413–9422.

[7] Habibi, M. H., Shojaee, E., Nichol, G. S. 2012. Synthesis, spectroscopic characterization and crystal structure of novel NNNN-donor μ-bis(bidentate) tetraaza acyclic Schiff base ligands. *Spectrochima Acta A*, 98: 396–404.

[8] Ulusoy, M., Birel, O., Sahin, O., Buyukgungor, O., Cetinkya, B. 2012. Structural, spectral, electrochemical and catalytic reactivity studies of a series of N2O2 chelated palladium(II) complexes. *Polyhedron* 38:141–148.

[9] Billman, J. H., Tai, K. M. 1958. Reduction of Schiff bases. II. Benzhydrylamines and structurally related compounds. *Journal of Organic Chemistry* 23:535–539.

[10] Valcarcel, M., Laque de Castro, M. D. 1994. Flow-Throgh Biochemical Sensors, Elsevier, Amsterdam.

[11] Spichiger-Kelle, U. 1998. Chemical Sensors and Biosensors for Medical and Biological Applications, Wiley-VCH, Weinheim.

[12] Gebreyesus, S. T., Khan, M. A. 2015. An overview on metal complexes of selected Schiff-bases with their electrochemical and sensor aspects. *Journal of Chemistry and Chemical Sciences*, 5(1):19–27.

[13] Khan, S., Chen, X., Almahri, A., Allehyani, E. S., Alhumaydhi, F. A., Ibrahim, M. M., Ali, Sh. 2021. Recent developments in fluorescent and colorimetric chemosensors based on Schiff bases for metallic cations detection: A review. *Journal of Environmental Chemical Engineering* 9:1063814.

[14] Berhanu, A. L., Gaurav, M. I., Malik, A. K., Aulakh, J. S., Kumar, V., Kim, K. H. 2019. A review of the applications of Schiff bases as optical chemical sensors. *Trends in Analytical Chemistry* 116:74–91.

[15] Khan, E., Khan, S., Gul, Z., Muhammad, M., 2020. Medicinal importance, coordination chemistry with selected metals (Cu, Ag, Au) and chemosensing of thiourea derivatives. a review. *Critical Review in Analytical Chemistry* 1–23.

[16] Kaur, B., Kaur, N., Kumar, S. 2018. Colorimetric metal ion sensors: A comprehensive review of the years 2011–2016. *Coordination Chemistry Reviews* 358:13–69.

[17] Lin, Q., Yang, Q., Sun, B., Wei, T., Zhang, Y. 2014. A novel highly selective "turn-on" fluorescence sensor for silver ions based on Schiff base. *Chinese Journal of Chemistry* 32:1255–1258.

[18] Chen, Z., Zhou, H., Gu, W., Liu, T., Xie, Z. 2019. A medium controlled fluorescent enhancement probe for Ag+ and Cu2+ derived from pyrene containing Schiff base, *Journal of Photochemistry and Photobiology A: Chemistry* 379:5–10.

[19] Aksuner, N., Henden, E., Yilmaz, I., Cukurovali, A., 2009. A highly sensitive and selective fluorescent sensor for the determination of copper(II) based on a schiff base, *Dyes Pigment.* 83:211–217.

[20] Shellaiah, M., Rajan, Y. C., Balu, P., Murugan, A., 2015. A pyrene based Schiff base probe for selective fluorescent turn-on detection of Hg 2+ ions with live cell application, *New Journal of Chemistry* 39:2523–2531.

[21] Singhal, D., Gupta, N., Singh, A. K. 2015. Chromogenic "naked eye" and fluorogenic "turn on" sensor for mercury metal ion using thiophene-based Schiff base, *RSC Advances* 5:65731–65738.

[22] Feng, L., Shi, W., Ma, J., Chen, Y., Kui, F., Hui, Y., Xie, Z. 2016. A novel thiosemicarbazone Schiff base derivative with aggregation-induced emission enhancement characteristics and its application in Hg2+ detection. *Sensors Actuators B Chemistry* 237:563–569.

[23] Su, Q., Niu, Q., Sun, T., Li, T. 2016. A simple fluorescence turn-on chemosensor based on Schiff-base for Hg2+-selective detection, *Tetrahedron Letters* 57:4297–4301.

[24] Li, Y., Shi, W., Ma, J., Wang, X., Kong, X., Zhang, Y., Feng, L., Hui, Y., Xie, Z. 2017. A novel optical probe for Hg2+ in aqueous media based on mono-thiosemicarbazone Schiff base. *Journal of Photochemistry and Photobiology A Chemistry* 338:1–7.

[25] Yan, J., Fan, L., Qin, J. C., Li, C. R., Yang, Z. Y. 2016. A novel and resumable Schiff-base fluorescent chemosensor for Zn(II). *Tetrahedron Letters* 57:2910–2914.

[26] Li, Y., Li, K., He, J. 2016. A "turn-on" fluorescent chemosensor for the detection of Zn(II) in aqueous solution at neutral pH and its application in live cells imaging. *Talanta* 153:381–385. https://doi.org/10.1016/j.talanta.2016.03.040.

[27] Das, B., Jana, A., Das Mahapatra, A., Chattopadhyay, D., Dhara, A., Mabhai, S., Dey, S. 2019. Fluorescein derived Schiff base as fluorimetric zinc (II) sensor via 'turn on' response and its application in livecell imaging. *Spectrochimica Acta Part A: Molecular and Biomolecular* 212:222–231.

[28] Chowdhury, B., Karar, M., Paul, S., Joshi, M., Choudhury, A. R., Biswas, B. 2018. Salen type ligand as a selective and sensitive nickel(II) ion chemosensor: A combined investigation with experimental and theoretical modelling. *Spectrochimica Acta Part A: Molecular and Biomolecular* 276:560–566.

[29] Rahimi, M., Amini, A., Behmadi, H. 2020. Novel symmetric Schiff-base benzobisthiazolesalicylidene derivative with fluorescence turn-on behavior for detecting Pb2+ ion. *Journal of Photochemistry and Photobiology A Chemistry* 388:112190.

[30] Wan, X., Ke, H., Tang, J., Yang, G. 2019. Acid environment-improved fluorescence sensing performance: A quinoline Schiff base-containing sensor for Cd2+ with high sensitivity and selectivity. *Talanta.* 199:8–13.

[31] Yan, J., Fan, L., Qin, J. C., Li, C. R., Yang, Z. Y. 2016. A novel chromone schiff-base fluorescent chemosensorfor Cd(II) based on C–N Isomerization. *Journal of Fluorine Chemistry.* 26:1059–1065.

[32] Qin, J. C., Yang, Z. Y. 2015. Bis-Schiff base as a donor-acceptor fluorescent probe: Recognition of Al3+ ions in near 100% aqueous solution. *Journal of Photochemistry and Photobiology A: Chemistry* 303–304:99–104.

[33] Bhorge, Y. R., Chou, T. L., Chen, Y. Z., Yen, Y. P. 2015. New coumarin-based dual chromogenic probe: Naked eye detection of copper and silver ions. *Sensors and Actuators B: Chemica* 220:1139–1144.

[34] Vashisht, D., Sharma, S., Kumar, R., Saini, V., Saini, V., Ibhadon, A., Sahoo, S. C., Sharma, S., Mehta, S. K., Kataria, R. 2020. Dehydroacetic acid derived Schiff base as selective and sensitive colorimetric chemosensor for the detection of Cu(II) ions in aqueous medium. *Microchemical Journal* 155:104705.

[35] Wang, L., Ye, D., Cao, D. 2012. A novel coumarin Schiff-base as a Ni(II) ion colorimetric sensor. Spectrochimica Acta Part A: Molecular and *Biomolecular.* 90:40–44.

[36] Peralta-Domínguez, D., Rodríguez, M., Ramos-Ortíz, G., Maldonado, J. L., Meneses-Nava, M. A., Barbosa-García, O., Santillan, R., Farf´an, N. 2015. A Schiff base derivative from cinnamaldehyde for colorimetric detection of Ni2+ in water. *Sensors and Actuators B: Chemical* 207:511–517.

[37] Ghorai, A., Mondal, J., Saha, R., Bhattacharya, S., Patr,a G. K. 2016. A highly sensitive reversible fluorescent-colorimetric azino bis-Schiff base sensor for rapid detection of Pb^{2+} in aqueous media. *Analytical Methods.* 8:2032–2040.

[38] Wu, G., Li, M., Zhu, J., Lai, K. W. C., Tong, Q., Lu, F. 2016. A highly sensitive and selective turn-on fluorescent probe for Pb(II) ions based on a coumarin-quinoline platform. *RSC Advances* 6:100696–100699.

[39] Rout, K., Manna, A., Sahu, M., Mondal, J., Singh, S. K., Patra, G. K. 2019. Triazole-based novel bis Schiff base colorimetric and fluorescent turn-on dual chemosensor for Cu^{2+} and Pb^{2+} application to living cell imaging. *RSC Advances* 944:25919–25931.

[40] Andrews, S., Silviya, S., Jeyanthi, D., Dev, E. S., Jebaraj, J. W., Balakrishnan, C. 2020. Biocompatible alkyne arms containing Schiff base fluorescence indicator for dual detection of Cd II and Pb II at physiological pH and its application to live cell imaging. *Analyst* 145:4576–4586.

[41] Bhanja, A., Mishra, S., Saha, K., Sinha, C. 2017. A fluorescence "turn-on" chemodosimeter for the specific detection of Pd^{2+} by a rhodamine appended Schiff base and its application in live cell imaging. *Dalton Transactions* 46:9245–9252.

[42] Adak, A. K., Purkait, R., Manna, S. K., Ghosh, B. C., Pathak, S., Sinha, C. 2019. Fluorescence sensing and intracellular imaging of Pd2+ ions by a novel coumarinyl-rhodamine Schiff base. *New J. Chemistry* 43:3899–3906.

[43] Li, B., Tian, J., Zhang, D., Tian, F. 2017. A novel colorimetric fluorescence sensor for Fe 3+ based on quinoline Schiff base. *Luminescence.* 32:1567–1573.

[44] Ambrosi, G., Borgogelli, G. E., Formica, M., Fusi, V., Giorgi, L., Micheloni, M., Rampazzo, E., Sgarzi, M., Zaccheroni, N., Prodi, L. 2015. PluS Nanoparticles as a tool to control the metal complex stoichiometry of a new thio-aza macrocyclic chemosensor for Ag(I) and Hg(II) in water. *Sens. Actuators B Chem.* 207:1035–1044.

[45] Amatori, S., Ambrosi, G., Borgogelli, E., Fanelli, M., Formica, M., Fusi, V., Giorgi, L., Macedi, E., Micheloni, M., Paoli, P., Rossi, P., Tassoni, A. 2014. Modulating the sensor response to halide using NBD-based azamacrocycles. *Inorganic Chemistry* 53:4560–4569.

[46] Amatori, S., Ambrosi, G., Fanelli, M., Formica, M., Fusi, V., Giorgi, L., Macedi, E., Micheloni, M., Paoli, P., Pontellini, R., Rossi, P., Varrese, M. A.. 2012. Multi-use NBD-based tetra-amino macrocycle: Fluorescent probe for metals and anions and live cell marker. *Chemistry – A European Journal* 18:4274–4284.

[47] Prodi, L., Bolletta, F., Montalti, M., Zaccheroni, N., Savage, P. B., Bradshaw, J. S., Izatt, R. M. 1998. A fluorescent sensor for magnesium ions. *Tetrahedron Letters* 39:5451–5454.

[48] Lvova, L., Caroleo, F., Garau, A., Lippolis, V., Giorgi, L., Fusi, V., Zaccheroni, N., Lombardo, M., Prodi, L., Di Natale, C., Paolesse, R. 2018. A fluorescent sensor array based on heteroatomic macrocyclic fluorophores for the detection of polluting species in natural water samples. *Frontiers Chemistry* 6:258.

[49] Ambrosi, G., Paz Clares, M., Pont, I., Formica, M., Fusi, V., Ricci, A., Paoli, P., Rossi, P., García-Espa˜na, E., Incl´an, M. 2020. Zn2+ and Cu2+ complexes of a fluorescent scorpiandtype oxadiazole

azamacrocyclic ligand: crystal structures, solution studies and optical properties. *Dalton Transactions* 49:1897–1906.
[50] Formica, M., Ambrosi, G., Fusi, V., Giorgi, L., Arca, M., Garau, A., Pintus, A., Lippolis, V. 2018. CdII/ZnII discrimination using 2,5-diphenyl[1,3,4]oxadiazole based fluorescent chemosensors. *New Journal Chemistry* 42:7869–7883.
[51] Canonico, B., Giorgi, L., Nasoni, M. G., Montanari, M., Ambrosi, G., Formica, M., Ciacci, C., Ambrogini, P., Papa, S., Fusi, V., Luchetti, F. 2021. Synthesis and biological characterization of a new fluorescent probe for vesicular trafficking based on polyazamacrocycle derivative. *Journal of Biological Chemistry.* 402:1225–1237.
[52] Conti, L., Flore, N., Formica, M., Giorgi, L., Pagliai, M., Mancini, L., Fusi, V., Valtancoli, B., Giorgi, C. 2021. Glyphosate and AMPA binding by two polyaminophenolic ligands and their dinuclear Zn(II) complexes. *Inorganica Chimica Acta* 519:120261.
[53] Ambrosi, G., Fanelli, M., Paoli, P., Formica, M., Paderni, D., Rossi, P., Micheloni, M., Giorgi, L., Fusi, V. 2020. Zn(II) detection and biological activity of a macrocycle containing a bis(oxadiazole)pyridine derivative as fluorophore. *Dalton Transactions* 49:7496–7506.
[54] Wang, L., Zhang, C., He, H., Zhu, H., Guo, S., Zhou, S., Wang, S., Zhao, J. R., Zhang, J. 2020. Cellulose-based colorimetric sensor with N, S sites for Ag+ detection. *International Journal of Biological Macromolecules* 163:593–602.
[55] Kumar, R., Bawa, R., Gahlyan, P., Dalela, M., Jindal, K., Jha, P. K., Tomar, M., Gupta, V. 2019. Pyrene appended bis-triazolylated 1,4-dihydropyridine as a selective fluorogenic sensor for Cu2+. *Dye. Pigment.* 161:162–171.
[56] Kumar, A., Chae, P. S., Kumar, S. 2020. A dual-responsive anthrapyridone-triazole-based probe for selective detection of Ni2+ and Cu2+: A mimetic system for molecular logic gates based on color change. *Dyes Pigments.* 174:108092.
[57] Li, Zhou Y., Li, Y. F., Kong, X. F., Zou, C. X., Weng, C. 2013. Colorimetric and fluorescent chemosensor for citrate based on a rhodamine and Pb2+ complex in aqueous solution. *Analytica Chimica Acta.* 774:79–84.
[58] Mahata, S., Bhattacharya, A., Kumar, J. P., Mandal, B. B., Manivannan, V. 2020. Naked-eye detection of Pd2+ ion using a highly selective fluorescent heterocyclic probe by "turn-off" response and in-vitro live cell imaging. *Journal of Photochemistry and Photobiology* 394:112441.
[59] Gupta, V. K., Singh, A. K., Gupta, B. 2006. A cerium (III) selective polyvinyl chloride membrane sensor based on a Schiff base complex of N.N'-bis[2-(salicylideneamino) ethyl]ethane-1.2-diamine. *Anaytica Chimica Acta.* 575:198–204.
[60] Wang, F. C., Chai, Y. Q., Yuan, R. 2008. Thiocyanate-selective electrode based on nsalicylidene-benzylamineato copper(II) complex. *Russian Journal of Electrochemistry*, 44(3):272–277.
[61] Ganjali, M. R., Rasoolipour, S., Rezapour, M., Norouzi, P., Adib, M. 2008. Synthesis of thiophene-2-carbaldehyde-(7-methyl-1,3-benzothiazol-2-yl)hydrazone and its application as an ionophore in the construction of a novel thulium (III) selective membrane sensor. *Electrochemistry Communications* 7:989–994.
[62] Chowdhury, D. A., Ogata, T., Kamata, S., Ohashi, K. 1996. Samarium(III)-selective electrode using neutral bis(thiaalkylxanthato)alkanes. *Analytical Chemistry* 68:366–370.
[63] Ganjali, M. R., Rasoolipour, S., Rezapour, M., Norouzi, P., Tajarodi, A., Hanifehpour, Y. 2005. Novel ytterbium(III) selective membrane sensor based on N-(2-pyridyl)-N-(2-methoxyphenyl)-thiourea as an excellent carrier and its application to determination of fluoride in mouthwash preparation samples. *Electroanalysis* 17:1534–1539.
[64] Ganjali, M. R., Ghesmi, A., Hosseini, M., Pourjavid, M. R., Rezapour, M., Shamsipur, M., Salavati-Niasari, M. 2005. Novel terbium(III) sensor based on a new bispyrrolidene Schiff's base. *Sensors Actuators B.* 105:334–339.
[65] Ito, T., Coto, C. 2001. Ion-selective solvent polymeric membrane electrodes based on 1- phenyl-3-methyl-4-acyl-5-pyrazolones for trivalent lanthanoid ions. *Journal of Trace and Microprobe Techniques* 19:601–613.
[66] Ganjali, M. R., Rahimi, M., Maddah, B., Moghimi, A., Borhany, S. 2004. An Eu(III) sensor based on N,N-diethyl-N-(4-hydroxy-6-methylpyridin-2-yl)guanidine. *Analytical Sciences* 20:1427–1431.
[67] Ganjali, M. R., Rezapour, M., Pourjavid, M. R., Haghgoo, S. 2004. PPT level detection of samarium(III) with a coated graphite sensor based on an antibiotic. *Analytical Sciences* 20:1007–1011.

[68] Ganjali, M. R., Norouzi, P., Daftari, A., Faridbod, F., Salavati-Niasari, M. 2007. Fabrication of a highly selective Eu(III) membrane sensor based on a new S-N hexadenatates Schiff's base. *Sensors and Actuators B* 120:673–678.
[69] Aksuner, N., Henden, E., Yilmaz, I., Cukurovali, A. 2009. A high sensitive and selective fluorescent sensor for the determination of copper (II) based on a Schiff base. *Dyes and Pigments*, 83: 211–217.
[70] Mashhadizadeh, M. H., Sheikhshoaie, I., Saeid-Nia, S. 2003. Nickel (II)-selective membrane potentiometric sensor using a recently synthesized Schiff base as neutral carrier. *Sensor and Actuators* B. 94:241–146.
[71] Gupta, A. K, Singh, A. K., Sameena, S., Gupta, B. 2006. A cobalt (II)-selective PVC membrane based on a Schiff base complex of N.N'-bis(salicylidene)-3.4- diaminotoulene. *Analytica Chimica Acta*. 566:5–10.
[72] Ardakani, M. Md. 2010. Highly selective perchlorate membrane electrode based on synthesized Schiff base complex. *Iran. J. Chem. Eng*. 29(3):123–132.
[73] Fu, Q., Qian, S., Li, N., Xia, Q., Ji, Y. 2012. Characterization of a new Zn(II)- selective electrode based on Schiff-base as ionophore. *International Journal of Electrochemical Science* 7:6799–6806.
[74] Sindhu, S. K., Kumar, S., Singh, L. R. 2010. Fabrication of novel Cu(II) selective PVC electrode and its estimation in herbal drugs. *International Journal of Pharma World Research* 1(1).
[75] Chandra, S., Sharma, K., Kumar, A., Tomar, P. K. 2010. Cu(II) selective PVC membrane electrode based on zinc complex of Acetophenonethiosemicarbazone (ZATSC) as an ionophore. *Der Pharma Chemica*. 2(6):256–266.
[76] Bandi, K. R., Singh, A. K., Upadhyay, A. 2013. Electroanalytical and naked eye determination Cu(II) ion in various environmental samples using 5-amino-1.3.4-thiadiazole-2-thio based Schiff bases. *Materials Science and Engineering C*. 34:149–157.
[77] Bandi K. R., Singh A. K., Upadhyay A. 2013. Biologically active Schiff bases as potentiometric sensor for the selective determination of Nd(III) ion. *Electrochimica Acta* 105:654–664.
[78] Ayman, A., Abdel, A., 2013. A novel highly sensitive and selective optical sensor based on a symmetric tetradentate Schiff-base embedded in PVC polymeric film for determination of Zn(II) ion in real samples. *Journal of Luminescence*. 143:663–669.
[79] Zamani, H. A., Ganjali, M. R., Faridbod, F., Salavati-Niasari, M. 2012. Heptadentate Schiff-base based PVC membrane sensor for Fe(III) ion determination in water samples. *Materials Science and Engineering C*. 32:564–568.
[80] Yuan, X. J., Wang, R. Y., Mao, C. B., Wu, L., Chu, C. Q., Yao, R., Gao, Z. Y., Wu, B. L., Zhang, H. Y.. 2012. New Pb (II)-selective membrane electrode based on new Schiff base complex. *Inorganic Chemistry Communications*. 15:29–32.
[81] Ghanei-Moltlagh, M., Ali Taher, M., Ahmadi, K., Sheikhshoaie, I. 2011. Iodide selective membrane electrodes based on a Molybdenum a Molybdenum-Salen as a neutral carrier. *Materials Science and Engineering C*. 31:1625–1631.
[82] Ghanei-Motlagh, M., Taher, M. A., Saheb, V., Fayazi, M., Sheikhshoaie, I. 2011. Theoretical and practical investigations of copper ion selective electrode with polymeric membrane based on N.N'-(2.2'-dimethylpropane-1.3-diyl)-bis(dihydroxyacetophenone). *Electrochimica Acta*. 56:5376–5385.
[83] Shokrollahi, A., Abbaspour, A., Ghaedi, M., Naghashian Haghighi, A., Kianfar, A. H., Ranjbar, M. 2011. Construction of new Cu(II) coated wire ion selective electrode based on 2-((2-(2-(2-(2-hydroxy-5-methyoxybenzylideneamino)phenyl) disufanyl)phenylimino)methyl)-4-methoxphenol Schiff base. *Talanta*, 84:34–41.
[84] Hosseini, M., Dehghan Abkenar, S., Ganjali, M. R., Faridbod, F. 2011. Determination of zinc (II) ions in waste water samples by a novel zinc sensor based on a new synthesized Schiff's base. *Materials Science and Engineering C*. 31:428–433.
[85] Zamani, H. A., Zabihi, M. S., Rohani, M., Zangeneh-Asadabdi, A., Reza Ganjali, M, Faridbod, F., Meghdadi, S. 2011. Quantitative monitoring of terbium ion by a Tb(III) selective electrode based on a new Schiff's base. *Materials Science and Engineering C*. 31:409–413.
[86] Zamani, H. A., Rohani, M., Zangeneh-Asadabadi, A., Zabihi, M. S., Ganjali, M. R., Salavati-Niasar, M. 2010. A novel lutetium (III) membrane sensor based on a new symmetric S-N Schiff's base for Lu (III) analysis in real sample. *Material and Engineering C*. 30: 917–920.
[87] Kamal,A.,Kumar,N.,Bhalla,V.,Kumar,M.,KumarMahajan,R.2014.Rhodaminedimethyliminocinnamyl based electrochemical sensors for selective detection of iron(II). *Sensors andActuators B*. 190:127–133.

[88] Awad Alo, T., Mohamed, G. G., Azzam, E. M. S., Abd-elaal, A. A. 2014. Determination of Ce(III) in environmental polluted samples. *Sensor and Actuators B*. 191:192–203.
[89] Gupta,, V. K., Singh A. K., Mergu, N. 2012. A new beryllium ion-selective membrane electrode based on dibenzo(perhydrotriazino)aza-14-crown-4-ether. *Analytica Chimica Acta*. 749: 44–50.
[90] Gupta, V. K., Singh, A. K., Kumawat, L. K. 2013. A novel gadolinium ion-selective membrane electrode based on 2-(4-phenyl-1.3-thiazol-2-yliminomethyl)phenol. *Electrochimica Acta*. 95:132–138.
[91] Gupta, V. K., Jain, R., Hamdan, A. J., Agarwal, S., Bharti, A. K. 2010. A novel ion selective sensor for promethium determination. *Analytica Chimica Acta*. 681:27–32.
[92] Gupta, V. K., Singh, A. K., Mergu, N. 2014. Antipytine based Schiff bases as turn-on fluorescent sensors for Al(III) ion. *Electrochimica Acta* 117:405–412.
[93] Gupta, V. K., Singh, A. K., Ganjali, M. R., Norouzi, P., Faridbod, F., Mergu, N. 2013. Comparative study of colorimetric sensors based on newly synthesized Schiff bases. *Sensors and Actuators B: Chemical*. 182: 642–651.
[94] Gupta, V. K., Jain, A. K., Pal, M. K., Bharti, A. K. 2012. Comparative study of fluoride selective PVC based electrochemical sensors. *Electrochimica Acta*. 80:316–325.
[95] Bandi, K. B., Singh, A. K, Jain, V. A. K., Gupta, K. 2011. Electroanalytical studies on cobalt(II) ion-selective sensor of polymeric membrane electrode and coated graphite electrode based on n2o2 salen ligand. *Electroanalysis*. 23(12):2839–2850.

11 State of the Art and Future Perspective of Schiff Base Metal Complexes in Biological Treatment

Archana Gautam[1*] *and Monica Tyagi*[2]
[1]Department of Chemistry, S.G.T.B. Khalsa College, University of Delhi, Delhi, India
[2]Department of of Applied Science, GN Group of Institutes, Greater Noida, Uttar Pradesh, India
Email address: archanagautam_00@yahoo.com

CONTENTS

11.1 GENERAL INTRODUCTION OF SCHIFF BASES

In 1864, Hugo Schiff, a German scientist synthesized a new class of organic compounds which are named as Schiff bases in his honor. Schiff bases are prepared by the condensation of a primary amine with an aldehyde or a ketone under suitable conditions. Chemically, the oxygen of the carbonyl group (>C=O) present in aldehyde or ketone is replaced by nitrogen with the elimination of a water molecule, resulting in the formation of azomethine or imine (>C=N) group or Schiff bases. A number of metals with various oxidation states can be stabilized by these Schiff bases ligands, thereby regulating the performance of these metals in a wide variety of catalytic transformations. The ligands containing Sp2 hybridized nitrogen atoms in which N-atom is a part of the aromatic system are extensively important in coordination chemistry [1] and are accountable for broad-spectrum properties. The synthesis of Schiff base compounds is quite simple and they have high structural flexibility due to the presence of azomethine or imine (>C=N-) group which make them excellent chelating agents.

The method for the preparation of Schiff bases is shown in Figure 11.1. The reaction include the nucleophilic addition of aliphatic or aromatic primary amines (RNH_2 or $ArNH_2$) to carbonyl compounds (>C=O) ketones or aldehydes.

DOI: 10.1201/9781003291459-11

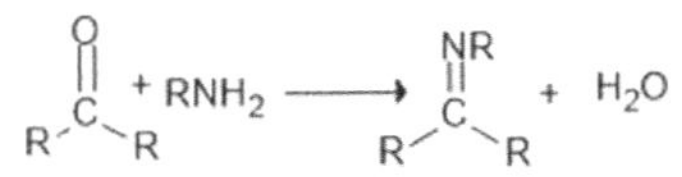

FIGURE 11.1 Structure of Schiff's base.

Generally, aldehydes react faster than ketones to form Schiff bases in condensation reactions. This is because aldehydes are sterically less hindered as compared to ketones. Schiff bases are capable of forming highly stable complexes with transition metals producing bidentate, tridentate, tetradentate, hexadentate, or polydentate ligands in common.

Schiff base is a topic of current and rising interest because it has anticancer, antibacterial, antiviral, antifungal, and other biological properties. Chemically macrocyclic moieties are of significant interest due to their versatility as ligands, the availability of various potential donor atoms. They can produce mononuclear or polynuclear complexes, some of which are physiologically relevant; in particular, first row transition metal complexes containing such ligands exhibit a diverse set of biological features [2–11].

The Schiff bases have received a lot of attention during the last decade due to their ability to selectively encapsulate transition metal ions [12–15]. Macrocyclic as well as acyclic Schiff bases, in particular, have been postulated to be ideal systems for the production of mono and polynuclear complexes with transition metals and lanthanides [16, 17].

The applications of Schiff base ligands and their metal complexes depend on nature and structure of the reactants (dicarbonyl and diamine compounds). The corresponding metal ions also play an important role in a number of activities, due to their electronic, magnetic, and electrochemical properties.

Schiff base macrocyclic transition metal complexes are important because of their biological relevance, the stable structure [18], and different other spectral characteristics [19]. Schiff base macrocyclic ligands display remarkable biological properties such as antitumor [20], antibacterial [21], antifungal activities [22]. Coordination of these compounds with transition metal ions often enhances their activitiy [23]. Cobalt complexes containing nitrogen donor ligands have been utilised to simulate biological systems such as vitamin B12 coenzyme [24, 25]. As transition metals, they are also of relevance in the context of dioxygen activation.

Coordination complexes that oxidatively cleave DNA under physiological conditions are gaining attention in the progress of artificial nucleases [26]. Due to the importance of the copper ion in enzymatic catalysis, efforts are being made to explore the chemistry of redox active copper complexes as chemical nucleases [27, 28].

Schiff base macrocyclic ligands and their metal complexes constitute a significant class of antibiotics [29] that are active against both Gram-negative and Gram-positive bacteria. Since the discovery of its first representative, Streptomycin, the mechanism of their action has been intensively investigated [30, 31].The continued interest in finding such chelating agents is mainly because they are used in labeling monoclonal antibodies [32] with radioactive metal for cancer diagnosis therapy. They show great potential as highly efficient catalysts in asymmetric transformation [33]. During the past decades much attention has been focused on DNA binding with inert transition metal complexes and monitored by spectral studies [34]. Some important biological activities of Schiff base and their metal complexes are shown in Figure 11.2.

11.2 ANTICANCER ACTIVITY

Wongsuwan et al. (2021) [35] studied *in vitro* anticancer activity of Fe(II) and Fe(III) complexes of the Schiff base ligand, N-(8-quinolyl)-X-salicylaldimine (qsal), against the A549 human lung adenocarcinoma cell line. Fe(III) complex as shown in Figure 11.3 exhibits highest anticancer

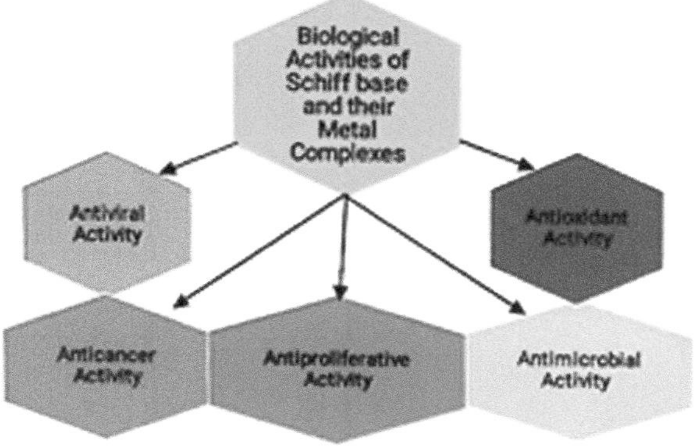

FIGURE 11.2 Biological activities of Schiff's base.

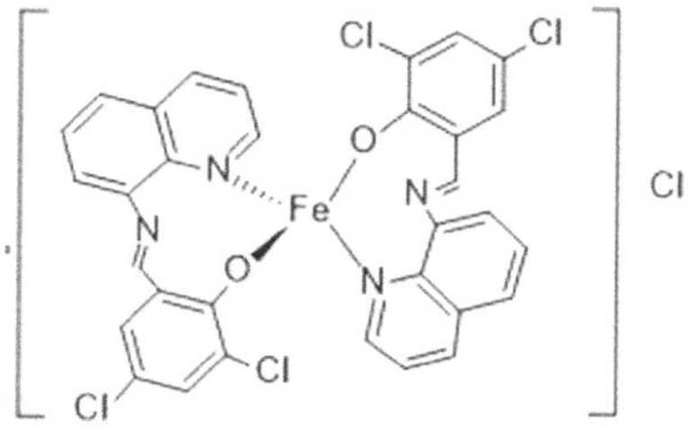

FIGURE 11.3 Fe(III) complexed with shiff's base.

activity (IC50 = 10 μM) which is better than two common anticancer agents, Etoposide and Cisplatin.

Chen et al. (2021) [36] have studied three Schiff base ligands [Zn(La)$_2$] (4a), [Zn(Lb)$_2$] (4b), and [Cu(Lc)$_2$] (4c), of Zn(II) and Cu(II) complexes as shown in Figure 11.4 and examined for antiproliferative activity against T-24, HepG2 and SK-OV-3 tumor cell lines. The complexes showed varied cytotoxicity against the three cell lines, with IC50 values ranging from 9.00 and 46.72 μM against 6.828–7.719 μM of cisplatin. They were found to be non-cytotoxic on normal liver cell line HL-7702. Studies on DNA binding found that complexes connected to DNA predominantly through weak binding interactions, which have been corroborated by spectral and other analysis. The complex 3b was found to be capable of blocking the HepG2 cells cycle at the S phase and caused apoptosis by the mitochondria-related apoptotic pathway.

Abu-Dief et al. (2021) [37] studied 1-((3,5-di-tert-butyl-2-hydroxybenzylidene) amino) naphthalen-2-ol-5-sodium sulfonate (DSHN) complexes with Zn(II) and VO(II). The studies revealed that both the complexes shown in Figure 11.5, DSHNZn (5a) and DSHNVO (5b), exhibit high antiproliferative activity against HCT-116, MCF-7, and HepG2 cancer cell lines with IC50 = 30.20, 13.90 and 21.10 μM for DSHNZn, respectively, and IC50 = 31.90, 15.30 and 22.80 μM for DSHNVO, respectively, against IC50 = 13.30, 4.12 and 7.50μM of vinblastine. Further, the binding nature of the complexes was examined with calf thymus DNA (CTDNA), indicating that the binding occurs just through the intercalation or replacement mode.

Liao et al. (2021) [38] studied a Schiff base Co(III) complex of N,N'-bis(20-hydroxyphenyl acetone)o-ethanediamine as shown in Figure 11.6, and evaluated for cytotoxicity against HeLa, human colon cancer cells (LoVo), A549 by MTT assay using cisplatin as reference, the cisplatin-resistant cell line (A549/cis). Further, the cytotoxicity against the normal cell lines LO2 was also

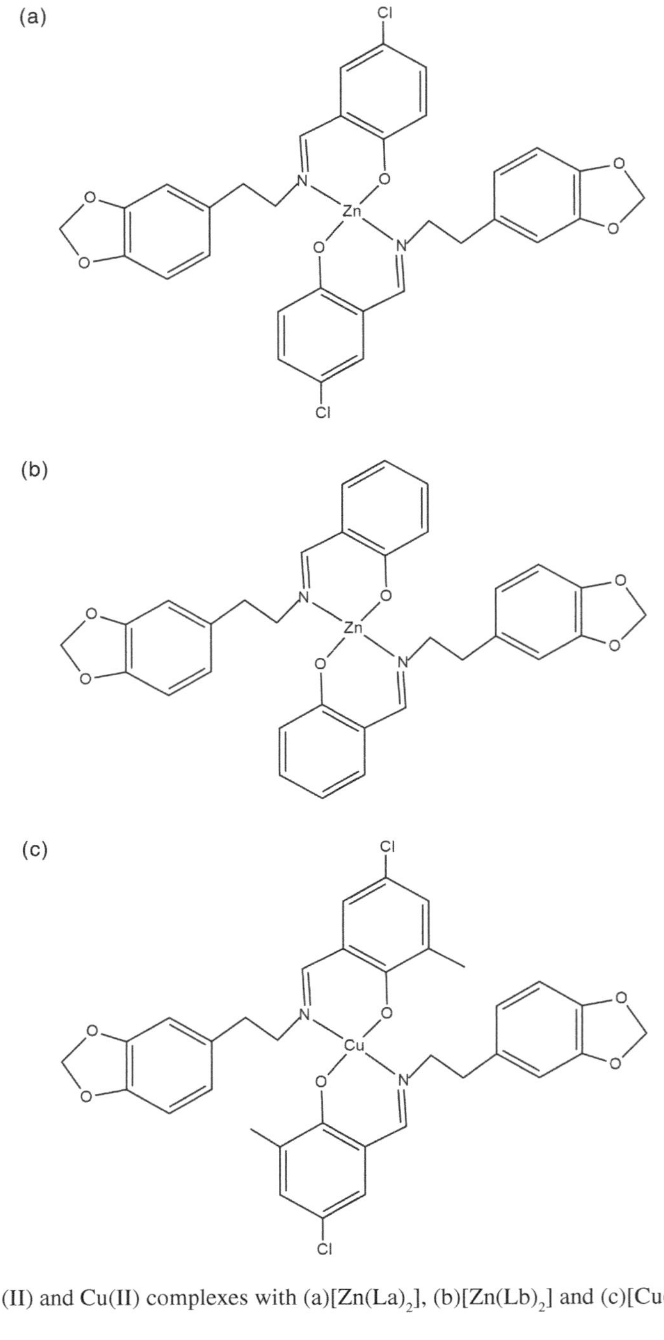

FIGURE 11.4 Zn (II) and Cu(II) complexes with (a)[Zn$(La)_2$], (b)[Zn$(Lb)_2$] and (c)[Cu$(Lc)_2$].

evaluated. Complex M3 efficiently inhibited the growth of all the tested cancer cells in 48 h treatment in a dose-dependent manner. The IC50 against HeLa was found to be close to that of cisplatin (IC50 = 12.40 μM versus 9.74 <μM). The activity of the complex against the cisplatin-resistant cell line A549/cis was found to be higher than the reference (IC50 = 18.03 μM versus 44.79 μM). However, the complex shown in Figure 11.6 was found to be less toxic than cisplatin (IC50 = 6.27

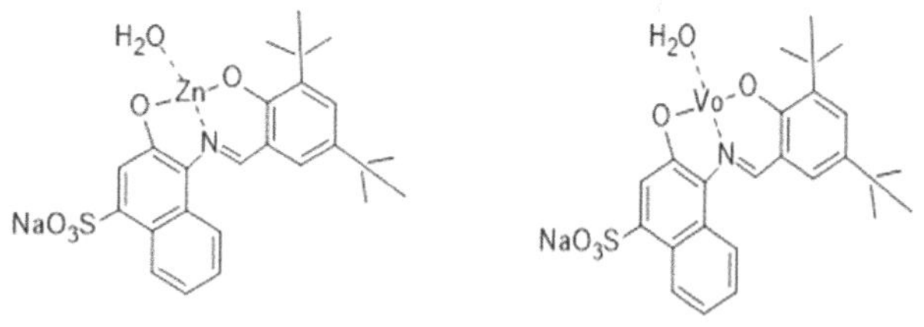

FIGURE 11.5 DSHN complexed with (a) Zn(II) and (b) VO(II).

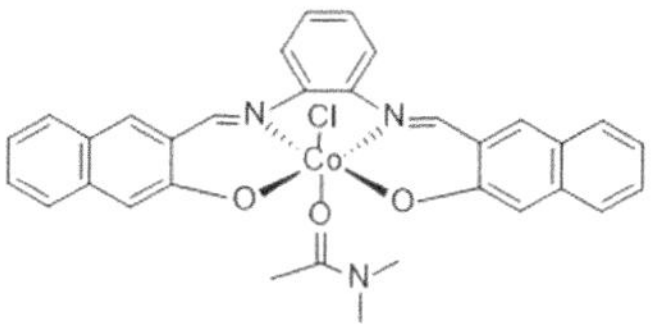

FIGURE 11.6 Co(III) complexed with N,N'-bis(20-hydroxyphenyl acetone)o-ethanediamine.

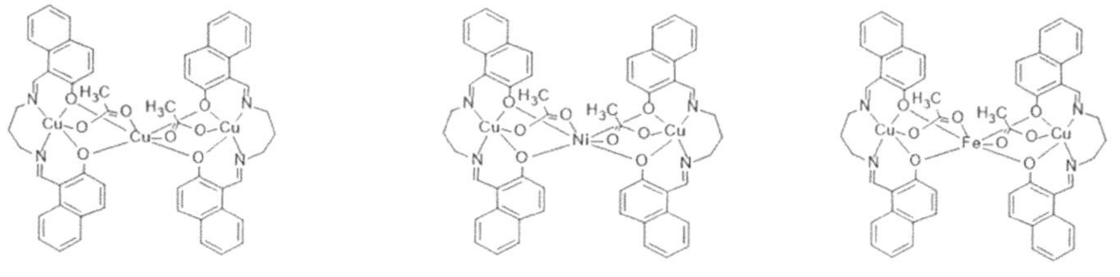

FIGURE 11.7 Schiff base transition metal complexes of (a)Cu, (b)Ni and (c)Fe with N,N-bis[(2-hydroxy-1-naphthalenyl)methylene]-propane-1,3-diamine.

μM versus 2.61 μM) but in due course it inhibited cell proliferation by blocking the synthesis of DNA and acting on HeLa cells nuclear division.

Shi et al. (2020) [39] have studied several Schiff base transition metal complexes, N,N-bis [(2-hydroxy-1-naphthalenyl)methylene]-propane-1,3-diamine, and determined them for *in vitro* anticancer activities against some human cancer cell lines by the MTT assay and compared them with cisplatin. The study revealed that the complexes shown in Figure 11.7, Cu (7a), Ni (7b) and Fe (7c), have higher antitumor activities than cisplatin. Further, it was revealed from flow cytometry that (7c) complex have highest inhibitory effects (IC50 below 0.5 μM for T-24 human bladder cancer cell line) and almost non-toxic to normal cell HL-7702.

Hassan et al. (2020) [40] synthesized and screened Mn(II), Co(II), Ni(II), Cu(II), Zn(II) and Zr(IV) metal complexes of a tridentate Schiff base ligand, (E)-1-(2-hydroxy-3-methoxybenzylidene)-3-phenylurea. The cytotoxic activity of synthesized compounds were carried out against human colon carcinoma (HCT-116), found to have IC50 value 6.9 μg/mL as well as breast carcinoma cells (MCF-7), found to have IC50 value 7.22 μg/mL, using cisplatin as standard drug. All the complexes were found to be more active in comparision to free ligand especially, Mn (8a), Cu (8b), and Zn (8c) complexes as shown in Figure 11.8, and have high cytotoxic activity.

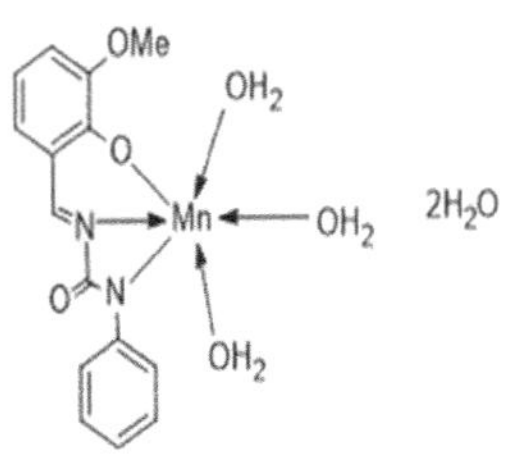

8a

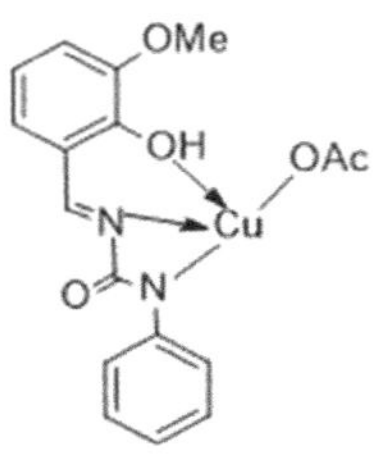

8b

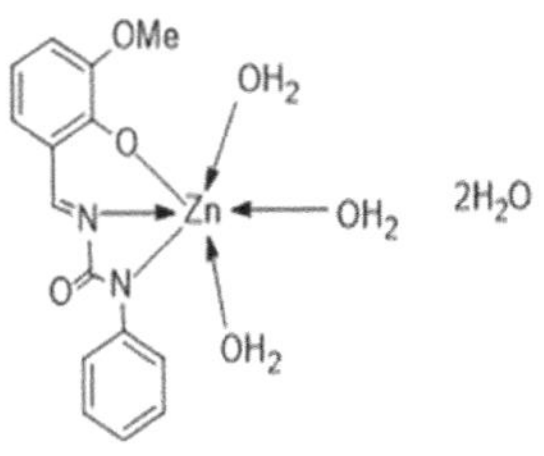

8c

FIGURE 11.8 (E)-1-(2-hydroxy-3-methoxybenzylidene)-3-phenylurea complexed with (a)Mn, (b)Cu, and (c)Zn.

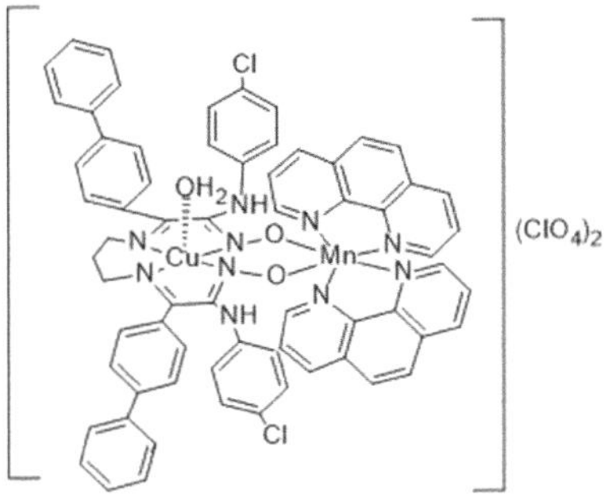

FIGURE 11.9 Structure of Schiff's base complex [Cu(1-(biphenyl)-2-hydroxyimino-2-(4-chloroanilino)-1-ethanone)(H_2O)Mn(phen)$_2$](ClO4)$_2$.

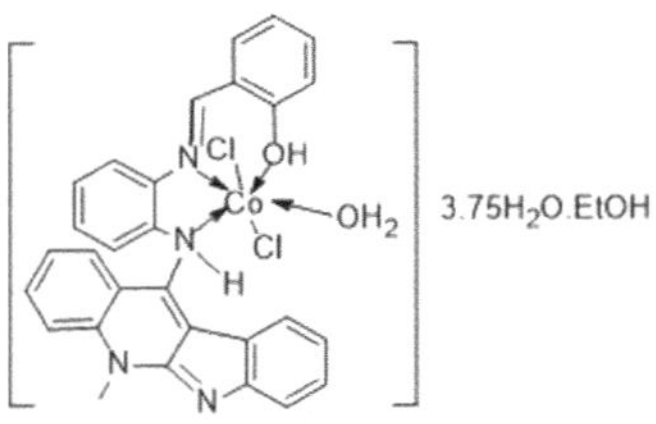

FIGURE 11.10 Structure of Cu(II) complex.

Al-Serwi et al. (2020) investigated the in vitro efficacy of the Schiff base tetradentate Cu(II)Mn(II) complex against the (SCC) squamous cell carcinoma cell line. As a comparison to cisplatin, oral-derived (GMSCs) gingival mesenchymalstem cells were employed as the control. The IC50 value, 1 µg/mL, was obtained for cisplatin. The Schiff base complex [Cu(1-(biphenyl)-2-hydroxyimino-2-(4-chloroanilino)-1-ethanone)(H2O)Mn(phen)2](ClO4)2 shown in Figure 11.9 was found to have IC50 value, 250 µM on both the cell lines by MTT assay.

Emam et al. (2017) [42] studied some Schiff bases complexes with Co(II), Ni(II) and Cu(II) for antitumor activity against colorectal adenocarcinoma (HT29) cells. The Co(II) complex shown in Figure 11.10 showed superior anticancer activity as compared to free ligand. It inhibited cancer cell growth by about 83.22% after 72 h, as examined from 3-(4,5-dimethylthiazol-2-yl)-2,5-diphenyltetrazolium bromide (MTT) assay.

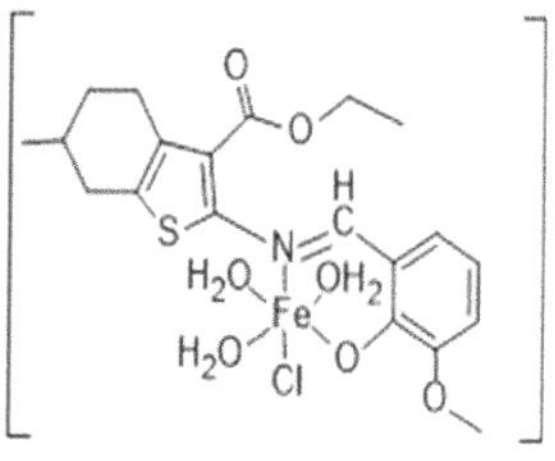

11(a)

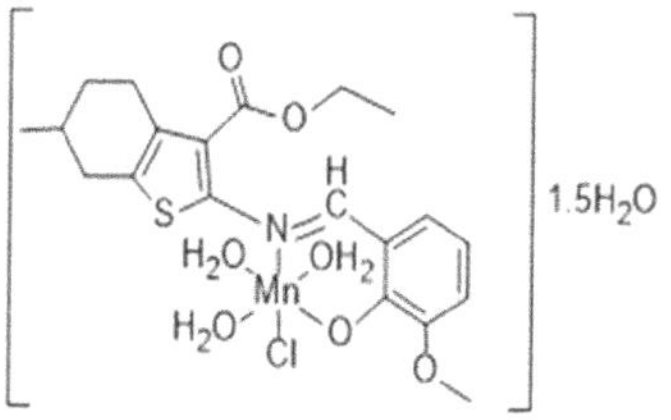

11(b)

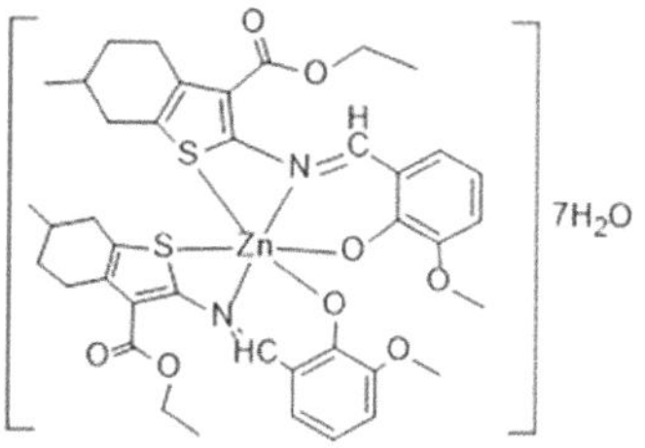

11(c)

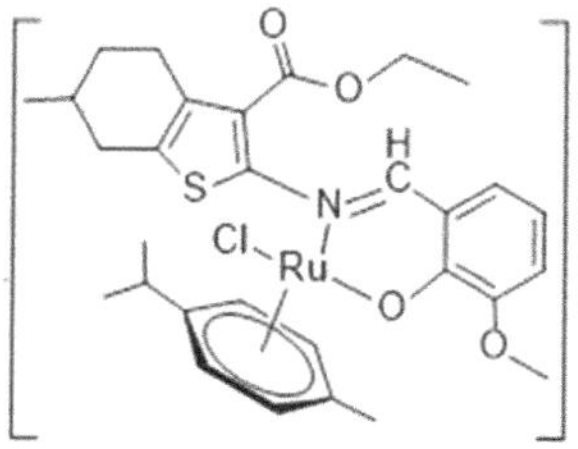

11(d)

FIGURE 11.11 Ethyl-2-(2-hydroxy-3-methoxybenzylideneamino)-6-methyl-4,5,6-tetrahydro benzothiophene-3-carboxylate complexed with (a) Fe(II), (b) Mn(II), (c) Zn(II) and (d) Ru(II).

11.3 ANTIOXIDANT ACTIVITY

Turan et al. (2018) [43] synthesized and analysed *in vitro* the antioxidant properties of schiff base ligand, ethyl-2-(2-hydroxy-3-methoxybenzylideneami no)-6-methyl-4, 5, 6-tetrahydro benzo[b]thiophene-3-carboxylate and its Fe(II) (11a), Mn(II) (11b), Zn(II) (11c), and Ru(II) (11d) complexes as shown in Figure 11.11 by using FRAP and CUPRAC assay, DPPH free radical scavenging, and ABTS cation radical scavenging. BHA, BHT, and ascorbic acid were used as standard antioxidants for comparision. The radical scavenging activity of antioxidant compounds was extensively evaluated by using the ABTS assay, according to which antioxidant activity of a compound depends on the reduction of $ABTS^{•+}$ cation radicals [44,45]. The study found that Ru(II) complexes had higher antioxidant activity than the parent ligand and Mn(II), Fe(II), and Zn(II) complexes.

11.4 ANTIMICROBIAL ACTIVITY

Gautam et al. [46] synthesized and characterised Ni(II), Pd(II), Pt(IV), and Cu(II) complexes of a schiff base azamacrocyclic ligand, 5,7,12,14-tetramethyl-1,2,4,8,10,11-hexaazacyclotetradeca-4,7,11,14-tetraene-3,9-dione. *In vitro* testing was performed on the ligand and its complexes against a variety of bacteria and pathogenic fungi. The Agar Plate Technique was used to perform preliminary fungitoxicity screening at various doses in vitro [47, 48]. Test fungi included Aspergillus niger, Aspergillus glaucus, and Aspergillus flavus. As a commercial fungicide, chlorothalonil was employed, and DMSO was used as as a control. The ligand and its metal complexes were evaluated for antibacterial activity against Sarcina lutea (Gram-positive) and Escherchia coli (Gram-negative) using the disc diffusion technique [49, 50]. Streptomycin was used as a standard. According to antimicrobial screening results, metal chelates have more inhibitory effects than free ligand. Cu complexes were discovered to have the strongest antibacterial action. The enhanced activity of metal chelates is explained on the basis of Chelation [51].

The bioactive Schiff base 2-hydroxy-benzoic acid(3,4-dihydro-2Hnaphthalen-1-ylidene)-hydrazide and its Co(II), Ni(II), and Zn(II) complexes were synthesised and characterised by Chandra et al. (2017) [52]. The compounds were tested against Gram-positive bacteria such as *Staphylococcus aureus* and *Bacillus subtilis*, as well as Gram-negative bacteria such as *Escherichia coli*, *Xanthomonas campestris*, and *Pseudomonas aeruginosa*. The minimum inhibitory concentration (MIC) of the produced compounds was determined by Andrews' technique [53]. As a control, Ciprofloxacin was employed. All of the compounds in the tested series showed potent antibacterial action against Gram-negative bacteria (*E. coli*, *X. campestris* and *P. aeruginosa*). The compounds in this series, however, proved ineffective against Gram-positive bacteria (*S. aureus*, *B. subtilis*). Ni(II) and Zn(II) were discovered to have the highest inhibitory action against bacterium compounds.

In the same year, Chandra et al. [54] synthesised and characterised a schiff base ligand derived from 4-aminoantipyrine and its Ni(II) and Cu(II) complexes. The antifungal activity of ligand and synthesized compounds was tested to determine the minimum inhibitory concentration (MIC) using serial dilution technique [55, 56]. The anti-fungal activity of Phoma sorghina, Aspergillus niger, and Fusarium oxys-porum microorganisms were investigated. The standard drug was bavistin, and DMSO served as control. The biological activity data revealed that the majority of metal complexes had moderate activity against all pathogens examined. When Schiff base ligands were coordinated with various metal ions, their activity increased. This increase in ligand activity upon complexation may be explained by Overtone's idea and Chelation theory [57–60].

Chandra et al. (2009) [61] synthesized and characterized Cr(III) 12(a), Mn(II) 12(b) and Co(II) 12(c) complexes with a novel 5,7,12,14-tetraphenyl-1,2,4,8,10,11-hexaazacyclotetra decane-3,9-dione. The ligand and its complexes, shown in Figure 11.12, were tested against a variety of bacteria and plant pathogenic fungi in vitro. The antifungal activity was measured *in vitro* using the agar plate technique [47,48]. Aspergillus niger, Aspergillus glaucus, and Aspergillus flavus were the fungus strains chosen for *in vitro* antifungal activity. The commercial fungicide was chlorothalonil. The disc diffusion technique [49] was used to investigate the antibacterial activity of the ligand and its metal complexes against Sarcina lutea (gram-positive) and Escherchia coli (gram-negative). The results of antimicrobial screening demonstrate that metal chelates have a greater inhibitory impact than free ligand. The chelation theory [62] can explain the enhanced action of metal chelates.

Chandra et al. (2007) [63] synthesized and characterized cobalt(II) 13(a), nickel(II) 13(b) and copper(II) 13(c) complexes, as shown in Figure 11.13, of a tetradentate ligand 1,3,7,9-tetraaza-4,6,10,12-tetraphenyl-2,8-dithiacyclodo decane. The synthesized compounds were tested against several species of bacteria and fungi using the technique described previously [64–67]. In both antibacterial and antifungal analyses, ligand-free metal ions inhibit somewhat more inhibition capacity

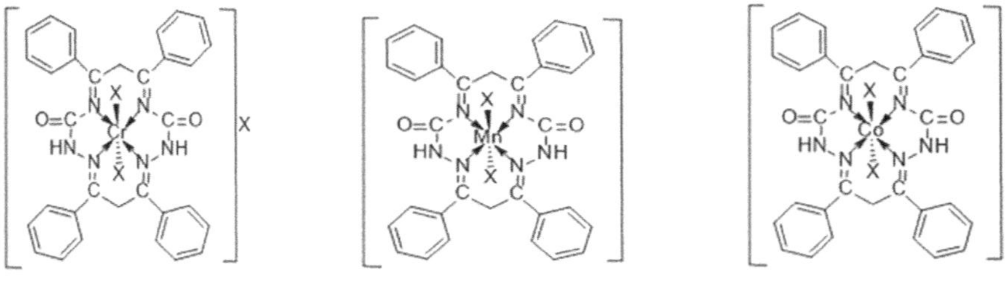

12(a) 12(b) 12(c)

FIGURE 11.12 5,7,12,14-tetraphenyl-1,2,4,8,10,11-hexaazacyclotetra decane-3,9-dione ligand complexed with (a) Cr(III), (b) Mn(II) and (c) Co(II).

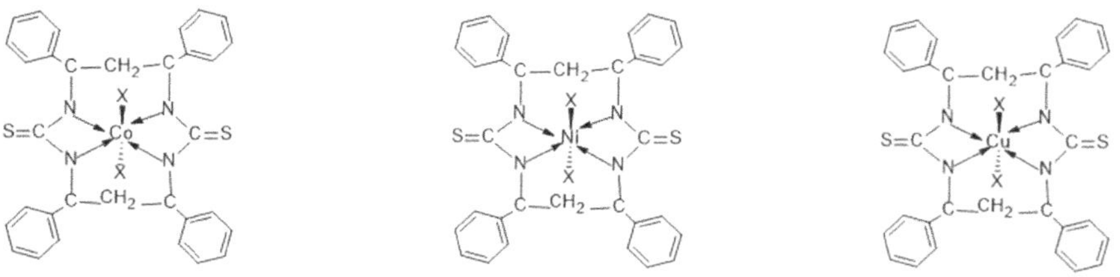

13(a) 13(b) 13(c)

FIGURE 11.13 1,3,7,9-tetraaza- 4,6,10,12-tetraphenyl-2,8-dithiacyclododecane ligand complexed with (a) Co(II), (b) Ni(II) and (c) Cu(II).

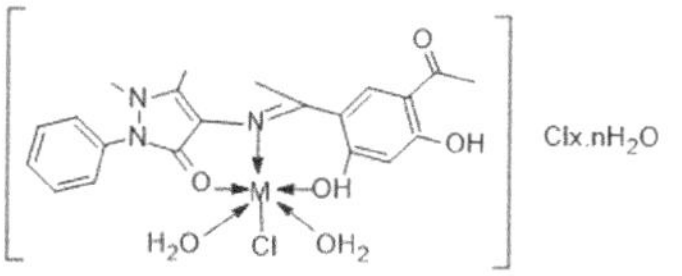

Where, M= Cr(III), X=n=2, Mn(II), X=1, n=2, Fe(III), X=2, n=0.

14(a)

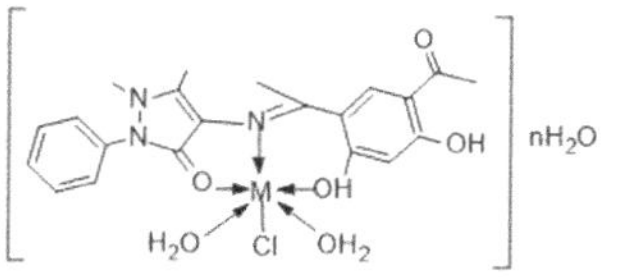

Where, M= Cd(II), n=0 and Zn(II), n=2.

14(b)

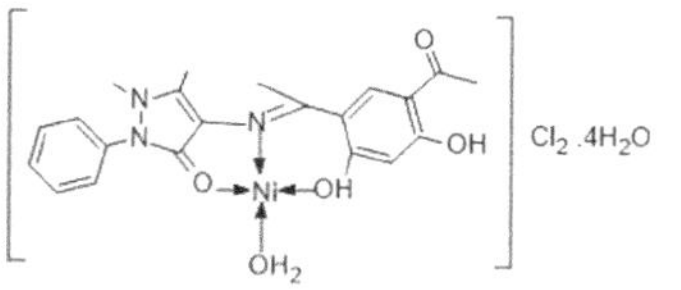

14(c)

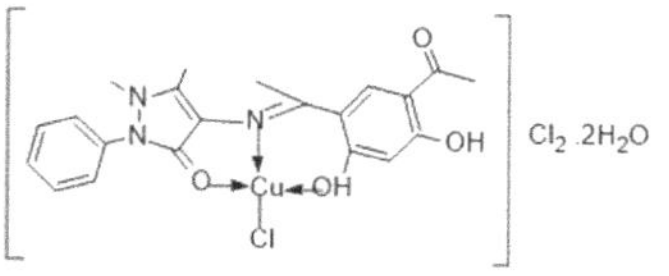

14(d)

FIGURE 11.14 4-((1-(5-acetyl-2,4-dihydroxyphenyl)ethylidene)amino)-1,5-dimethyl-2-phenyl-1H-pyrazol-3(2H)-one complexed with Cr(III), Mn(II), Fe(III) (a), Cd(II), Zn(II) (b), Ni(II) (c) and Cu(II) (d).

as compared to ligand but far less than complexes. The compounds were tested as growth inhibitors against *Sarcina lutea* (gram-positive) and *Escherchia coli* (gram-negative) bacteria using the disc diffusion technique [64,65]. The antibacterial screening study showed that Cu(II) complexes inhibit far more bacteria, whereas Co(II) and Ni(II) complexes showed activity roughly equivalent to free ligand. *Aspergillus niger* and *Aspergillus glaucus* were chosen as test fungi for antifungal screening by agar plate technique [66,67]. The results showed that Cu(II) and Ni(II) complexes show almost identical inhibition: however Co(II) complexes show weak inhibition (roughly equivalent to ligand) against both the tested fungi.

Gehad et al. (2021) [68] synthesised and studied a novel Schiff base ligand, 4-((1-(5-acetyl-2,4-dihydroxyphenyl)ethylidene)amino)-1,5-dimethyl-2-phenyl-1H-pyrazol-3(2H)-one, as well as its Cr(III), Mn(II), Fe(III) 14(a), Cd(II), Zn(II) 14(b), Ni(II) 14(c) and Cu(II) 14(d) as shown in Figure 11,14. They were evaluated *in vitro* for antibacterial and antifungal activity against some bacterial and fungal species. The synthesised ligand and its complexes were tested *in vitro* for anti-breast cancer activity, and a molecular docking research was conducted to investigate the various binding mechanisms of active compounds against the SARS-CoV-2 (COVID-19). MOE 2008 was used to test promising drugs using molecular docking at the protein sites of new corona viruses, and the project was designed to employ molecular docking without validation using MD simulations.

According to a recent coronavirus research, a novel tridentate Schiff base ligand, 4-((1-(5-acetyl-2,4-dihydroxyphenyl)ethylidene)amino)-1,5-dimethyl-2-phenyl-1H-pyrazol-3(2H)-one, can be used to combat it. Its Cr(III) complex has a lower binding energy as compared to the ligand, implying that antivirals can be effective. In the future, this discovery may lead to new corona virus treatments [69].

11.5 ANTIVIRAL ACTIVITY

In recent years, scientists have been paying attention to evolve non-platinum complexes as anticancer agents for cilinical use in the treatment of cancer. More et al (2019) [70] reported some isatin Schiff base ligands showing antiviral activities against Moloney leukemia virus [71] and vaccinia [72] as well as rhino virus [73], and SARS virus [74]. Ronen et al. reported the mode of inhibition of (MIBDET) N-methylisatin-b-40,40-diethylthiosemicarbazone on the production of (MLV) Moloney leukemia virus. It is suggested from the study that the inhibition of MLV occurs by blocking the translation of viral RNA in place of interfering with viral RNA transcription by M-IBDET [71]. Chen et al. [74] have studied the inhibition activities of Isatin derivatives against pneumonia that severe acute respiratory syndrome corona virus (SARS CoV).

The study concluded that N-substituted Isatin derivatives inhibited SARS CoV 3C-like protease (3CLpro) at low levels. FRET (Fluorescence resonance energy transfer) was used to evaluate inhibitory activity, which was then verified using high performance liquid chromatography (HPLC). The compound displayed in Figure 11.15 was determined to be more potent than selective inhibit SARS CoV3CLpro among all the investigated derivatives.

Isatin-thiosemicarbazone (methisazone) is an important preventive medication against a variety of viral infections [75]. Abbas et al. [76] synthesized and tested various 5-fluoroisatin derivatives for antiviral activity, vesicular stomatitis virus (VSV) replication, and cytotoxicity against Vero clone CCL-81 cell lines. Among all the investigated compounds N1- (5-fluoro-2-oxoindolin-3-ylidene) N-allyl and -N4-(p-fluoro-(or p-dimethylamino) benzylidene) thiocarbo hydrazone (or n-butyl)-2-(5-fluoro-2-oxoindolin-3-ylidene)hydrazine carbo-thioamide were demonstrated as good antiviral agents; the compounds in Figure 11.16, which showed remarkable inhibitory activity as antiviral agents.

Mojzych et al. [77] studied and explored the biological activities of Schiff bases of hydrazone, oximes, thiosemicarbazones, and semicarbazones derived from 5-acyl-1,2,4-triazines compound 17(a) and showed significant antiviral activities and 5-acyl-3-methylsulfamyl-1,2,4-triazine and its

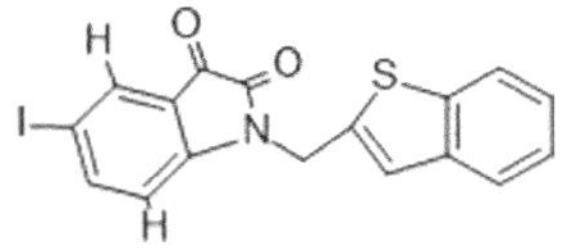

FIGURE 11.15 N-substituted Isatin derivative with SARS CoV 3C-like protease inhibiting properties.

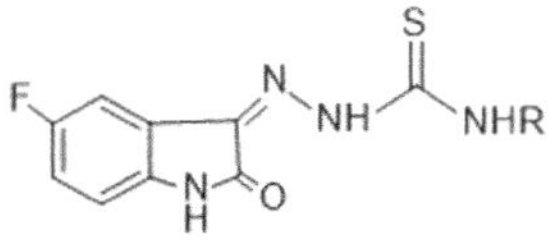

FIGURE 11.16 Structure of 5-fluoroisatin derivative with antiviral activity.

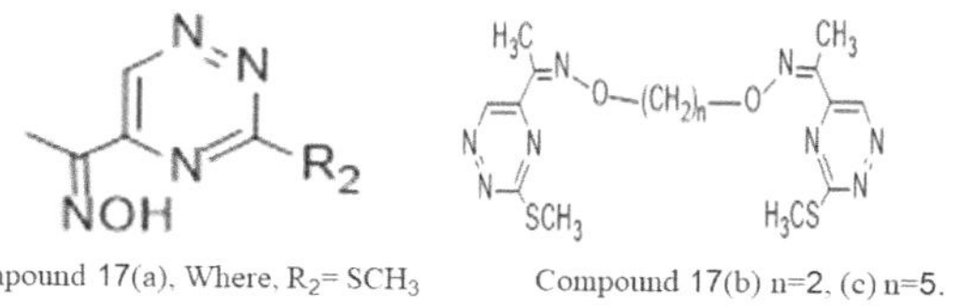

FIGURE 11.17 Structure of schiff's bases (a) 5-acyl-1,2,4-triazines compound, (b) 5-acyl-3-methylsulfamyl-1,2,4-triazine.

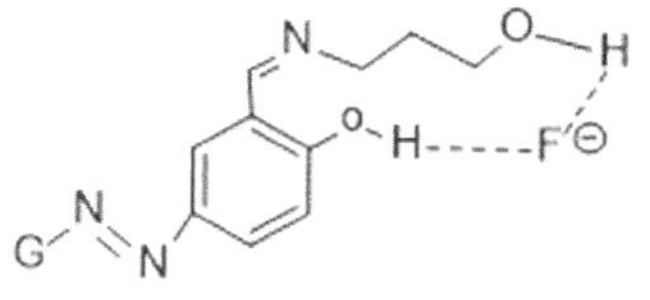

FIGURE 11.18 Structure of such azo Schiff base chemosensor.

derivatives, and compound 17(b) as shown in Figure 11.17 exhibited remarkable antiviral activity against Coxsackie Virus B4 in Vero cell culture and in HeLa cell culture.

Some novel antiviral agents have been designed using Schiff bases derived from salicylaldehyde and 1-amino-3-hydroxyguanidine tosylate. Wang et al. [78] altered several structural properties of Schiff bases of hydroxyamino guanidines (SB-HAG) to create new substituted salicylaldehyde Schiff bases of HAG (SSB-HAG) derivatives. All of these compounds were examined for the first time against coronavirus and mouse hepatitis virus (MHV) infection. When compared to TCIDw, compound [1-[(30-allyl-20-hydroxybenzylidene)amino]-3-hydroxyguanidine] was found to be around 376 times more active in comparision to hydroxylguanidine and about 564 times more efficient as compared to hydroxy amino guanidines (HAG) against growth of mouse hepatitis virus (MHV).

11.6 OTHER MISCELLANEOUS APPLICATIONS AND FUTURE SCOPE

In contrast to biological applications, Schiff base metal complexes also have various abiological applications in different fields such as in food industry, dye industry, agrochemicals, catalysis, analytical chemistry, polymer chemistry, and so on.

11.6.1 Schiff Base as Sensors

A sensor is a device capable of sensing changes in its surroundings and transmitting the data to a computer processor. Schiff bases may be used to efficiently monitor harmful compounds in the environment; for example, the presence of Cr3+ and organophosphates in the environment can be detected, measured, and eliminated [79]. They also form basic units in many dyes. Azo dye-based Schiff bases are used in detection and quantification of ions as effective chemosensors [80–83]. Chemosensors are abiotic molecules that have the capacity to attach to an analyte selectively and reversibly; azo Schiff bases have been characterized as fluoride chemosensors. Schiff base fluorescence sensors, on the other hand, are gradually becoming the preferred choice due to their great sensitivity even at extremely low analyte concentrations. Some novel azo Schiff base chemosensors have been recently studied for the detection of various ions such as Hg^{2+}, Cd^{2+}, and fluoride ions in the presence of competing cations/anions. The structures of such azo Schiff base chemosensors with high selectivity and sensitivity are shown in Figure 11.18.

Electroanalytical methods can be used to investigate laboratory, clinical, and environmental samples. Schiff bases serve as ionophores and are composed of organic polymers and converted into membranes to be used as ion-selective potentiometric sensors. Schiff bases can use electrical properties by proton transfer to form the structure of a new type of molecular conductor [84]. Conductive polymers can be synthesised using imine derivatives. ZnO Schiff base complexes are explored as semi-conductors by means of fabricating field-effect transistors to electronic performance [79].

Due to their photochromic properties, Schiff base compounds are used as photostabilizers, dyes for solar collectors and solar filters. They are also exerted in optical sound recording technology [85]. Schiff bases are used as electronic materials, in optical switches and other photonic components [86] because of their optical nonlinearity. Pyridine-based schiff bases are reported to exhibit photovoltaic characteristics in solar cell applications as current-voltage measurements. Schiff bases are employed in optical computers to monitor and regulate radiation intensity, in imaging systems, as organic materials in reversible optical memories, and as photo detectors in biological systems [85, 87]. Schiff bases can be used in gas chromatography, as stationary phase due to their high thermal stability [88]. Schiff bases compounds effectively inhibit corrosion of mild steel, aluminium, copper, zinc in acidic medium [89]; this is because of the presence of the azomethine group, the aromatic ring electron cloud, and electronegative nitrogen atom.

11.6.2 Schiff Base in Energy Applications

Presently, the main focus of the researchers all over the world is to encounter the challenges of energy demands for sustainable development. In this field, renewable sources of energy such as solar and wind energy are used to meet with our immediate energy demands. In addition, some Schiff base complexes of Pt^{2+}, Zn^{2+} and organometallic compounds are used to generate, convert, and utilize energy. This is because of their distinct electronic system that can be easily modified by changing their chemical structure to evolve an appropriate substance to meet with the requirement of energy applications.

11.7 CONCLUSION

Schiff base compounds have received considerable attention due to their broad-spectrum applications. These compounds and their metal complexes, on the other hand, are being explored extensively for their biological importance. They show multiple biological activities, such as antimicrobial, antitumor, antimalarial, antiviral, antioxidant, anti-inflammatory, and so on. Furthermore, it is an interesting topic of research that consistently furnishes knowledge about newly synthesized compounds and yet requires further investigation for future perspectives. This chapter highlights a broad study on the biological activity of Schiff base and their metal complexes, along with some miscellaneous applications.

REFERENCES

1. Gaikw, K.V. and Yadav, M.U. 2016. Metal complexes of Schiff bases. Scholarly Research Journal for Interdisciplinary Studies 3/24: 2225–34. http://www.srjis.com/pages/pdfFiles/146771229523.VIJAY%20GAIKWADpaper22.pdf
2. Singh, K.N. and Singh, S.B. 2001. Complexes of 1-isonicotinoyl-4-benzoyl-3 thiosemicarbazide with manganese (II), iron (III), chromium (III), cobalt (II), nickel (II), copper (II) and zinc (II). Transition Metal Chemistry 26, no. 4–5: 487–95. https://link.springer.com/article/10.1023/A:1011092801141
3. Hanna, W.G. and Maowad, M.M. 2001. Synthesis, characterization and antimicrobial activity of cobalt(II), nickel(II)and copper(II) complexes with new asymmetrical Schiff base ligands derived from 7-formyanil-substituted diamine-sulphoxine and acetylacetone. Transition Metal Chemistry 26, no. 6 (December): 644–51. https://link.springer.com/article/10.1023/A:1012066612090

4. Ali, A. et al. 2003. Novel pyrazolo[3,4-d]pyrimidine-based inhibitors of staphlococcus aureus DNA polymerase III: Design, synthesis, and biological evaluation. Journal of Medicinal Chemistry 46, no. 10 (March): 1824–30. DOI: 10.1021/jm020483c
5. Chandra, S. and Gupta, L.K. 2004. EPR and electronic spectral studies on Co(II), Ni(II) and Cu(II) complexes with a new tetradentate [N4] macrocyclic ligand and their biological activity. Spectrochim. Acta A 60, no. 7 (June): 1563–71 https://doi.org/10.1016/j.saa.2003.08.023
6. .Chandra, S. and Sangeetika 2004. Spectroscopic, redox and biological activities of transition metal complexes with ONS donor macrocyclic ligands derived from semicarbazide and thiodiglycolic acid. Spectrochim Acta A 60, no. 8–9 (July): 2153–62. https://doi.org/10.1016/j.saa.2003.09.027
7. Chandra, S. and Gupta, L.K. 2005. Spectroscopic studies on Co(II), Ni(II) and Cu(II) complexes with a new macrocyclic ligand: 2,9-dipropyl-3,10- dimethyl-1,4,8,11-tetraaza-5,7:12,14-dibenzocyclote ttradeca-1,3,8,10-tetraene Acta A 61, no. 6 (April): 1181–88. https://doi.org/10.1016/j.saa.2004.06.039
8. Chandra, S. and Gupta, L.K. 2005. Mass, EPR, IR and electronic spectroscopic studies on newly synthesized macrocyclic ligand and its transition metal complexes. Spectrochim. Acta A 62, no. 4–5 (December): 1125–30. https://doi.org/10.1016/j.saa.2005.03.029
9. Chandra, S., Gupta, R., Gupta, N. and Bawa, S. 2006. Biological relevant macrocyclic complexes of Copper: spectral, magnetic, thermal and antibacterial approach. Transition Metal Chemistry 31, no. 2 (March): 147–51. https://link.springer.com/article/10.1007/s11243-005-6194-5
10. Sharma, R., Agarwal, S., Rawat, S. and Nagar, M. 2006. Synthesis, characterization and antibacterial activity of some transition metal cis-3,7-dimethyl-2,6-octadiensemicarbazone complexes. Transition Metal Chemistry 31, no. 2 (March): 201–06. https://link.springer.com/article/10.1007/s11243-005-6374-3
11. Chandra, S. and Gupta, L.K. 2007. Modern spectroscopic and biological approach in the characterization of novel 14-membered [N4] macrocyclic ligand and its transition metal complexes. Transition Metal Chemistry 32, no. 2 (February): 240–45. https://link.springer.com/article/10.1007/s11243-006-0155-5
12. Karlin, K.D. and Zubieta, J. 1986. Copper coordination chemistry and biochemistry, Biochemical and Inorganic Perspectives, Adenine Press, Guilderland, NY, Vol .1 & 2 https://doi.org/10.1002/ange.1984 0960632
13. Srinivasan, S. and Athappan, P. 2001. Synthesis, spectral and redox properties of metal complexes of macrocyclic tetraaza chiral Schiff bases. Transition Metal Chemistry 26, no. 4–5 (August): 588–93. https://link.springer.com/article/10.1023/A:1011007429295
14. Chin, K.O.A., Morrow, J.R., Lake, C.H. and Churchill, M.R. 1994. Synthesis and solution properties of lanthanum(III), europium(III), and lutetium(III) THP complexes and an x-ray diffraction study of a crystal containing four stereoisomers of a europium(III) THP. Inorganic Chemistry 33, no. 4 (February): 656–64. https://doi.org/10.1021/ic00082a008
15. Rudkevich, D.M. et al. 1995. Bifunctional recognition: Simultaneous transport of cations and anions through a supported liquid membrane. J. Am Chem. Soc.117, no. 22: 6124–25. https://ris.utwente.nl/ws/portalfiles/portal/187762524/Rudkevich1995bifunctional.pdf
16. Guerriero, P., Tamburini, S., and Vigato, P.A.1995. From mononuclear to polynuclear macrocyclic or macroacyclic complexes. Coordination Chemistry Reviews 139 (February): 17–243. https://doi.org/10.1016/0010-8545(93)01105-7
17. Chandra, S., Tyagi, M., Rani, S. and Kumar, S. 2010. Lanthanide complexes derived from hexadentate macrocyclic ligand: Synthesis, spectroscopic and thermal investigation. Spectrochim. Acta A, 75, no. 2 (February): 835–40. https://doi.org/10.1016/j.saa.2009.12.009
18. Tyagi, M. and Chandra, S. 2014. Synthesis and spectroscopic studies of biologically active tetraazamacrocyclic complexes of Mn(II), Co(II), Ni(II), Pd(II) and Pt(II). Journal of Saudi Chemical Society 18, no. 1 (January): 53–58. http://dx.doi.org/10.1016/j.jscs.2011.05.013
19. Lampeka, Y.D. and Gavrish, S.P. 2000. Spectral characteristics of the copper(III) and nickel(III) coordination compounds with open-chain and macrocyclic dioxotetraamines. Polyhedron, 19, no. 26–27 (December): 2533–38. https://doi.org/10.1016/S0277-5387(00)00557-X
20. Das, M. and Livingstone, S.E. 1976. Metal chelates of dithiocarbazic acid and its derivatives. IX. Metal chelates of ten new Schiff bases derived from S-methyldithiocarbazate. Inorganic Chimica Acta 19: 5–10. https://doi.org/10.1016/S0020-1693(00)91065-X

21. Mohan, M., Agarwal, A. and Jha, N.K. 1988. Synthesis, characterization, and antitumor properties of some metal complexes of 2,6-diacetylpyridine bis (N4-azacyclic thiosemicarbazones). Journal of Inorganic Biochemistry 34, no. 1 (September): 41–54. https://doi.org/10.1016/0162-0134(88)85016-5
22. Tyagi, M., Chandra, S. and Choudhary, S.K. 2001. Tetraaza macrocyclic complexes: Synthesis, spectral and antifungal studies. J. Chem Pharma. Res., 3, no. 1: 56..jocpr.com/articles/tetraaza-macrocyclic-complexes-synthesis-spectral-and-antifungal-studies.pdf
23. Singh, N.K. and Kushawaha, S.K. 2001. Complexes of N-phenyl-N′-2 furan thio carbo hydrazide with oxovanadium (IV), manganese (III), iron (III), cobalt (II), nickel (II), copper (II) and zinc (II) Transition Metal Chemistry 26, no. 1–2 (February): 140–46. https://link.springer.com/article/10.1023/A:1007109116827
24. Hodgkin, D.G. et al. 1955. The crystal structure of the hexacarboxylic acid derived from B_{12} and the molecular structure of the vitamin. Nature 176: 325–28. https://doi.org/10.1038/176325a0
25. Battersby, A.R. 2000. Tetrapyrroles: The pigments of life. Nat Prod Rep. 17, no. 6: 507–26. https://doi.org/10.1039/%20B002635M
26. Thomas, A.M., Neelkanta, G., Mahadevan, S. Nethaji, M. and Chakravarty, A.R. 2002. Synthesis, crystal structure, and nuclease activity of oxalato-bridged dicopper(II) complexes with planar n-donor heterocyclic bases. European Journal of Inorganic Chemistry 2002, no. 10 (October): 2720–26. https://doi.org/10.1002/1099-0682(200210)2002:10%3C2720::AID-EJIC2720%3E3.0.CO;2-I
27. Lamour, E., Routierj, S. and Bernier, J.L. 1999. Oxidation of Cu^{II} to Cu^{III}, free radical production, and DNA cleavage by hydroxy-salen–Copper complexes. isomeric effects studied by ESR and electrochemistry. J. Am. Chem. Soc. 121, no. 9: 1862–69. https://doi.org/10.1021/ja982221z
28. Pittie, M., Horn, J.D.V., Brion, D., Burrows, C.J. and Meunier, B. 2000. Targeting the DNA cleavage activity of copper phenanthroline and clip-phen to A·T tracts via linkage to a poly-n-methylpyrrole. Bioconjugate Chem. 11, no. 6: 892–900. https://doi.org/10.1021/bc000050t
29. Szezepanic, W., Bal, W., Getner, K. and Bojczuk, M.J. 2002. Copper (II) binding by kanamycin A and hydrogen peroxide activation by resulting complexes. New J. Chem., 26, no. 10 (September): 1507–14. https://doi.org/10.1039/B203812A
30. Packer, L., Mccord, J.M. and Fridovich, I. 2002. Superoxide dismutase. Methods in Enzymology. J Biol Chem. 331341: 349.
31. Falter, F., Vicens, Q. and Westhof., E. 1999. Aminoglycoside–RNA interactions. Curr. Opin. Chem. Biol. 3, no. 6 (December): 694–704. https://doi.org/10.1016/s1367-5931(99)00028-9
32. Mishra, A.K. and Chatal, J.F. 2001. Synthesis of macrocyclic bifunctional chelating agents: 1,4,7-tris(carboxymethyl)-10-(2-aminoethyl)-1,4,7,10 tetraazacyclododecane and 1,4,8-tris(carboxymethyl)-11-(2-aminoethyl)-1,4,8,11-tetraazacyclotetradecane. New J. Chem. 25, no. 2: 336–39. https://doi.org/10.1039/B006978G
33. Tiwari, T. 2019. Yttrium-90 radioembolization: Current clinical practice and review of the recent literature. Journal of Radiology Nursing 38, no. 2 (June): 86–91. https://doi.org/10.1016/j.jradnu.2019.03.004
34. Zou, X.H. et al. 1999. Mono- and bi-nuclear ruthenium(II) complexes containing a new asymmetric ligand 3-(pyrazin-2-yl)-as-triazino[5,6-f]1,10-phenanthroline: synthesis, characterization and DNA-binding properties. J. Chem. Soc. Dalton Trans.: 1423–28. https://doi.org/10.1039/A900064J
35. Wongsuwan, S. et al. 2021. Synthesis, characterization and anticancer activity of Fe (II) and Fe(III) complexes containing N-(8-quinolyl) salicylaldimine Schiff base ligands. J. Biol. Inorg. Chem. 26, no. 2–3 (May): 327–39. https://doi.org/10.1007/s00775-021-01857-9
36. Chen, S.Y. et al. 2021. New cytotoxic zinc(II) and copper(II) complexes of Schiff base ligands derived from homopiperonylamine and halogenated salicylaldehyde. Inorganica Chimica Acta, 516, no. 1 (February): 120171. http://doi.org/10.1016/j.ica.2020.120171
37. Abu-Dief, A.M. et al. 2021. Targeting ctDNA binding and elaborated in-vitro assessments concerning novel Schiff base complexes: Synthesis, characterization, DFT and detailed in-silico confirmation. J. Mol. Liq. 322 (January): 114977. https://doi.org/10.1016/j.molliq.2020.114977
38. Liao, W.H. et al. 2021. A novel Schiff base cobalt(III) complex induces a synergistic effect on cervical cancer cells by arresting early apoptosis stage. Biometals 34, no. 2 (April): 277–89.
39. Shi, S. et al. 2020. Synthesis and antitumor activities of transition metal complexes of a bis-Schiff base of 2-hydroxy-1-naphthalenecarboxaldehyde. Journal of Inorganic Biochemistry 210 (September): 111173. https://doi.org/10.1016/j.jinorgbio.2020.111173

40. Hassan, A.M., Heakal, B.H., Younis, A., Abdelmoaz, M.A. and Abdrabou, M.M. 2020. Conventional and microwave-assisted synthesis, antimicrobial and antitumor studies of tridentate schiff base derived from o-vanillin and phenyl urea and its complexes. Adv. J. Chem. Sect. A. 3: S621–S38. https://dx.doi.org/10.22034/ajca.2020.105996
41. Al-Serwi, R.H. et al. 2020. Enhancement of cisplatin cytotoxicity by Cu(II)–Mn(II) schiff base tetradentate complex in human oral squamous cell carcinoma. Molecules 25, no. 20: 4688. https://doi.org/10.3390/molecules25204688
42. Emam, S., El Sayed, I., Ayad, M. and Hathout, H. 2017. Synthesis, characterization and anticancer activity of new Schiff bases bearing neocryptolepine. J. Mol. Struct. 1146: 600–19. http://dx.doi.org/10.3390/molecules25204688
43. Turan N. and Buldurun K. 2018. Catalytic activity of ruthenium(II) complex. European Journal of Chemistry 9, no. 1: 22–29. https://doi.org/10.5155/eurjchem.9.1.22-29.1671
44. Turan, N. and Şekerci, M. 2009. Synthesis and spectral studies of novel Co(II), Ni(II), Cu(II), Cd(II), and Fe(II) metal complexes with N-[5′-Amino-2,2′-bis(1,3,4-thiadiazole)-5-yl]-2-hydroxybenzaldehyde imine (HL). Spectroscopy Letters 42, no. 5 (July): 258–67. http://dx.doi.org/10.1080/00387010902897521
45. Turan, N., Savci, A., Buldurun, K., Alan, Y. and Adiguzel, R. 2016. Synthesis and chemical structure elucidation of two Schiff base ligands, their iron (II) and zinc (II) complexes, and antiradical, antimicrobial, antioxidant properties. Letters in Organic Chemistry 13, no. 5: 343–51. http://dx.doi.org/10.2174/1570178613666160422161855
46. Gautam A. and Chandra S. 2019. Spectral and biological studies of Ni(II),Pd(II),Pt(IV) and Cu(II) complexes with novel azamacrocyclic Ligand. Asian Journal of Chemistry 31, no. 2: 396–402. https://doi.org/10.14233/ajchem.2019.21706
47. Agarwal, R.K. and Prasad, S. 2005. Synthesis, spectroscopic and physicochemical characterization and biological activity of co(II) and Ni(II) coordination compounds with 4-aminoantipyrine thiosemicarbazone. Bioinorg Chem Appl. 3: 271–88. https://doi.org/10.1155/BCA.2005.271
48. Chandra, S. and Verma, S. 2008. Synthesis, spectral and antifungal studies of transition metal complexes with sixteen membered hexadentate macrocyclic ligand. Russ J. Coord. Chem. 34, no. 7 (July): 499–503. https://doi.org/10.1134/S107032840807004X
49. Anacona, J.R. and Sillva, G.D. 2005. Synthesis and antibacterial activity of cefotaxime metal complexes. J. Chil. Chem. Soc. 50, no. 2: 447–50. www.scielo.cl/scielo.php?script=sci_arttext&pid=S0717-97072005000200001&lng=en&nrm=iso&tlng=en
50. Chandra, S., Gautam, A. and Tyagi, M. 2009. Synthesis, structural characterization and antibacterial studies of a tetradentate macrocyclic ligand and its Co(II), Ni(II) and Cu(II) complexes. Russ. J. Coord. Chem., 35, no. 1 (January): 25–29. https://doi.org/10.1134/S1070328409010060
51. Chandra, S. and Tyagi, M. 2008. Ni(II), Pd(II), and Pt(II) complexes with ligand containing thiosemicarbazone and semicarbazone moiety: synthesis, characterization and biological investigation. J. Serb. Chem. Soc. 73, no. 7: 727–34. https://doi.org/10.2298/JSC0807727C
52. Tyagi, M., Tyagi, P. and Chandra, S. 2017. Synthesis, spectral and antibacterial activity of Co(II), Ni(II) and Zn(II) complexes with 2-hydroxy-benzoic acid(3,4-dihydro-2-Hnaphthalen-1-ylidene)-hydrazide. Applied Organometallic Chemistry 31, no. 12: e3880. https://doi.org/10.1002/aoc.3880
53. Andrews, J.M. 2001. Determination of minimum inhibitory concentration. J Antimicrob. Chemother 48, no. 1 (July): 5–16. https://doi.org/10.1093/jac/48.suppl_1.5
54. Tyagi, M., Chandra, S. and Tyagi, P. 2017. Synthesis, characterization and anti-fungal evaluation of Ni(II) and Cu(II) complexes with a derivative of 4-aminoantipyrine. Journal of Taibah University for Science 11, no. 1 (January): 110–20. https://doi.org/10.1016/j.jtusci.2015.11.003
55. NCCLS, Method for Dilution Antimicrobial Susceptibility Test for Bacteria that Grow Aerobically. Approved Standards, 5th ed., National Committee for Clinical Standards, Villanova, PA, 2000.
56. Tyagi, P., Chandra, S. Saraswat, B.S. and Sharma, D. 2015. Design, spectral characterization, DFT and biological studies of transition metal complexes of Schiff base derived from 2-aminobenzamide, pyrrole and furan aldehyde. Spectrochim Acta A. 143 (May): 1–11. https://doi.org/10.1016/j.saa.2015.02.027
57. Raman. N., Selvan, A. and Manishankar, P. 2010. Spectral, magnetic, biocidal screening, DNA binding and photocleavage studies of mononuclear Cu(II) and Zn(II) metal complexes of tricoordinate heterocyclic Schiff base ligands of pyrazolone and semicarbazide/thiosemicarbazide based derivatives. Spectrochim. Acta A 76, no. 2 (July): 161–73. https://doi.org/10.1016/j.saa.2010.03.007

58. Tyagi, M., Chandra, S. and Tyagi, P. 2014. Mn(II) and Cu(II) complexes of a bidentate Schiff's base ligand: Spectral, thermal, molecular modeling and mycological studies. Spectrochim. Acta A 117 (January): 1–8. https://doi.org/10.1016/j.saa.2013.07.074
59. Chandra, S. and Agrawal, S. 2014. Spectroscopic characterization of Lanthanoid derived from a hexadentate macrocyclic ligand: Study on antifungal capacity of Complexes, Spectrochim. Acta A 124 (April): 564–70. https://doi.org/10.1016/j.saa.2014.01.042
60. Chandra, S., Bargujar, S., Nirwal, R., Qanungo, K. and S.K. Sharma. 2013. Synthesis, spectral characterization, molecular modeling, thermal study and biological evaluation of transition metal complexes of a bidentate Schiff base ligand. Spectrochim. Acta A 113 (September): 164–70. https://doi.org/10.1016/j.saa.2013.04.114
61. Chandra, S. and Gautam, A. 2009. Spectroscopic and biological approach in the characterization of Cr(III), Mn(II) and Co(II) complexes with a novel hexaaza macrocyclic ligand derived from semicarbazide. J. Serb. Chem. Soc. 74, no. 12: 1413–22. https://doi.org/10.2298/JSC0912413C
62. Sengupta, S. K., Pandey, O. P., Srivastava, B. K. and Sharma, V. K. 1998. Synthesis, structural and biochemical aspects of titanocene and zirconocene chelates of acetylferrocenyl thiosemicarbazones. Transition Metal Chemistry 23, no. 2 (August): 349–53. https://doi.org/10.1023/A:1006986131435
63. Chandra, S., Gautam, A. and M. Tyagi. 2007. Synthesis and spectroscopic characterization of transition metal complxes of 12-membered tetraaza [N4] macrocyclic ligand. Transition Metal Chemistry 32, no. 8 (August): 1079–84. https://doi.org/10.1007/s11243-007-0289-0
64. Chandra, S. and Gupta, L.K. 2004. EPR and electronic spectral studies on Co(II), Ni(II) and Cu(II) complexes with a new tetradentate [N_4] macrocyclic ligand and their biological activity. Spectrochim. Acta A 60, no. 7 (June): 1563–71. https://doi.org/10.1016/j.saa.2003.08.023
65. Chandra, S., Gupta, L.K. and Sangeetika. 2004. Synthesis, physiochemical and electrochemical studies of Mn(II), Co(II), Ni(II) and Cu(II) complexes with N-donor tetradentate[N_4] macrocycle ligand derived from ethyl cinnamate. Synth. React. Inorg. Met. Org. Chem. 34, no. 9 (November): 1591–604. https://doi.org/10.1081/SIM-200026596
66. Bansal, A. and Singh, R.V. 2001. Template synthesis, spectroscopic studies and biological screening of macrocyclic complexes of lead(II). Indian J. Chem. 40A (September): 989–93. http://nopr.niscair.res.in/bitstream/123456789/18565/1/IJCA%2040A(9)%20989-993.pdf
67. Hanna, W.G. and Moawad, M.M. 2001. Synthesis, characterization and antimicrobial activity of cobalt(II), nickel(II)and copper(II) complexes with new asymmetrical Schiff base ligands derived from 7-formyanil-substituted diamine-sulphoxine and acetylacetone. Transition Metal Chemistry 26, no. 6 (December): 644–51. https://doi.org/10.1023/A:1012066612090
68. Gehad G., Mohamed, M.M., Omar and Yasmin M. Ahmed. 2021. Metal complexes of tridentate Schiff base: Synthesis, characterization, biological activity and molecular docking studies with COVID-19 protein receptor. Z. Anorg Allg.Chem. 647, no. 23–24 (December): 2201–18. https://doi.org/10.1002/zaac.202100245
69. Ahmed, Y. M., Omar, M.M. and Mohamed, G.G. 2021. Synthesis, spectroscopic characterization, and thermal studies of novel Schiff base complexes: Theoretical simulation studies on coronavirus (COVID-19) using molecular docking. Journal of the Iranian Chemical Society. https://doi.org/10.1002/zaac.202100245
70. More, M.S., Joshi, P.G., Mishra, Y.K. and P.K. Khanna. 2019. Metal complexes driven from Schiff bases and semicarbazones for biomedical and allied applications: A review. Materials Today Chemistry 14, 100195: 1–22. doi: 10.1016/j.mtchem.2019.100195
71. Ronen, D., Sherman L., Barnun, S. and Teitz, Y. 1987. N-methylisatin-beta-4',4'-diethylthiosemicarbazone, an inhibitor of Moloney leukemia virus protein production: Characterization and in vitro translation of viral mRNA. Antimicrob. Agents Chemother. 31: 1798–802. doi: 10.1128/aac.31.11.1798
72. Cooper, J.A., Moss, B. and Katz., E. 1979. Inhibition of vaccinia virus late protein synthesis by isatin-β-thiosemicarbazone: Characterization and in vitro translation of viral mRNA. Virology. 96: 381–392. doi: 10.1016/0042-6822(79)90096-5
73. Webber, S.E. et al. 1996. Design, synthesis, and evaluation of nonpeptidic inhibitors of human rhinovirus 3C protease. J. Med. Chem. 39: 5072–82. https://doi.org/10.1021/jm960603e

74. Chen, L.R. et al. 2005. Synthesis and evaluation of isatin derivatives as effective SARS coronavirus 3CL protease inhibitors. Bioorg. Med. Chem. Lett. 15: 3058–3062. DOI: 10.1016/j.bmcl.2005.04.027
75. Richman, D.D. 2006. Antiviral drug resistance. Antiviral Research 71(2–3): 117–21. DOI: 10.1016/j.antiviral.2006.03.004
76. Abbas, S.Y. et al. 2013. Synthesis, characterization, and antiviral activity of novel fluorinated isatin derivatives. Chem. 14: 1725–33. DOI: 10.1007/s00706-013-1034-3.
77. Mojzych, M. and Sebela, M. 2015. Synthesis and biological activity evaluation of Schiff bases of 5-Acyl-1, 2, 4-Triazine. J. Chem. Soc. Pak. 37(2): 300–5. www.researchgate.net/publication/281961593
78. Wang, P.H., Keck, J.G., Lien, E.J. and Lai, M.M.C. 1990. Design, synthesis, testing, and quantitative structure-activity relationship analysis of substituted salicylaldehyde Schiff bases of 1-amino-3-hydroxyguanidine tosylate as new antiviral agents against coronavirus. J. Med. Chem. 33(2), 608–14. DOI: 10.1021/jm00164a023
79. Khan M.I., Gul Saima and Khan M.A., Chapter: Schiff Bases and their matallic derivatives: Highly versatile molecules with biological and abiological perspective. 2019. DOI: 10.5772/intechopen.80799
80. Asnake L.B., et al. 2019. A review of the applications of Schiff bases as optical chemical sensors. Trends in Analytical Chemistry 116: 74–91. DOI:10.1016/J.TRAC.2019.04.025
81. Lawrence, J.F. and Frei, R.W. Chemical Derivatization in Liquid Chromatography, vol. 7, Elsevier, 2000. eBook ISBN: 9780080858074
82. Nourifard, F. and Payehghadr, M. 2016. Conductometric studies and application of new Schiff base ligand as carbon paste electrode modifier for mercury and cadmium determination. Int. J. Environ. Anal. Chem. 96: 552–67. https://doi.org/10.1080/03067319.2016.1172219
83. Gupta, V.K., Singh, A.K. and B. Gupta. 2006. A cerium (III) selective polyvinyl chloride membrane sensor based on a Schiff base complex of N, N′-bis [2-(salicylideneamino) ethyl] ethane-1, 2-Diamine. Anal. Chim. Acta 575: 198–204. doi: 10.1016/j.aca.2006.05.090.
84. Amany, M.A. and Ibrahim. 1992. Electrical and thermal behaviour of some Schiff bases and their charge transfer complexes with acidic acceptors. Thermochim. Acta. 197, 1: 211–17. DOI:10.1016/0040-6031(92)87052-C
85. Tanaka, K., Shimoura, R., and Caira, M.R. 2010. Synthesis, crystal structures and photochromic properties of novel chiral Schiff base macrocycles. Tetrahedron Lett. 51, 2: 449–52. 10.1016/j.tetlet.2009.11.062
86. Jia, J.H., Tao, X.M., Li, Y.J. and Sheng, W.J. 2011. Synthesis and third-order optical nonlinearities of ferrocenyl Schiff base. Chem. Phys. Lett. 514, 1–3: 114–18. https://doi.org/10.1021/ic0258633
87. Pistolis, G., Gegiou, D. and Hadjoudis, E. 1996. Effect of cyclodextrin complexation on thermochromic Schiff bases. J. Photochem. Photobiol. A. Chem. 93, 2–3: 179–84. https://doi.org/10.1016/1010-6030(95)04182-6
88. Atta, A.M., Shaker, N.O. and Maysour, N.E. 2006 Influence of the molecular structure on the chemical resistivity and thermal stability of cured Schiff base epoxy resins. Prog. Org. Coat. 56: 100–10. 10.1016/j.porgcoat.2005.12.004
89. Emregül, K.C., Düzgün, E. and Atakol O. 2006. The application of some polydentate Schiff base compounds containing aminic nitrogens as corrosion inhibitors for mild steel in acidic media. Corr. Sci. 48: 3243–60. https://doi.org/10.1016/j.corsci.2005.11.016

12 Molecular Docking and Drug Design of Schiff Base Metal Complexes

Ramadan M. Ramadan[1] *and Amani F.H. Noureldeen*[2]
[1] Chemistry Department, Faculty of Science, Ain Shams University, Cairo, Egypt
[2] Biochemistry Department, Faculty of Science, Ain Shams University, Cairo, Egypt
E-mail: r_m_ramadan@yahoo.com; amaninoureldeen@yahoo.com

CONTENTS

12.1 INTRODUCTION

Molecular interactions between biological targets (receptors) and substrates, such as protein-protein, protein-drug, enzyme-drug and DNA-drug, display essential roles in numerous biological processes. These interactions usually drive to formation of stable macromolecular target-substrate complexes, which may perform essential biological functions. The structure and binding sites of the biologically active target are a requirement to recognize the binding modes and the binding affinities between the interacting compounds. The difficulty and expenses of elucidation of the complex structures by experimental methods prompted researchers to consider molecular docking computation as a vital approach to understanding the target-substrate interactions (Halperin et al., 2002; Taylor et al., 2002; Sousa et al., 2006; Huang and Zou, 2010; Meng et al., 2011; Khan et al., 2020). Molecular docking processes are theoretical calculation methods performed for drug design and drug discovery procedures. It helps in suggesting the drug interaction model as well as to recognize the action of the new drug interactions against a selected biological receptor (Gupta et al., 2018; Márquez et al., 2020). Thus, docking is a computer simulation method and is used to identify the conformations of the formed complex of the target-ligand (Figure 12.1). The target could be any molecule such as a protein, a DNA or an enzyme. The docked ligands either are naturally (protein /nucleic acid) or synthesized derivatives. The precise detection of the binding modes between the docked ligand and target is substantially important in the designing of the methods of the structure-based and ligand-based drugs. The analyses of docking results are important in detecting the conformational

DOI: 10.1201/9781003291459-12

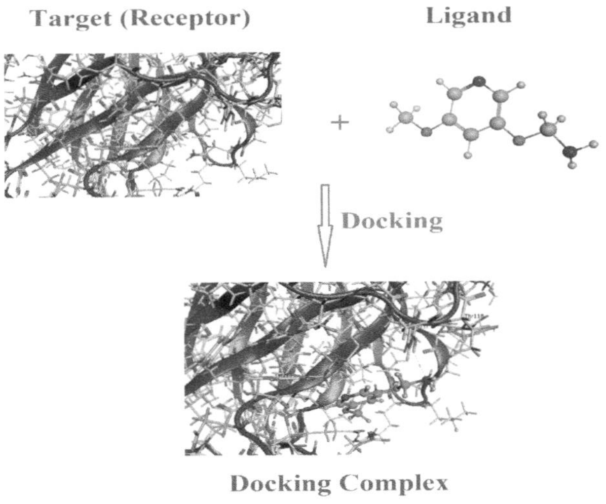

FIGURE 12.1 Elements in molecular docking.

changes, which are linked with amino acid parts in the binding locations to host the hydrophobic docked molecules (inhibitors). Thus, the most significant findings of the computer programs software packages for molecular docking are the virtual screening output that can decide whenever the selected hopeful molecules will go further *in vitro* and *in vivo* screening.

12.2 CONCEPT OF MOLECULAR DOCKING

The molecular docking concept is a study investigating how two or more molecules, such as synthesized compounds or drugs, could fit together with a bioactive macromolecule like an enzyme or protein (guest-host concept) (Kirkpatrick, 2004). Thus, it is a technique for molecular modeling, which is performed to indicate how the target interacts with ligands of small compounds. The ability of a biological macromolecule such as an enzyme or a nucleic acid to form a supramolecular complex array from the interaction with molecules is a leading part in the dynamics of a target that might inhibit or enhance its biological action. Therefore, action of compounds inside the binding pockets (*cavities on the surface or in the interior of a protein molecule that have suitable properties for binding a ligand*) of the target is described by molecular docking. This technique performs operations to give the correct poses that show how ligands can lie in the binding pockets of a target. In addition, it predicts the binding affinity (*the affinity of a molecule, such as a drug, to bind a target, and defines how tightly a ligand binds to a particular protein*) between the ligand and the target. When molecular docking is carried out, many probable poses of a ligand-target complex are generated. These binding poses are estimated from the scoring function that determines the resulted poses. (Scoring functions are mathematical functions, which are used to indicate the affinity of binding between two species when they are docked.) Negative values of score functions are usually generated from the docking operations. The higher negative values mean that the docked ligands have more affinity of binding toward the receptor. The molecular docking precision, thus, relies on the accuracy of the selected score functions. Therefore, scoring functions calculate the binding modes and sites of the binding of ligand as well as they predict the binding affinity between the target and ligand (Li et al., 2019). Various molecular docking processes can be carried out, such as protein-DNA docking, protein or DNA-synthesized molecules docking and protein-protein docking (Ewing and Kuntz, 1997).

However, small molecules are used to identify the preferences of charge and shape of the binding sites of macromolecular target, which are considered for both ligand-based and structure-based drug design cycles. The virtual screening has succeeded in developing large chemical databases in order to suggest the experimentally active derivatives that can serve as a practical and efficient alternative to screening processes with high-throughput. Methodological procedures that applied in the virtual screening like molecular docking and functional scoring offered rapid and accurate identification of compounds as well as predicting the native binding conformations (Satyanarayanajois, 2011).

There are many available commercial programs (such as Molecular Operating Environment (MOE), GOLD and AutoDock Vina software packages) which can perform and predict the possible poses and scores that molecules may potentially bind and inhibit DNA or proteins. Docking is proceeded by locating rigid molecules or fragments (ligands) into active sites (pockets) of the target. This can be achieved by using different mathematical methods such as geometric hashing, clique-searching or pose clustering. First, the docking starts by generating the compositions of all possible orientations and conformations of the macromolecule target that paired with the docked ligand. Then, the scoring function gives a value, which indicates the preferred interactions (Ewing and Kuntz, 1997). Selection of the desired structures of the biological macromolecular targets, obtained from the web site of Protein Data Bank, are used to identify and optimize their active sites (pockets). The Protein Data Bank, PDB, is a web site offering large database for 3D structures of the biological active macromolecules (proteins or nucleic acids). These data are mainly obtained from nuclear magnetic resonance (NMR) spectroscopy, X-ray (Xrd) crystal structures as well as electron microscopy (EM). These data are free of charge and can be accessed from the Internet through the websites of the following organizations: RCSB (Research Collaboratory for Structural Bioinformatics), PDBe (Protein Data Bank in Europe), PDBj (Protein Data Bank in Japan) and BMRB (Biological Magnetic Resonance Data Bank).

12.3 DRUG DESIGN AND DRUG DISCOVERY

Drugs are vital materials for life in this world. They are effective substances for the preventing, diagnosis and diseases treatment as well as improving the health and characteristics of human life. Discovery of novel bioactive chemotherapeutic drugs with new functionalities able to overcome severe emerging diseases are the most considerable current researches in pharmaceutic drug development. Drug discovery is an approach where potential newfangled therapeutical derivatives have been characterized by using a combination of experimental, clinical and computational techniques (Zhong and Zhou, 2014; Xiao et al., 2015; Zhou and Zhong, 2017). However, the development of a new drug from a primary idea to the final product is a complicated process that can take many years and cost a lot of money and effort.

Drug design is an innovation process to find novel medications depending on knowledge of biologically active targets. That means drug design includes the invention of molecularly designed compounds, which are integrated in charge and shape with a specific target, and can bind and interact with each other. Modern drug design is usually dependent on computer modeling methods and bioinformatics processes. Drug discovery also identifies the screening hits and then optimizes them to raise their affinity, selectivity, potency, metabolic stability and oral bioavailability. Thus, if the compound achieves all these requirements, the process of drug development will begin before starting clinical trials. Drug development and drug discovery include pre-clinical investigations of *in vitro* and *in vivo* studies on cell lines and animal models as well as the clinical tests on human volunteers. The final step is then gaining official approval in order to market the drug (Luu et al., 2013). Figure 12.2 outlines the different steps in drug design and accreditation, while the drug discovery cycle is given in Figure 12.3 (Hughes et al., 2011).

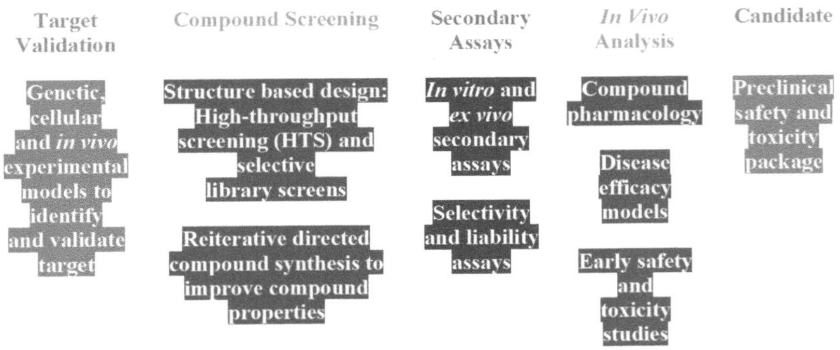

FIGURE 12.2 The different steps for drug design and accreditation.

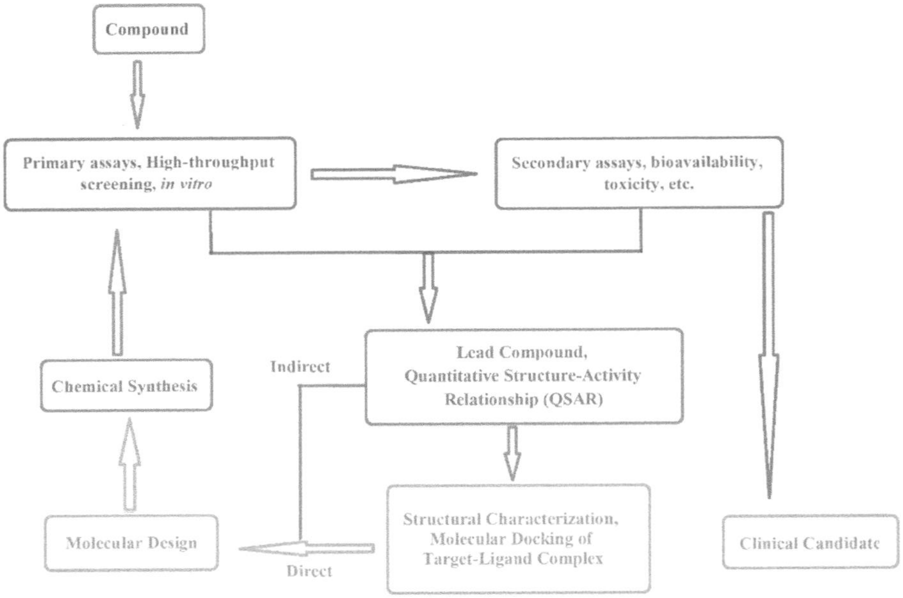

FIGURE 12.3 Drug discovery cycle.

12.4 MOLECULAR DOCKING: AN APPROPRIATE PROCESS FOR DRUG DESIGN AND DRUG DISCOVERY

Molecular docking recently grasped the attention to understand how chemical compounds are interacted with macromolecular receptor as well as its related applications in drug discovery and development. The use of the molecular docking approach to distinguish between the structures, demands is the requisite for the efficiency in binding between ligand and target. In addition, molecular docking employs predicting the potential conformations of the formed ligand-receptor complex. Thus, it is commonly performed for designing and optimization of compounds that have potential therapeutic use (Pinzi and Rastelli, 2019). For example, several methods for the design of a structure-based drug were applied using molecular docking to distinguish the non-peptide inhibition of the enzymes of the cysteine and serine protease groups (Ring et al., 1993). Moreover, computational programs were carried out to predict the binding affinity in the development of malarial plasmepsin inhibitors. Homology modeling and molecular docking were used to identify the inhibitor design and their combination with the binding free energy calculations (Bjelic et al., 2007).

The processes of designing methods for structure-based drug depend on information given from the 3D structures for the targets of interest. It also allows the ranking of the databases of these molecules according to the electronic and structural complementarity of them to those targets. Therefore, molecular docking is a very effective structure-based *in silico* (computer or via computer simulation) method that helps in predicting the interactions that occur between the molecules and the bioactive targets. The structure-based drug design is a refined process that proceeds via many cycles prior to an optimized lead drug being considered in the clinical trials. Figure 12.4 demonstrates simplified steps of structure-based drug design process.

The first four steps in Figure 12.4 involve the selection of the target protein or nucleic acid, purified, and its structure is determined using either X-ray or NMR studies. Identification of the target binding sites is then performed followed by docking and the score of the suggested lead compounds or fragments of compounds to illustrate the selected binding zones of the target structure. According to the electrostatic and steric interactions (such as H-bonding, hydrophobic interactions, etc.), the docked drugs are scored and ranked with receptor binding sites. Biochemical assays of the best promising compound are tested. Determination of structure of the target complex for *in vitro* micromolar and nanomolar inhibitions is again carried out using either X-ray or NMR studies. Obviously, there are other additional steps undertaken during the process of structure-based drug design like the resynthesizing of the optimized drug, determination of the structure of the target-ligand species, followed by further lead compound re-optimization. Usually, the final optimized compound (obtained after many working steps on the drug design process) shows remarkable improvement in the binding and the specificity for a target. Following these steps, the lead drug will be ready to proceed to phase I clinical trials.

Recently, the entire world has been choked by a flare-up of a pandemic corona virus disease (COVID-19) that has caused a bad situation globally and affected the lives of millions of people. No effective medication until now is approved to treat the infected patients. Hundreds of clinical trials are still trying to get an effective drug that is considered a safe and efficient antiviral to combat COVID-19 and can be approved by US FDA agency. The development of new drugs needs many years to be available for distribution on the market. Thus, there is an urgent requirement to discover effective drugs to prevent and cure such deadly diseases. Therefore, both the designing methods for structure-based and ligand-based drug are commonly used in the drug discovery process against any discovered severe disease. This can be established with the aid of molecular modeling, molecular docking and virtual screening. Structural identification of suitable targets for SARS-CoV-2 may then help the researchers to use the ligand- and structure-based virtual identification of inhibitors to discover some lead drugs against COVID-19 (Goel et al., 2021; Gurung et al., 2021). Figure 12.5 demonstrates the shape of the coronavirus COVID-19 and some of the molecules (Favipiravir, Oseltamivir, Galidesivir, Disulfiram and Bromhexine) that are currently investigated as potential drugs for the clinical trials.

12.5 SCHIFF BASE METAL COMPLEXES AS POTENTIAL THERAPEUTIC REAGENTS

The Schiff base derivatives are typically prepared by the condensation of a ketone or an aldehyde (compound containing –C=O group) with a primary amine (-NH_2 moiety) under different reaction conditions and drying solvents (see Figure 12.6) (Rezaeivala and Keypour, 2014; Jia and Li, 2015). The Schiff base may contain, besides the nitrogen atom donor of the azomethine (-C=N-) moiety, other nitrogen and oxygen donor atoms. Thus, the Schiff base compounds could be either monodentate with one nitrogen donor atom (V-shaped ligand) or multidentate derivative (bi, tri or tetradentate with combination of nitrogen and oxygen donors). The azomethine group (-C=N–) is

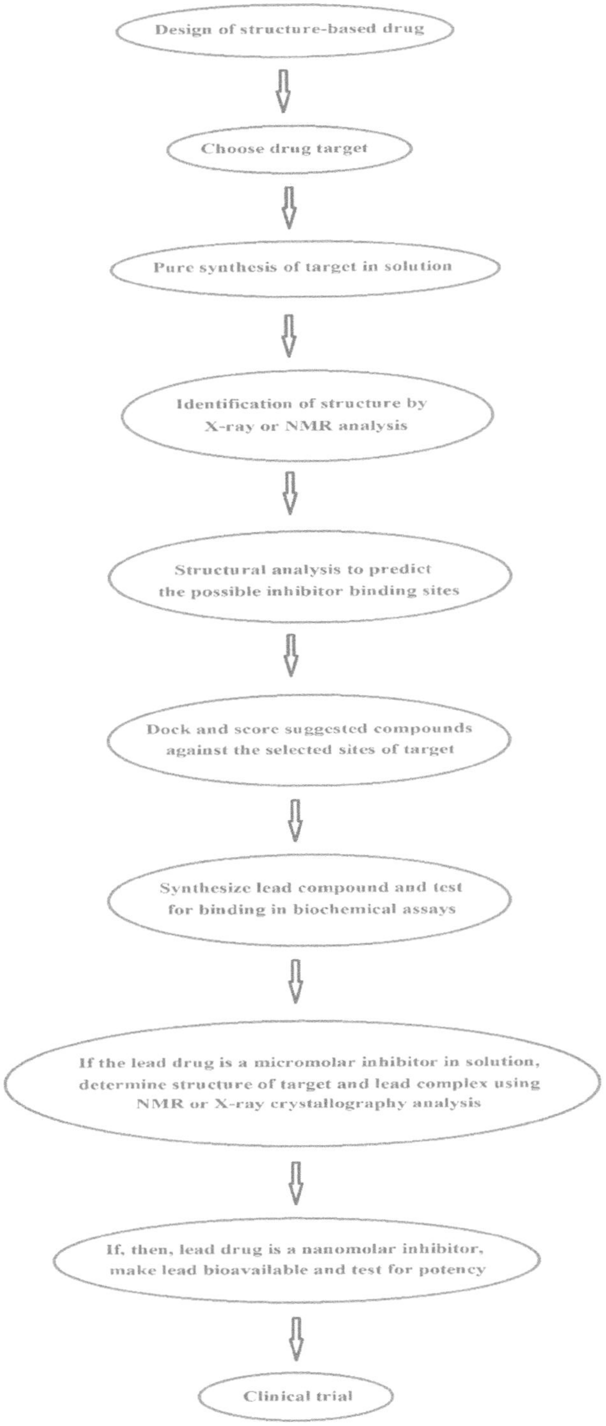

FIGURE 12.4 Steps of structure-based drug design process.

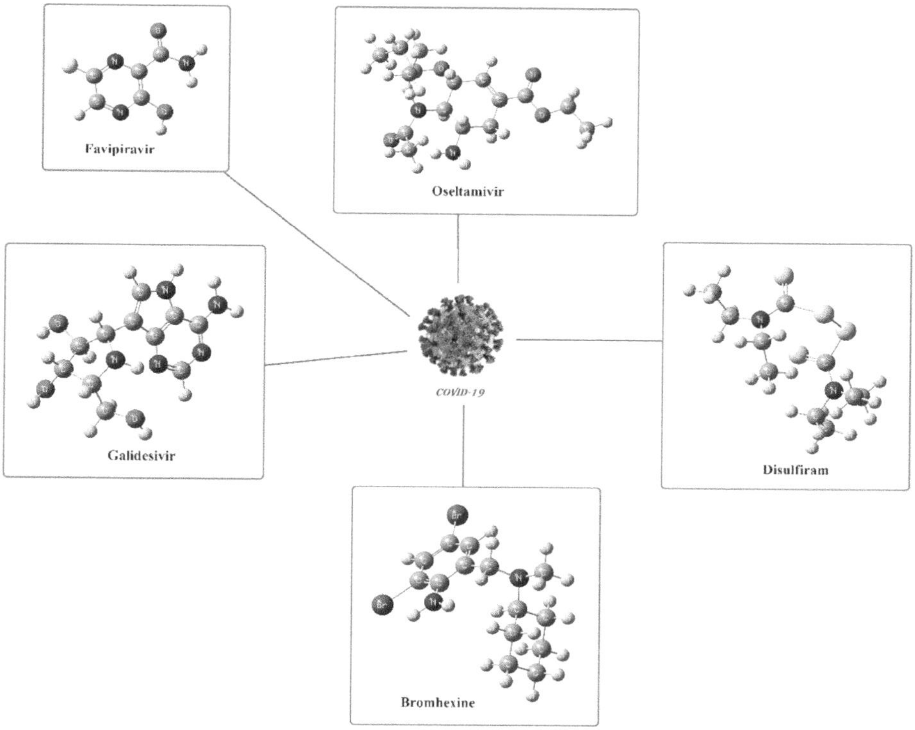

FIGURE 12.5 Some molecules currently investigated in the clinical trials for the coronavirus COVID-19 inhibition.

often accountable for the wide ranges of the biological behaviors of these compounds. Therefore, these derivatives of Schiff bases are a substantial class of organic compounds that play considerable roles in the development of coordination chemistry and form stable complexes with transition metals having different oxidation states (Abdel-Rahman et al., 2017; Zhu, 2019; El-Medani et al., 2020; Ramadan et al., 2021; a- Mohamed et al., 2021). The excellent chelating ability of Schiff base derivatives with the different transition metals is responsible for their enhanced biological activities. Because of their synthetic flexibility and structure stability, these derivatives are considered as versatile ligands in the field of coordination chemistry (Abu-Dief and Mohamed, 2015).

Numerous research on drugs associated with metal-based entities showed that complexes of Schiff bases are versatile pharmacophores and have valuable applications in medical sciences and medicinal chemistry. As therapeutic drugs, the Schiff base complexes have possible applications as medication for anti-inflammatory, antibiotic, antifungal, antiviral, antimicrobial, anticancer, analgesic effect, and so on. Recently, drug resistance represents a global threat in medical therapy because the most pathogenic organisms can develop their capability to deactivate these drug compounds. Therefore, urgent attention from chemistry and pharmaceutical researchers have been paid to handle such critical challenges of multi-drug resistance. Notably, the chemotherapeutic activities of the complexes of Schiff bases have grasped the interest of many academic and clinical investigators and already become the subject of numerous articles (Chaudhary et al., 2021).

The development of molecularly designed small compounds like antimicrobial and anticancer drugs concluded their high potential applications. Although there are large numbers of drugs that already exist to combat cancer and microbial infections, the wide existence of multi-drug resistance requires expansion of modified and potential derivatives having desired features and can void the

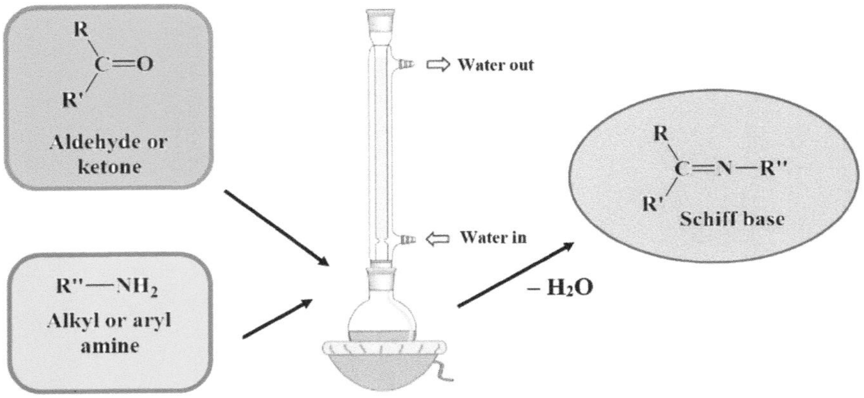

FIGURE 12.6 A typical reaction procedure for the formation of the Schiff base.

problem of drug resistance. The use of metal-organic drugs in anticancer chemotherapy proved to be a successful strategy. Such strategy has efficient potential uses for treating the multi-drug-resistant infections. Interest in binding the Schiff base complexes with biologically active molecules, such as genes and nucleic acids, was encouraged to understand the basis of such interaction modes. Consequently, development considerations for these species is achieved to be utilize as anti-inflammatory and anticancer agents (Kaplánek et al., 2015; Bingul et al., 2016; a- Mohamed et al., 2021; Ramadan et al., 2021). In particular, Schiff bases with hydrazide-hydrazone group [-(C=O)NHN=CH] as well as their transition metal complexes are indicated to have excellent applications as models in medicinal and biological sciences (Abdel Aziz et al., 2017; El-Medani et al., 2020).

The category of Schiff bases, as a type of organic compounds, showed a priority of interest in coordination chemistry because of their multiple features of metal chelating, biological activity and flexible modification, and in the tuning of their structures to implement in a particular biological application. The Schiff base-metallodrugs have been widely investigated to develop novel promising antimicrobial and anticancer chemotherapeutic reagents. Owing to the differences between those two targets, structurally modified and designed Schiff bases, especially those having heterocyclic moieties, are currently extensively investigated to obtain the desired drug that could target a particular disease (Malik et al., 2018; El-Medani et al., 2020; Ramadan et al., 2021).

Therapeutic drugs containing metal-based complexes offer versatile structural and electronic features with diversity ranges of oxidation states, coordination geometries, number and type of ligands. Thus, metallodrugs may be activated by ligand substitutions or oxidation-reduction reactions. They are also multi-targeting and required to be considered to confirm structure-activity relations. One of the most researched diseases is cancer, rated as the second leading cause of death in the world after coronary artery and stroke diseases. Cancer refers to any one of many diseases characterized by the development of abnormal cells. These abnormal cells rapidly divide and are able to destroy the normal body cells. Cisplatin, a square planar Pt(II) complex with a formula cis-$Pt(NH_3)_2(Cl)_2$, is one of the earliest used chemotherapy drugs, playing an important role in fighting many types of cancer like testicular, ovarian, bladder, cervical, and lung cancers. Since the discovery of cisplatin in 1965, other similar platinum-based drugs were developed and used for different cancer patients. Cisplatin can obstruct the DNA from replicating and fparticipating in protein synthesis. The cells of cancer start to recognize that there is something wrong and, thus, initiate apoptosis (a type of cell-suicide). However, cisplatin causes terrible side effects, including severe nausea, vomiting, critical anemia, and hair loss, and the sufferer is susceptible to infections as well as it destroys the patients' kidneys and livers. Thus, the problems of such severe symptoms of platinum-based drugs (cisplatin, oxaliplatin and carboplatin) as antitumor drugs encouraged the developing of novel metal complexes able to interact with DNA with potential anticancer activity but with lesser side effects. Therefore, researchers seek other metal-based drugs, especially those containing Schiff bases with different metals than platinum(II), such as platinum(IV), gold(III), gold(I), ruthenium (II), and ruthenium (III). Both gold(III) and platinum(II) are isoelectronic metal ions and their complexes adopt square planar structural arrangements. Gold(III) compounds were investigated for their anticancer activities. Contrary to cisplatin, the biological studies of these complexes indicated that they targeted proteins instead of DNA, the same as cisplatin. On the other hand, ruthenium complexes present as an interesting subject in the medicinal chemistry field as they display great importance as anticancer agents with low toxicity and selective anti-metastatic properties. Further, ruthenium complexes proved to be successful in penetrating tumor cells and bind effectively to the nucleic acids. Thus, ruthenium complexes presented great research subjects and they are tested for their effects versus the different cell lines of cancer. These complexes are noted to be promising substitutes instead of cisplatin and its derivatives. In particular, Ru-derivatives are generally less toxic than other platinum drugs and they are most efficient against drug resistance induced by other drugs. Ruthenium complexes of Schiff base ligands succeeded in showing positive achievement for biological applications as antioxidant reagents, and antimicrobial and anticancer potential medications. This is probably because of the octahedral coordination preference of both ruthenium(II) and ruthenium(III) species. Such three-dimensional frameworks may provide potential elevation of site selectivity to better binding to the biologically active macromolecular targets. Furthermore, these molecular structure arrangements may cause deactivation of tumor suppressor gene (p53) in cancer cells. In addition, they can overcome the bad chemotherapeutic results and the clinical drugs resistance (Singh and Barman, 2021). Additionally, copper is another expected contender to create and improve new anticancer agents. The metal complexes of Zn(II), Ti(IV), Ga(III) and vanadium (with different oxidation states) have also demonstrated to be promising biological active models (Shekhar et al., 2021).

12.6 MOLECULAR DOCKING OF SCHIFF BASE METAL COMPLEXES

Analysis of docking results in the molecular docking process is helpful in detecting the conformation alterations connected with the binding positions of the amino acid parts, which accommodate the docked hydrophobic inhibitors. A large number of articles and reviews reported the molecular docking of transition metal complexes with many macromolecular bioactive targets. Among these articles, molecular docking operations of the Schiff base metal complexes with vast number of macromolecular targets have been thoroughly investigated and reported. Many reported metal-Schiff base derivatives that have promising applications as potential drugs for the treatment of antimicrobial, antiviral and anticancer diseases. What follows are illustrated represented examples of the molecular docking of some metal complexes of Schiff base derivatives.

Biological investigations of complexes of some bivalent metal ions of molecularly designed Schiff base derivatives with a phenylacetohydrazide moiety, 2-hydroxybenzylidene-2-phenylacetohydrazid (**A**), 1-(2-hydroxyphenylethylidene)-2-phenylacetohydrazide (**B**), and 1-(hydroxynaphthalenylmethylene)-2-phenylacetohydrazide (**C**) (Figure 12.7) supported that notion they can interact with nucleic acid through intercalative mode. They also displayed different DNA binding potency. Molecular docking of these derivative was carried out on a B-DNA target (PDB: 1BNA) to estimate the changes in conformational structures supported with amino acid parts. These residues accommodated the docked molecules (inhibitors) in their binding positions and showed how such docked derivatives mainly fit inside the nucleic acid minor grooves. The docked derivatives' conformational structures were rated according to their H-bonding and hydrophobic interaction between molecules and DNA receptor as well as the values of their binding energies. The studies also indicated that the optimum docking poses fitted in DA, DC and DG regions. The docked species displayed good high negative score values for binding (from -5.07 to -6.65 kcal/mol), which indicated higher efficiency of the bioactive reagents. The binding interactions of the

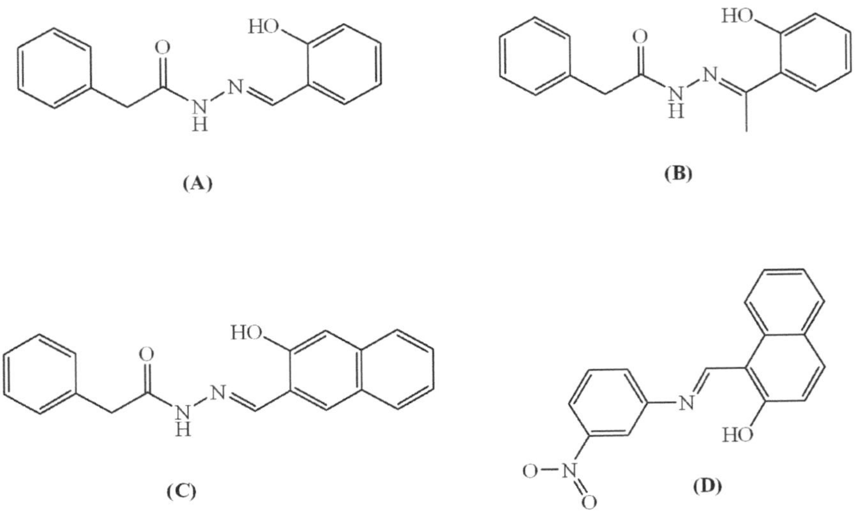

FIGURE 12.7 Structure of Schiff bases: (A) N'-(2-hydroxybenzylidene)-2-phenylacetohydrazid, (B) N-(1-(2-hydroxyphenyl)ethylidene)-2-phenylacetohydrazide, (C) N'-((1-hydroxynaphthalen-2-yl)methylene)-2-phenylacetohydrazide and (D) 1-((3-nitrophenylimino)methyl)naphthalen-2-olate.

[Co(**A**)$_2$], [Ni(**A**)$_2$], [Cu(**A**)$_2$], [Ni(**B**)$_2$], [Co(**C**)$_2$] and [Ni(**C**)$_2$] complexes resulted from hydrophobic interactions between the aromatic moieties of ligands and the amino acid residuals (DA-17, DA-18, DC-11, DC-15, DG3-12, DG-10 and DG-16). In addition, π-interaction with aromatic rings and hydrogen bonds was observed. The binuclear copper complex [Cu$_2$**B**$_2$] exhibited four hydrogen bonds to the DG-10 and DT-19 residue of DNA with the best binding score among the complexes (-6.65 kcal/mol). Therefore, it was concluded that these biologically active complexes were effectively interacted with the ready binding sites of the macromolecular receptor (El-Medani et al., 2020; Ramadan et al., 2021).

The molecular docking studies of cobalt and zinc (Co**D**$_2$ and Zn**D**$_2$) complexes of 1-((3 nitrophenylimino)methyl)naphthalen-2-olate Schiff base (**D**, Figure 12.7) were performed on the DNA-bound human topoisomerase II alpha (PDB ID: 4FM9) receptor to describe the potential inhibition and indicate the modes of binding in the enzyme pockets. Topoisomerases II are desired for managing the DNA super-helicity and chromosomes separation. They also perform as front-line receptors for varieties of therapeutic small molecules. The two studied compounds displayed particular binding affinity and good interaction with the binding pockets of topoisomerase II enzyme. CoL$_2$ bound with the receptor via three hydrogen bonds, two of them were between the nitro group of the nitroaniline moiety with GLN-938, and THR-934 residues, and one between the aromatic part of nitroaniline with LYS-798 part of the enzyme. Hydrophobic interactions were also observed with the residues MET-766 as well as weak π-anion interactions with the DA-8, DC-9 and DG-10 base pairs of DNA strand. The ZnL$_2$ complex showed only two H-bonds with the nitro group and the THR-934 and GLN-938. Further, the LYS-798 and MET-766 residues interacted hydrophobically, while two π-alkyl interactions with the aromatics adjacent to zinc were observed. In addition, weak π-anion and π- cation interaction was noted with DC-9, DA-8 and DG-10 nucleotides of the DNA strand (Rauf et al., 2020).

The molecular docking studies of the two V-shaped Schiff bases, **L1** and **L2**, along with ruthenium(II) complexes having the formulas [Ru(bpy)$_2$**L1**Cl]$^+$ and [Ru(bpy)$_2$**L2**Cl]$^+$, bpy = bipyridine (Figure 12.8), were carried out to recognize the interactions of the compound-DNA and their potential binding affinities and energy. The conformational structures were assessed according to the values of binding scores where hydrophobic interaction and H-bonding occurred among the docked compounds and 1BNA (DNA) receptor. The most favorable docking findings were in the GC guanine-cytosine zones, with a lesser magnitude in adenine area. The analysis supported the data of viscosity measurements and fluorescence quenching that confirmed the intercalation interaction mode of the DNA. The best binding poses obtained from the docking results of the two ligands and the two complexes are illustrated in Figure 12.8. The ligands and complexes displayed good

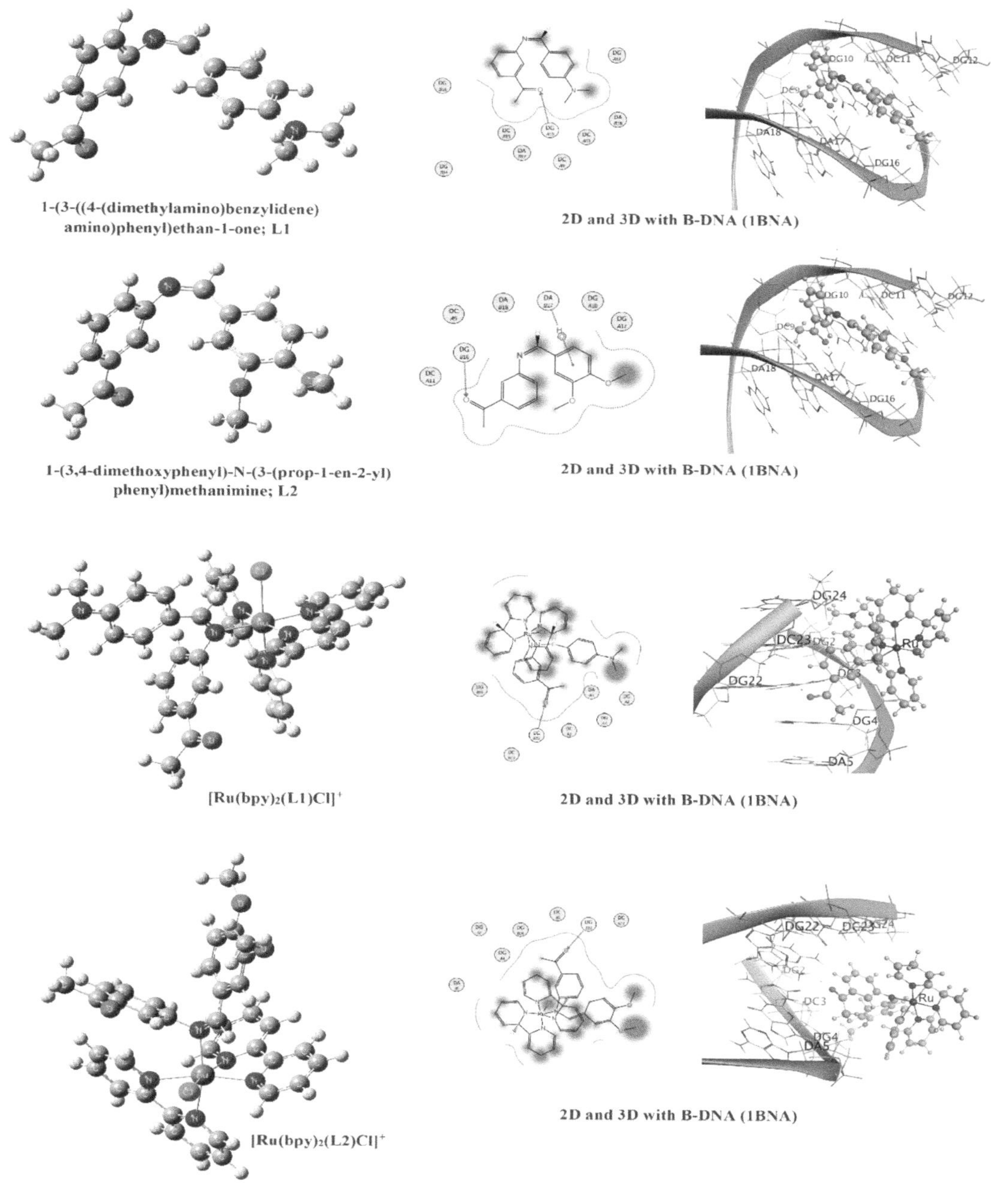

FIGURE 12.8 Molecular docking of L1, L2, and their Ru(II) complexes, [Ru(bpy)$_2$(L1)Cl]$^+$ and [Ru(bpy)$_2$(L1) Cl]$^+$, with a B-DNA (1BNA).

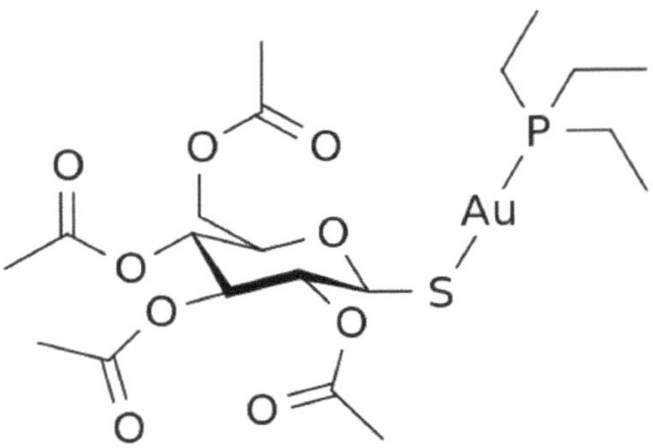

FIGURE 12.9 Structure of Auranofin: gold(I) complex.

high negative values for the binding scores (-4.60 to -5.70 kcal/mol). These findings indicated high affinity of binding properties between docked compounds and target. They also proved their elevated efficiency as biologically active compounds. The binding interactions of the amino acid moieties such as DG (B16), DG (A12), DC (A11), DC (B15) and DA (B18) occurred from hydrophobical interactions with the methyl and aromatic groups of the docked compounds. Furthermore, formation of an H-bond was formed between oxygen of the ketonic group of ligand with DA (B17), DG (B16) and DG (A10) residues. It was also noted that complexes interacted with nucleic acid from about the same zones like the ligands themselves (Figure 12.8). The two complexes showed hydrophobical interaction with the DA, DC and DG zones. In addition, the oxygen of the ketone part of ligands in complexes performed H-bonding through the DG (B22) region of target. Biological studies of the compounds were correlated with their molecular docking. The data exhibited different DNA binding potency and supported binding to DNA was through intercalative mode. It was concluded that these derivatives may thus be counted as potential reagents for therapeutically intervention in many diseases (a- Mohamed et al., 2021).

The drug auranofin is a trialkylphosphine gold(I) complex (Figure 12.9). It is an anti-arthritic agent orally taken and is prescribed for the therapeutic treatment of rheumatoid arthritis. It is also prescribed for persistent rheumatoid arthritis, which does not respond to conventional chemotherapy. It is well known that gold(I) complexes, like auranofin, selectively inhibit the thioredoxin reductase enzyme that is considered a main target of gold compounds. Thioredoxin reductase active sites include selenocysteine- and cysteine-amino acids that contribute to the mechanism of activity of enzymes. Interactions with the nucleophilic sulfur and selenium groups are believed to be the inhibition mode of action of gold complexes. Molecular docking analysis of the inhibition of the prominent anticancer drug, thioredoxin reductase enzyme, and a carrier for anticancer drug, human serum albumin, was carried out using some gold(I)-anticancer agents (synthesized by condensation, cyclization or addition reactions of Schiff base-aldehydes. Notably, these complexes exhibited good docking scores. The introduction of the Schiff base phenolic moieties was found to alter the sites of binding of gold compounds with the receptor. The docking strategy was performed on the human thioredoxin reductase-thioredoxin complex (PDB: 3QFB) and human serum albumin complexed with myristic acid (PDB: 1H9Z). The docking data revealed that the copper complexes possess powerful priority to bind to DII domain. The pocket of HSA (DII) is surrounded by the amino acid residues Asp-183, Asp-187, Ala-191, Ala-194, Glu-184, Glu-188, Lys-436, Lys-432, Lys-190, Arg-197, Asn-429, Val-433, Tyr-452, Val-462, Val-456, Val-455, Asn-458, and Gln-459. Some of those amino acid parts included hydrophobical interactions with the complexes as well as π–π interactions with Try-452 moiety (Babgi et al., 2021).

The molecular docking studies of the ruthenium complex $[Ru(Cl)_3(H_2O)(C_{13}H_{11}N_4O_2Cl)]$ was performed to explore complex-targets interacting sites, possible binding styles, and binding energies.

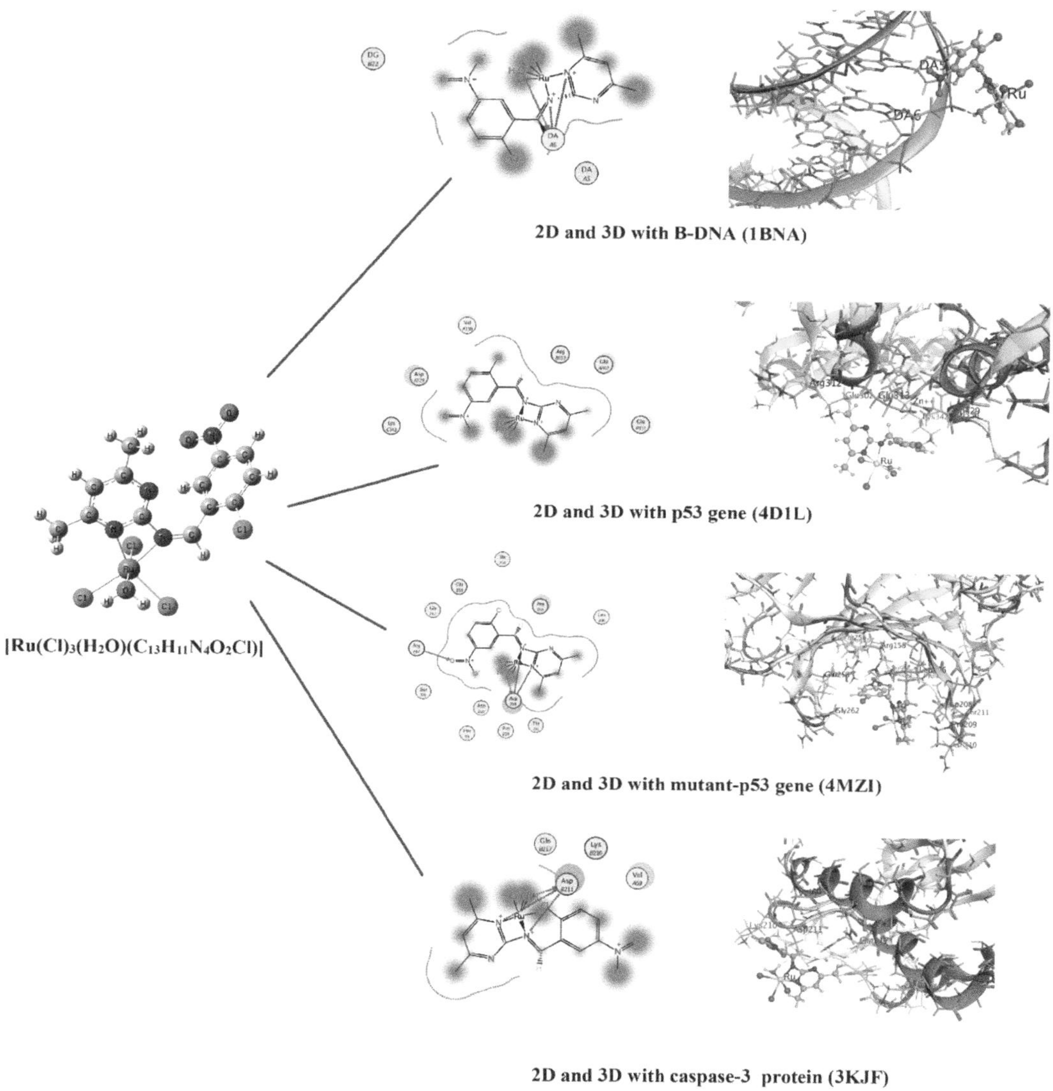

FIGURE 12.10 Molecular docking of $[Ru(Cl)_3(H_2O)(C_{13}H_{11}N_4O_2Cl)]$ complex with different targets.

Conformations of the docked complex were rated relative to the affinity of binding, hydrophobical interactions and H-bonding between the macromolecule target and the ruthenium complex. The used macromolecule targets were: B-DNA (1BNA), p53 tumor suppressor gene (4D1L), p53 tumor suppressor mutated gene (4MZI), caspase-3 protein (3KJF) and PTEN tumor suppressor (1D5R). The molecular docking analysis determines how the docked complex suitably fits into the minor groove of the macromolecule receptor and cover the different interaction modes with the target bases such as hydrophobic, ionic, and hydrogen bonding. Figure 12.10 illustrated the two- and three-dimensional (2D and 3D) molecular docking interaction representations of the ruthenium complex with the studied targets. The final score values of the interactions of ruthenium complex with targets, which are comparable to the binding affinity ranged from -2.52 to -4.12. These values illustrate that the investigated ruthenium complex has high efficiency as a bioactive compound. In the case of the 1BNA receptor, the binding interactions with the DNA resulted from hydrophobic interactions between amino acid parts, mainly DG 22 and DA 5, along with three hydrogen bonds between the DA 6 region and both the coordinated water and nitrogen atoms of the coordinated Schiff base ligand. On the other hand, according to the scoring values and binding interactions, the ruthenium complex showed different binding affinities with either p53 gene (4D1L) and the mutated p53 (4MZI). The binding score with the mutated p53 (S = -4.04) was found to be higher than the corresponding p53 gene itself (S = -2.52). Thus, it was concluded that the ruthenium complex has higher binding affinity toward the mutated p53. The interaction of the complex with p53 gene came only from hydrophobic interactions with Asp 329, Lys 342, Arg 312, Glu 302 and Glu 313 parts. On the other hand, in case of the mutated p53, the ruthenium complex formed three hydrogen bonds between the Asp 208 part of receptor along with hydrophobic interactions with several parts of the receptor such as Glu 258, Arg 267, Arg 158, Asn 210 and Thr 211 (Figure 12.10). Furthermore, the complex exhibited the best binding affinity (S = -4.12) with PTEN macrmolecule (1D5R) relative to the other investigated receptors. The most optimal docking was found to be hydrophobic interactions fitted in the Thr, Arg, Asn and Tyr parts of the receptor as well as three hydrogen bonds with the Asp 324 moiety. The docked ruthenium molecule also showed good binding score (S = -3.91) with caspase-3 protein (3KJF), as it formed hydrogen bonding with Asp 211 part and hydrophobic interactions with Lys 210 and Gln 217 moieties (Figure 12.10). Therefore, it was concluded that this ruthenium complex is regarded as a bioactive compound and has the ability to interact effectively with any available binding positions of the studied targets (Wahb, 2022).

Due to the sudden spread of the COVID-19 pandemic virus all over the world, which started in 2020, scientists and researchers have worked very hard to discover suitable drugs to cure such dangerous disease. Many organic and coordination compounds have been investigated in order to try treating this serious problem. Among those, many articles appeared describing the molecular docking and virtual screening of different Schiff bases along with their metal complexes and showed sufficient characteristics of inhibitions of COVID-19 disease. Molecular docking investigations were performed on metal ion complexes of Cr, Mn, Fe, Ni, Cu, Zn and Cd with a pyrazolone Schiff base, 4-((1-(5-acetyl-2,4-dihydroxyphenyl)ethylidene)amino)-1,5-dimethyl-2-phenylpyrazolone (Figure 12.11A) and the target SARS-CoV-2 main protease complexed to the UAW247 inhibitor, Protein Data Base ID: 6XBH (b- Mohamed et al., 2021). The analysis showed that the best predicted poses of the Schiff base ligand and its complexes have free binding energy (ΔG) ranged from -2.5 to -8.0 kcal/mol. According to the docked compound-target complexes, it was found that the docked compounds bound to target either by hydrogen donor with amino acid moieties of methionine (Met-49) and glutamic acid (Glu-166) or with hydrogen acceptor of the amino acid residues of glycine (Gly-143) and glutamine (Gln-189). The interaction energies of the complexes with the target revealed that the chromium complex had the higher binding interaction. This was due to formation of H-bonds and the amino acid parts, established by a π-H interaction with the Glu-189 residue. On the other hand, the molecular docking of 2-hydroxy-3,5-diiodobenzylidenedimethylcarbamohydrazonothiate base (L, Figure 12.11B), along with its copper(II) [Cu(L)(phen)] complex, phen = 1,10-phenanthroline, was performed on the COVID-19 main protease (M^{pro} protein) structure complexed with GC376 (PDB ID: 7C8U). The binding affinity of the docked molecules was found to be -7.14 and -6.18 kcal/mol for the ligand and the complex, respectively, which means strong binding interaction between the receptor and docked species. In addition, these binding affinities are comparable with some recognized antiviral medications. For example, the molecular docking analysis for remdesivir gave a binding affinity value equal to -6.352 kcal/mol, chloroquine gave -6.293 kcal/mol and hydroxychloroquine gave -5.573 kcal/mol with M^{pro} protein (Mohan and Choudhary, 2021). The docked molecules interact with the receptor via hydrogen bonding, π-H interaction and hydrophobic interactions with the residues SER-144, LEU-141, HIS-172, HIS-163, HIS-41, MET-165, THR-190 and GLU-166 for the Schiff base ligand. For the docked copper complex, interactions occurred with LEU-141, HIS- 41, LEU-262, VAL-171, VAL-297, THR-196, GLU-166 and ALA-193.

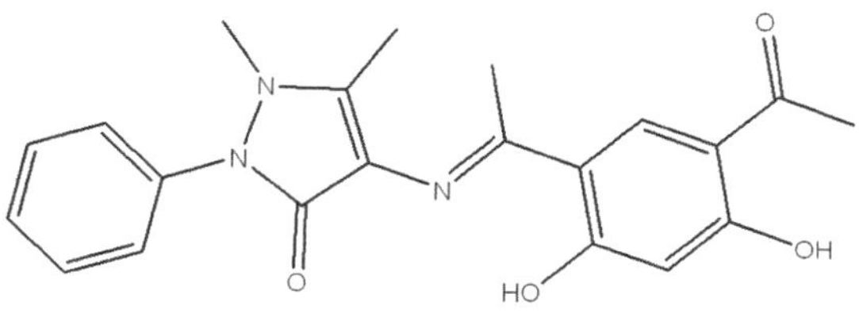

(A)

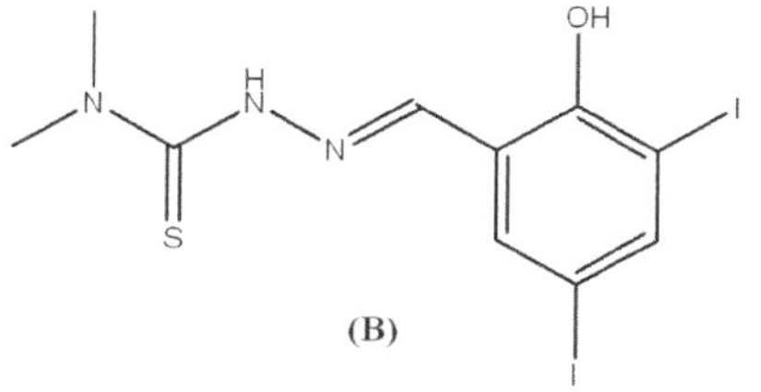

(B)

FIGURE 12.11 (A) Structure of 4-((1-(5-acetyl-2,4-dihydroxyphenyl)ethylidene)amino)-1,5-dimethyl-2-phenylpyrazolone Schiff base ligand. (B) Structure of 2-hydroxy-3,5-diiodobenzylidenedimethylcarbamohy drazonothioic.

12.7 EXPERIMENTAL STEPS FOR MOLECULAR DOCKING

Many reviewed articles and publications previously appeared to describe molecular docking methods, and to compare the performance of such docking tools (Halperin et al., 2002; Taylor et al., 2002; Sousa et al., 2006; Huang and Zou, 2010; Meng et al., 2011). To perform the molecular docking for a potential drug, four steps are adopted:

1- *Selection of the macromolecular target*: Structure of the bioactive macromolecular target must be experimentally elucidated by either NMR or X-ray crystallography (Xrd). Most of them are available and can be obtained from any site of protein databases (PDB). The model structure should have good quality and checked by using any available validation software. After the selection of the target, its structure has to be prepared by deleting the solvent molecules (often water) from its cavity, then stabilize charges and fill the missing residues. Generate the side chains depending on the available parameters. The target receptor is now in the stable state and considered as biologically active.
2- *Selection and preparation of ligand*: The word "ligand" is used to define the compound or drug, which is docked with the target. The kind of ligands demanded for docking depends on the required objective. Either it is downloaded from different available databases, or it is a synthetic molecule, where its structure is identified and sketched by means of drawing tool such as ChemDraw or Chemsketch. Often, it is the requisite to reduce the number of docked molecules by considering many parameters, such as molecular weight, net charge, solubility, polar surface area, toxicology properties, commercial availability, etc.
3- *Docking*: In this step, the drug or desired compound (ligand) is docked upon the target (receptor), then the different types of interactions (hydrophobic, H-bonding, π-aromatic, etc.) are reviewed. Generation of the scoring function will give scores due to the best poses of the docked ligand.
4- *Evaluation of docking data*: The docking algorithms will succeed in indicating the ligand-binding poses according to the amounts of root-mean-square deviation (RMSD). These values are defined as distances between observed atoms places of docked derivatives and those

indicated by algorithms. The best performance for the docked organic ligands is acceptable when the value of RMSD is less than 2Å. However, the RMSD value could be larger than 2Å if the docked ligand is a metal complex or in case of protein-protein docking. The system flexibility demonstrates challenges in search of corrected poses. Therefore, the coherent conformational numbers of degrees of freedom in the searches are considered to be a focal side in determining the searching efficiency.

12.8 MOLECULAR DOCKING TUTORIAL

Here, we will illustrate systematically detailed procedures for docking a ruthenium-Schiff base complex $[Ru(Cl)_3(H_2O)(C_{13}H_{11}N_4O_2Cl)]$ (Wahb, 2022) with a B-DNA target (PDB: 1BNA) using MOE software.

1- Select the macromolecular target such as a B-DNA (PDB: 1BNA) and download from the Protein Data Bank website. The target is prepared by deleting solvent molecules from the receptor cavity. Then, stabilize charges of target and fill the missing residues (Figure 12.12).
2- Draw the structure of the potential drug such as $[Ru(Cl)_3(H_2O)(C_{13}H_{11}N_4O_2Cl)]$ complex on ChemDraw in 2D pattern (Figure 12.13).

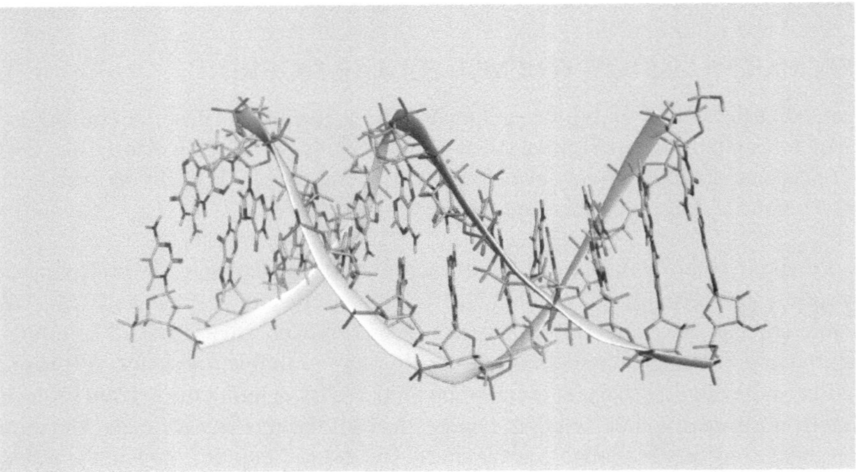

FIGURE 12.12 The preparation of B-DNA (1BNA) target.

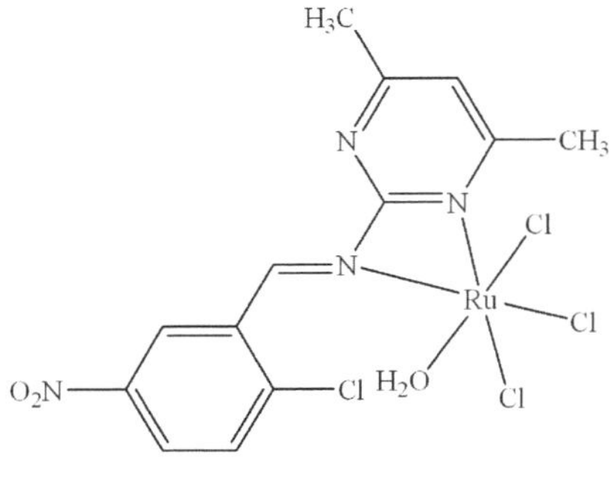

FIGURE 12.13 2D drawing of the Ru-complex.

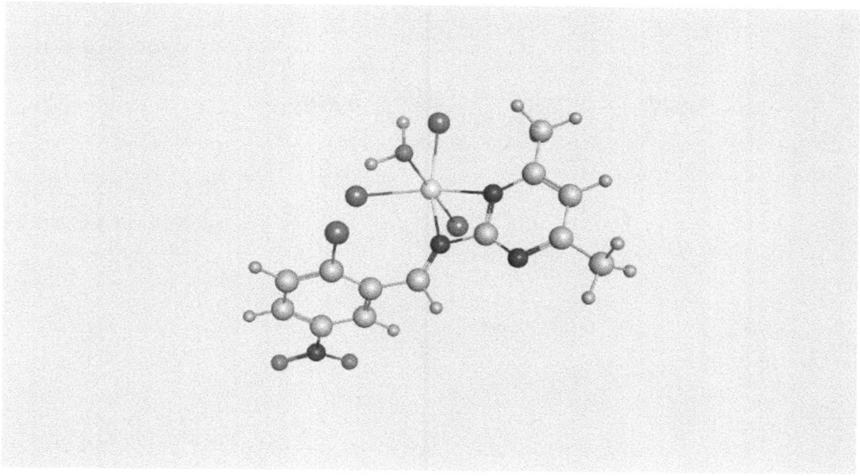

FIGURE 12.14 Optimized structure of the Ru-complex.

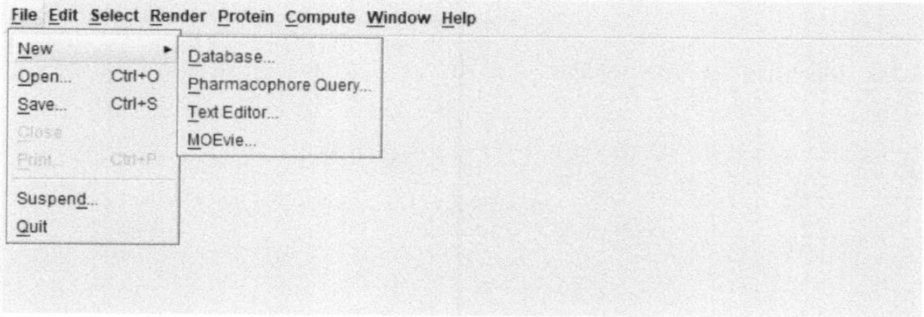

FIGURE 12.15 Open a database file.

3- Insert the structure of the ruthenium complex into the MOE main page. Carry out optimization and save the structure with a .mdb extension (Figure 12.14).
4- From the menu of the MOE screen, open **File →New →Database** (Figure 12.15). Name the file by appropriate name such as Ru-complex database. A pop-up page will appear with the choosing name.
5- From the **Database** page, click on **File →Import**. A pop-up page will ask to select the required file. Choose the .mdb file of the ruthenium complex from the list as shown in Figure 12.16.
6- Now, it is time to carry out the docking process. From the main page of MOE, click on the icon **Compute** then select **Dock**. On pressing on the icon **Dock**, a pop-up page will appear to adjust and select the docking protocol. Browse for the Dock data output, select the receptor (1bna.pdb), select the .mdb file of ligand (ruthenium complex) by clicking on **Browse**. Select rescoring 1 and 2 by choosing 10, 20 or 30 refinements (Figure 12.17).
7- After clicking **Run** on the docking protocol page, the main page of MOE will demonstrate the docking trails between ligand and target (Figure 12.18).
8- When the docking process is finished, a new pop-up page (**Database Viewer**) will appear illustrating the docking data, which include score values (S), the root-mean-square deviation (rmsd_refine) and the configuration, scoring and refining energies (Figure 12.19).

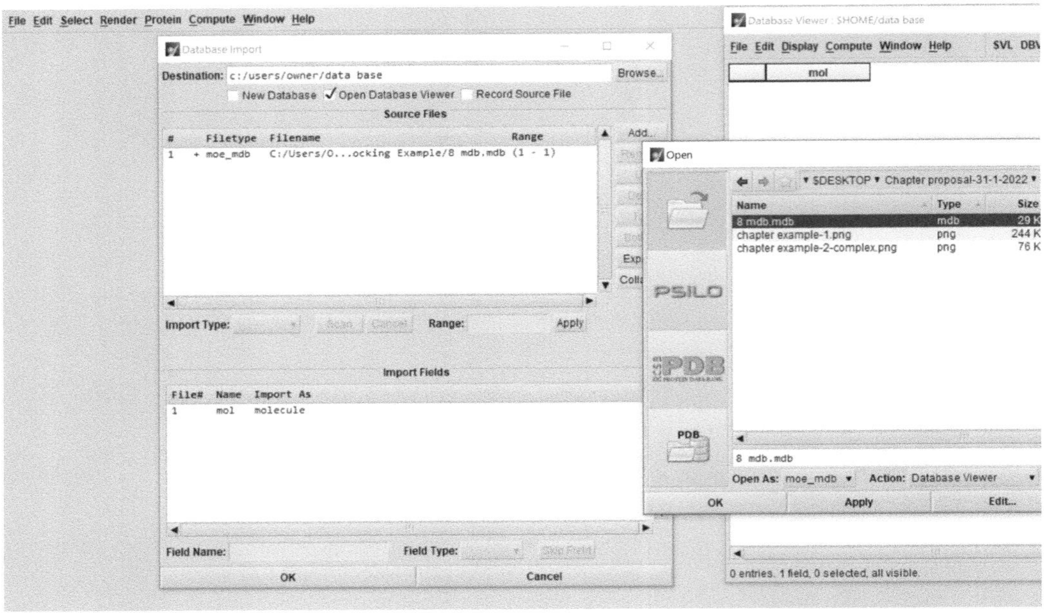

FIGURE 12.16 Importing the .mdb structure of the Ru-complex into the Database file.

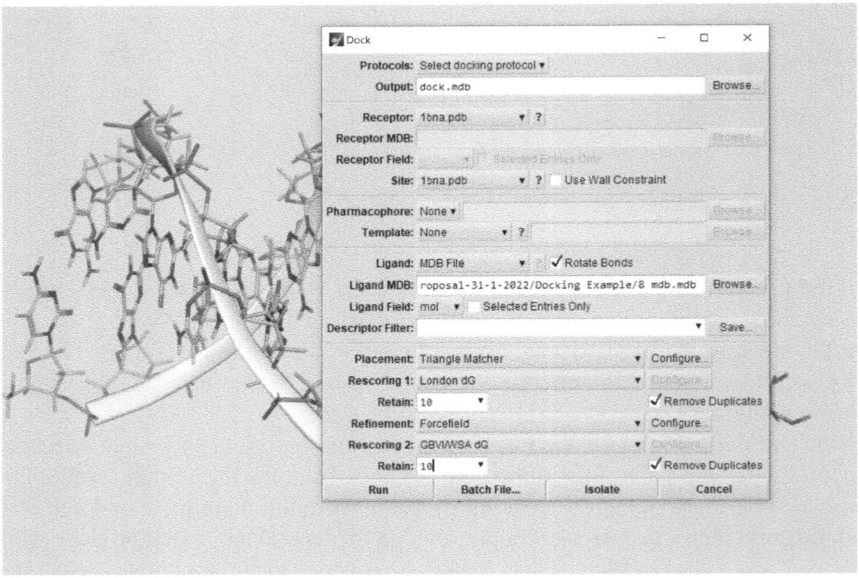

FIGURE 12.17 Selection of the docking protocol.

9- From the Database Viewer, click on **File** and select **Browse**. A new pop-up page illustrating the ligand structure will appear. You can choose from 1–10 poses as shown in the Database Viewer (Figure 12.20).

10- From the main page of MOE, click on **Compute** and then choose **Ligand Interactions** (Figure 12.21). A new pop-up page with the title "**Ligand Interactions**" appears and demonstrates the first two-dimensional docking pose (Figure 12.22). The illustration shows

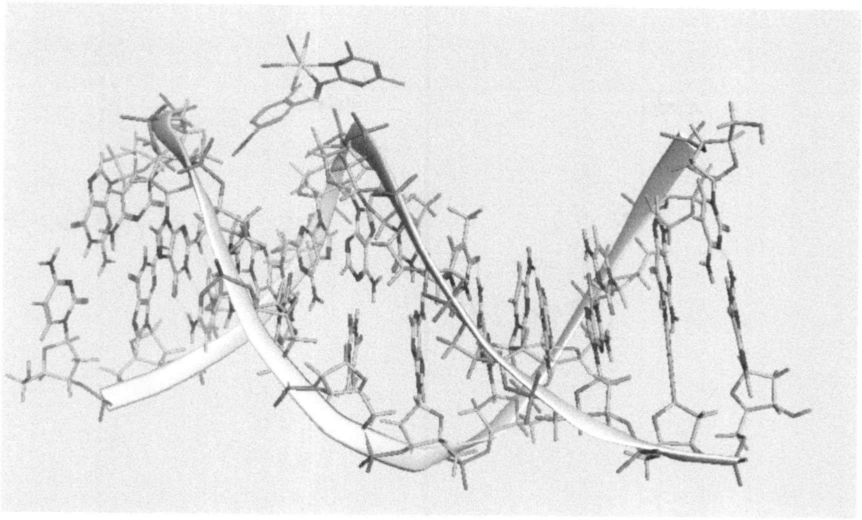

FIGURE 12.18 The docking processes.

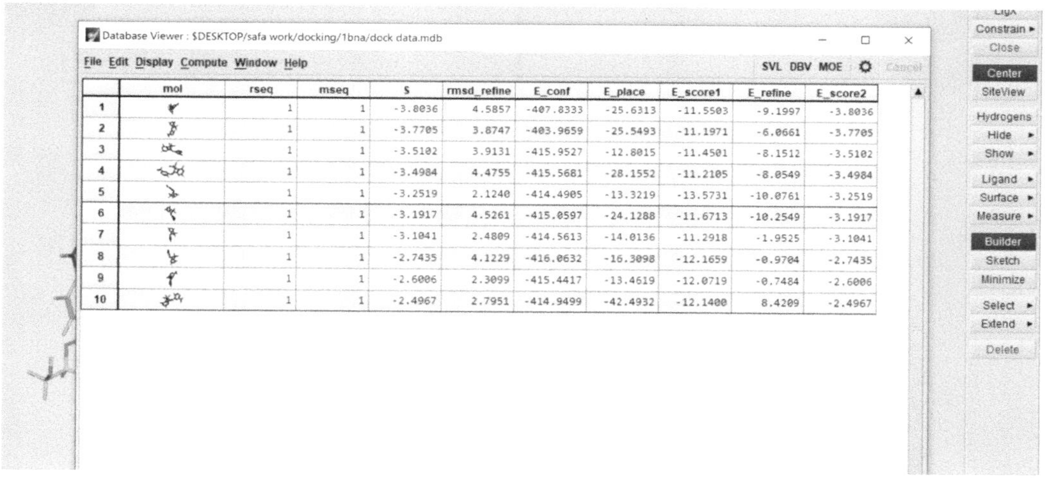

	mol	rseq	mseq	S	rmsd_refine	E_conf	E_place	E_score1	E_refine	E_score2
1		1	1	-3.8036	4.5857	-407.8333	-25.6313	-11.5503	-9.1997	-3.8036
2		1	1	-3.7705	3.8747	-403.9659	-25.5493	-11.1971	-6.0661	-3.7705
3		1	1	-3.5102	3.9131	-415.9527	-12.8015	-11.4501	-8.1512	-3.5102
4		1	1	-3.4984	4.4755	-415.5681	-28.1552	-11.2105	-8.0549	-3.4984
5		1	1	-3.2519	2.1240	-414.4905	-13.3219	-13.5731	-10.0761	-3.2519
6		1	1	-3.1917	4.5261	-415.0597	-24.1288	-11.6713	-10.2549	-3.1917
7		1	1	-3.1041	2.4809	-414.5613	-14.0136	-11.2918	-1.9525	-3.1041
8		1	1	-2.7435	4.1229	-416.0632	-16.3098	-12.1659	-0.9704	-2.7435
9		1	1	-2.6006	2.3099	-415.4417	-13.4619	-12.0719	-0.7484	-2.6006
10		1	1	-2.4967	2.7951	-414.9499	-42.4932	-12.1400	8.4209	-2.4967

FIGURE 12.19 The docking results from Database Viewer.

the type of interactions (hydrophobic or H-bonds) as well as the interacted fragments of the target such as DA6, DA5 and DG22. Browse between the different poses by clicking on **Apply** every time at the **Ligand Interactions** page. From the list, choose the best pose of docking according to the RMSD value and the number of hydrophobic and hydrogen-bond interactions. In this stage, some experience and information about the structure of ligand is required to select the best pose.

11- Now, we can save the two-dimensional structure of the chosen best pose (Figure 12.23). In addition, click on the **Report** icon on the page will give summarized data of the chosen pose, including the interacted atoms of the ligand and receptor, type of interaction, distances between interacted atoms and their interaction energies (Figure 12.24).

12- To get the three-dimensional docking pose, click on the icon **Isolate**. The chosen best docking pose will illustrate on the MOE main page. Color, background, type of bonds, etc.,

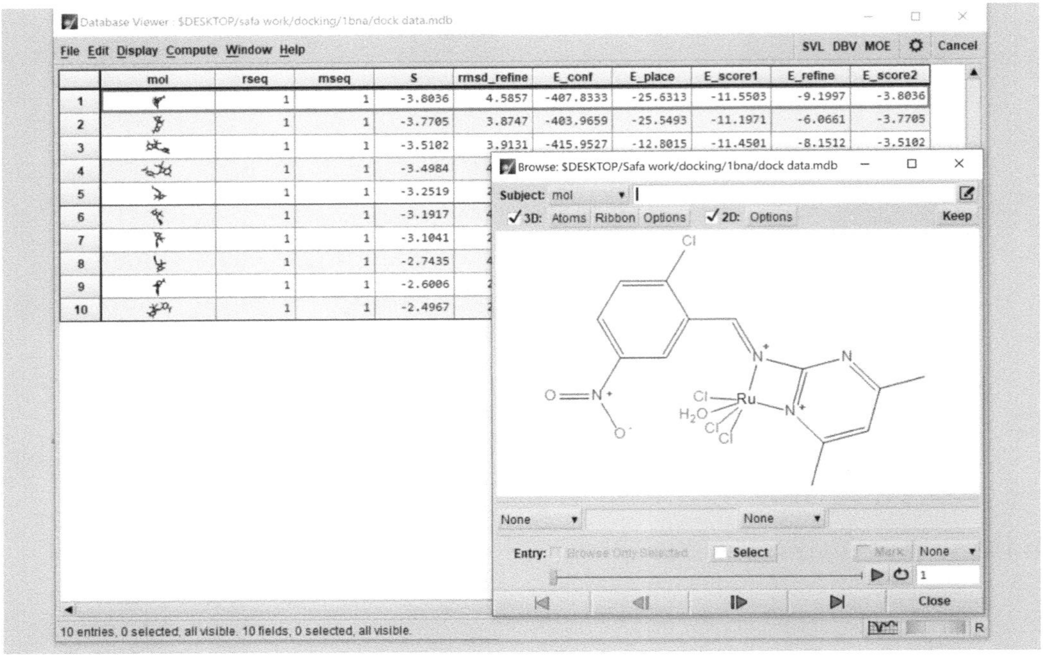

FIGURE 12.20 Browse for the ligand docking data.

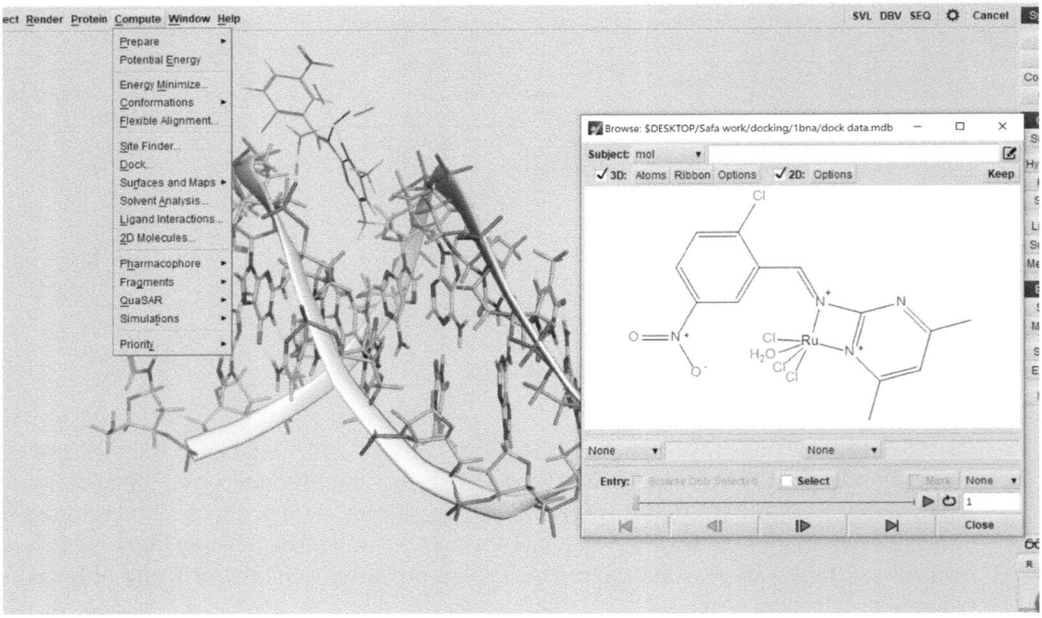

FIGURE 12.21 Select Compute →Ligand Interactions.

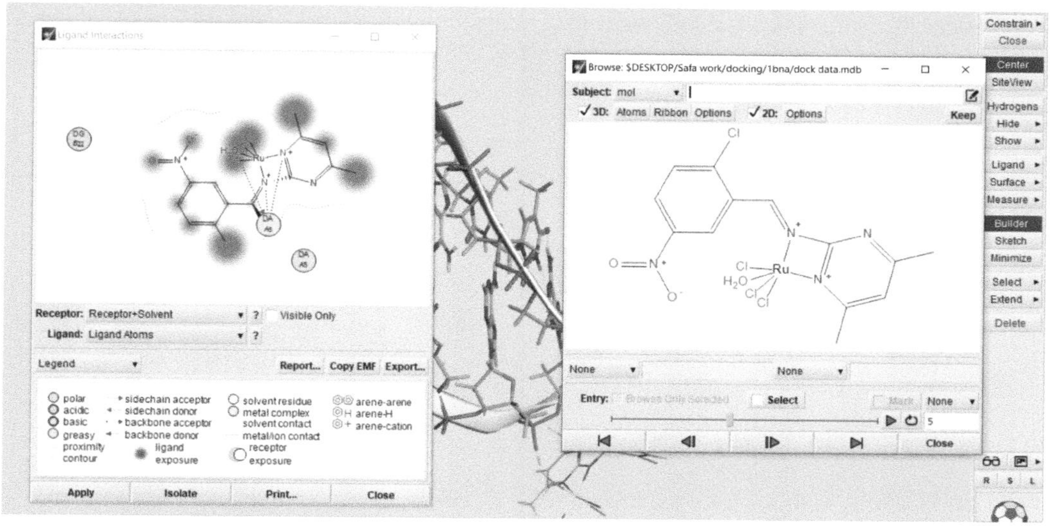

FIGURE 12.22 Ligand Interactions base demonstrating the two-dimensional docking pose.

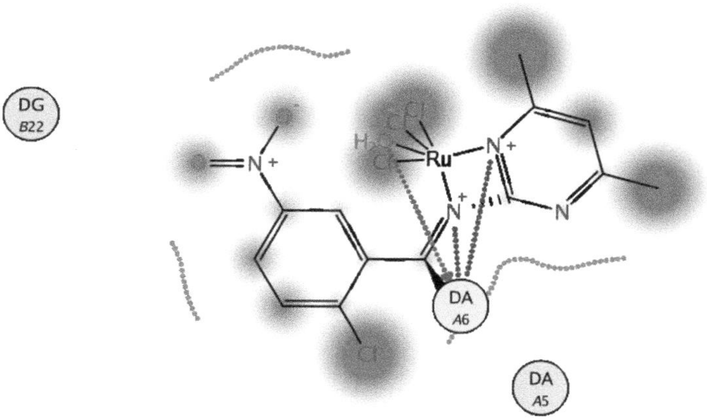

FIGURE 12.23 2D molecular docking interactions of the Ru-complex-target.

are selected in the desire to reach the final 3D view of the docked ligand-receptor complex (Figure 12.25).

13- Once you have gained all the information (type of interactions, score, and energies), you can decide to go further and continue with the structure-based drug design process (*vide supra*).

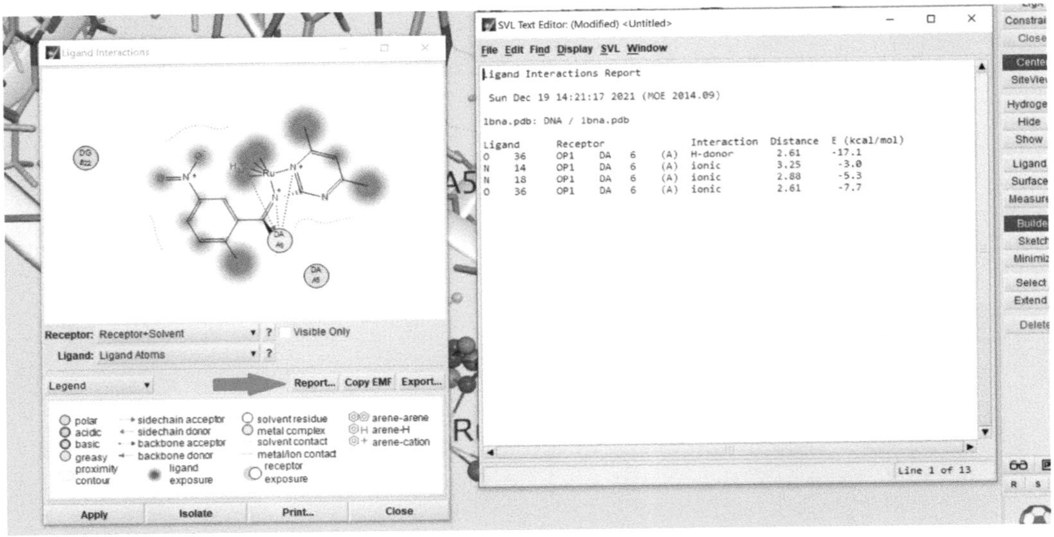

FIGURE 12.24 Report of the ligand interactions data.

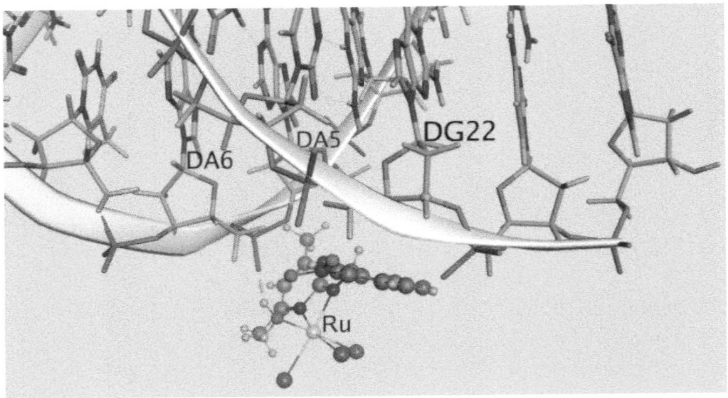

FIGURE 12.25 3D molecular docking interactions of the Ru-complex-target.

REFERENCES

Abdel-Rahman, L.H., Ismail, N.M., Ismael, M., Abu-Dief, A.M., and Ahmed, E.A. 2017. J. Mol. Struct. 1134: 851–862.

Abdel Aziz, A.A., Elantabli, F.M., Moustafa, H., and El-Medani, S.M. 2017. Spectroscopic, DNA binding ability, biological activity, DFT calculations and non linear optical properties (NLO) of novel Co(II), Cu(II), Zn(II), Cd(II) and Hg(II) complexes with ONS Schiff base. J. Mol. Struct. 1141: 563–576.

Abu-Dief, A.M. and Mohamed, I.M.A. 2015. A review on versatile applications of transition metal complexes incorporating Schiff Bases. Beni-Suef Univ. J. Basic Appl. Sci. 4: 119–133.

Babgi, B.A., Alsayari, J. Alenezi, H.M., Abdellatif, M.H., Eltayeb, N.E., Emwas, A.M., Jaremko, M., and Hussien, M.A. 2021. Alteration of anticancer and protein-binding properties of gold(I) alkynyl by phenolic schiff bases moieties. Pharmaceutics 13: 461. doi: 10.3390/pharmaceutics13040461.

Bingul, M., Tan, O., Gardner, C.R., Sutton, S.K., Arndt, G.M., Marshall, G.M., Cheung, B.B. Kumar, N., and Black D.S. 2016. Synthesis, characterization and anti-cancer activity of hydrazide derivatives incorporating a quinoline moiety. Molecules 21: 916. https://doi.org/10.3390/molecules21070916.

Bjelic, S., Nervall, M., Gutiérrez-de-Terán, H., Ersmark, K., Hallberg, A., and Aqvist, J. 2007. Computational inhibitor design against malaria plasmepsins. Cell Mol. Life Sci. 64: 2285–2305.
Chaudhary, N.K., Guragain, B., Chaudhary, S.K., and Mishra, P. 2021. Schiff base metal complexes as a potential therapeutic drug in medical science: A critical review. BIBECHANA 18: 214–230.
El-Medani, S.M., Makhlouf, A.A., Moustafa, H., Afifi, M.A., Haukka, M., and Ramadan, R.M. 2020. Spectroscopic, crystal structural, theoretical and biological studies of phenylacetohydrazide Schiff base derivatives and their copper complexes. J. Mol. Struct. 1208: 127860.
Ewing, J.A.T. and Kuntz, I.D. 1997. Critical evaluation of search algorithms for automated molecular docking and database screening. J. Comput. Chem. 18: 1175–1189.
Goel, B., Bhardwaj, N., Tripathi, N., and Jain, S.K. 2021. Drug discovery of small molecules for the treatment of COVID-19: A review on clinical studies. Mini Rev. Med. Chem. 21: 1431–1456.
Gupta, M., Sharma, R., and Kumar, A. 2018. Docking techniques in pharmacology: How much promising? Comput. Biol. Chem. 76: 210–217. doi:10.1016/j.compbiolchem.2018.06.005.
Gurung, A.B., Ali, M.A., Lee, J., Abul Farah, M., and Al-Anazi, K.M. 2021. An updated review of computer-aided drug design and its application to COVID-19. BioMed. Res. Int. 8853056. https://doi.org/10.1155/2021/8853056.
Halperin, I., Ma, B., Wolfson, H., and Nussinov, R. 2002. Principles of docking: An overview of search algorithms and a guide to scoring functions. Proteins 47(4): 409–443.
Huang, S.Y. and Zou, X. 2010. Advances and challenges in protein-ligand docking. Internat. J. Mol. Sci. 11: 3016–3034.
Hughes, J.P., Rees, S., Kalindjian, S.B., and Philpott, K.L. 2011. Principles of early drug discovery. British J. Pharmacol. 162: 1239–1249.
Jia, Y., and Li, J. 2015. Molecular assembly of Schiff base interactions: Construction and application. Chem. Rev. 115: 1597–1621.
Kaplánek, R., Havlík, M., Dolenský, B., Rak, J., Džubák, P., Konečný, P., Hajdúch, M., Králová, J., and Král, V. 2015. Synthesis and biological activity evaluation of hydrazone derivatives based on a Tröger's base skeleton. Bioorg. Chem. 23: 1651–1659.
Khan, S.U., Ahemad, N., Chuah, L.-H., Naidu, R., and Htar, T.T. 2020. Illustrated step by step protocol to perform molecular docking: Human estrogen receptor complex with 4-hydroxytamoxifen as a case study. Prog. Drug Discov. Biomed. Sci. 3. doi: 10.36877/pddbs.a0000054.
Kirkpatrick, P. 2004. Virtual screening: Gliding to success. Nat. Rev. Drug. Disc. 3: 299. https://doi.org/10.1038/nrd1364.
Li, J., Fu, A., and Zhang, L. 2019. An overview of scoring functions used for protein-ligand interactions in molecular docking. Interdiscip. Sci. Comput. Life Sci. 11: 320–328.
Luu, K.T., Kraynov, E., Kuang, B., Vicini, P., and Zhong, W.Z. 2013. Modeling, simulation, and translation framework for the preclinical development of monoclonal antibodies. AAPS J. 15: 551–558.
Malik, M.A., Dar, O.A., Gull, P., Wani, M.Y., and Hashmi, A.A. 2018. Heterocyclic Schiff base transition metal complexes in antimicrobial and anticancer chemotherapy. Med. Chem. Commun. 9: 409–436.
Márquez, E., Mora, J.R., Flores-Morales, V., Insuasty, D., and Calle, L. 2020. Modeling the antileukemia activity of ellipticine-related compounds: QSAR and molecular docking study. Molecules. 25: 24. doi:10.3390/ molecules25010024.
Meng, X-Y., Zhang, H-X., Mezei, M., and Cui, M. 2011. Molecular docking: A powerful approach for structure-based drug discovery. Curr. Comput. Aided Drug Design 7: 146–157.
a- Mohamed, S.E., Ramadan, R.M., Aboelhasan, A.E., and Abdel Aziz, A.A. 2021. Design, synthesis, biomedical investigation, DFT calculation and molecular docking of novel Ru(II)-mixed ligand complexes. J. Biomol. Struct. Dyn. 1–20. doi: 10.1080/07391102.2021.2017355.
b- Mohamed, G.G., Omar, M.M., and Ahmed, Y.M. 2021. Metal complexes of tridentate Schiff base: Synthesis, characterization, biological activity and molecular docking studies with COVID-19 protein receptor. Z. Anorg. Allg. Chem. 647: 1–19.
Mohan, B. and Choudhary, M. 2021. Synthesis, crystal structure, computational study and anti-virus effect of mixed ligand copper (II) complex with ONS donor Schiff base and 1,10-phenanthroline. J. Mol. Struct. 1246: 131246.
Pinzi, L. and Rastelli, G. 2019. Molecular docking: Shifting paradigms in drug discovery. Int. J. Mol. Sci. 20: 4331; doi:10.3390/ijms20184331.

Ramadan, R.M., El-Medani, S.M., Makhlouf, A.A., Moustafa, H., Afifi, M.A. Haukka, M., and Abdel Aziz, A.A. 2021. Spectroscopic, DFT, non-linear optical properties and *in vitro* biological studies of Co(II), Ni(II) and Cu(II) complexes of hydrazide Schiff base derivatives. Appl. Organometal. Chem. e6246.

Rauf, A., Shah, A., Munawar, K.S., Ali, S., Tahir, M.N., Javed, M., and Khan, A.M. 2020. Synthesis, physicochemical elucidation, biological screening and molecular docking studies of a Schiff base and its metal(II) complexes. Arab. J. Chem. 13: 1130–1141.

Rezaeivala, M. and Keypour, H. 2014. Schiff Base and non-Schiff Base macrocyclic ligands and complexes incorporating the pyridine moiety-The first 50 years. Coord. Chem. Rev. 280: 203–253.

Ring, C.S., Sun, E., McKerrow, J.H., Lee, G.K., Rosenthal, P.J., Kuntz, I.D., and Cohen, F.E. 1993. Structure-based inhibitor design by using protein models for the development of antiparasitic agents. Proc. Natl. Acad. Sci. USA. 90: 3583–3587.

Satyanarayanajois, S.D. 2011. Drug Design and Discovery (Methods and Protocols). Springer, New York.

Shekhar, S., Khan, A.M., Sharma, S. Sharma, B., and Sarkar, A. 2021. Schiff base metallodrugs in anti-microbial and anticancer chemotherapy applications: a comprehensive review. Emergent Mater. https://doi.org/10.1007/s42247-021-00234-1.

Singh, A. and Barman, P. 2021. Recent advances in Schiff base ruthenium metal complexes: Synthesis and applications. Topics in Current Chem. 379: 29. https://doi.org/10.1007/s41061-021-00342-w.

Sousa, S.F., Fernandes, P.A., and Ramos, M.J. 2006. Protein-ligand docking: Current status and future challenges. Proteins. 65: 15–26.

Taylor, R.D., Jewsbury, P.J., and Essex, J.W. 2002. A review of protein-small molecule docking methods. J. Comput. Aided Mol. Design. 16(3): 151–166.

Wahb, S. 2022. Anti-tumor activity of a novel ruthenium complex on progression of heptatocelluler carcinoma: *In vivo* and *in vitro* studies. PhD Thesis, Ain Shams University.

Xiao, X., Min, J.L., Lin, W.Z., and Liu, Z. 2015, Drug-Target: Predicting the interactions between drug compounds and target proteins in cellular networking via the benchmark dataset optimization approach. J. Biomol. Struct. Dyn. 33: 2221–2233.

Zhong, W.Z. and Zhou, S.F. 2014. Molecular science for drug development and biomedicine. Int. J. Mol. Sci. 15: 20072–20078.

Zhou, S.F., and Zhong, W.Z. 2017. Drug design and discovery: Principles and applications, molecules. 22: 279. doi:10.3390/molecules22020279.

Zhu, X.W. 2019. Russ. J. Coord. Chem. 45: 532–538.

Index